Data Management Using Stata: A Practical Handbook

Data Management Using Stata: A Practical Handbook

MICHAEL N. MITCHELL

A Stata Press Publication
StataCorp LP
College Station, Texas

Published by Stata Press, 4905 Lakeway Drive, College Station, Texas 77845
Typeset in LaTeX 2ε
Printed in the United States of America
10 9 8 7 6 5 4 3 2 1

ISBN-10: 1-59718-076-9
ISBN-13: 978-1-59718-076-4

Library of Congress Control Number: 2010926561

Acknowledgements

As I think about the process of creating this book, my heart fills with gratitude for all the help that I received in writing it.

I want to start by thanking the good, kind people at StataCorp and Stata Press for their help, encouragement, and support in writing this book. I am very grateful to Bill Rising. Much of what you read in this book has been shaped by his keen advice and astute suggestions. I am especially grateful to Vince Wiggins, who watered the seeds of this book, providing encouragement and, perhaps most important of all, faith in this endeavor. I also want to thank Deirdre Patterson for her excellent and meticulous editing, Annette Fett for creating such a clever and fun cover design, and Lisa Gilmore of the Stata Press production team for all the terrific things she does.

I also want to thank the members of the UCLA ATS Statistical Consulting Team, who generously shared their thoughts on this book. Xiao Chen, Phil Ender, Rose Medeiros, Brigid Brettesborn, and especially Christine Wells provided much needed feedback and thoughtful suggestions. I also want to thank Lynn Soban for her helpful review and much appreciated encouragement. I am also grateful to Frauke Kreuter for her very kind assistance in translating labels into German in chapter 4.

Finally, I want to thank all the terrific clients who asked me statistical consulting questions at UCLA. Working with you on your questions and problems taught me more than you could ever know.

Contents

Tables

Figures

Preface

There is a gap between raw data and statistical analysis. That gap, called data management, is often filled with a mix of pesky and strenuous tasks that stand between you and your data analysis. I find that data management usually involves some of the most challenging aspects of a data analysis project. I wanted to write a book showing how to use Stata to tackle these pesky and challenging data-management tasks.

One of the reasons I wanted to write such a book was to be able to show how useful Stata is for data management. Sometimes people think that Stata's strengths lie solely in its statistical capabilities. I have been using Stata and teaching it to others for over 10 years, and I continue to be impressed with the way that it combines power with ease of use for data management. For example, take the `reshape` command. This simple command makes it a snap to convert a wide file to a long file and vice versa (for examples, see section 8.3). Furthermore, `reshape` is partly based on the work of a Stata user, illustrating that Stata's power for data management is augmented by user-written programs that you can easily download (as illustrated in section 10.2).

Each section of this book generally stands on its own, showing you how you can do a particular data-management task in Stata. Take, for example, section 2.4, which shows how you can read a comma-delimited file into Stata. This is not a book you need to read cover to cover, and I would encourage you to jump around to the topics that are most relevant for you.

Data management is a big (and sometimes daunting) task. I have written this book in an informal fashion, like we were sitting down together at the computer and I was showing you some tips about data management. My aim with this book is to help you easily and quickly learn what you need to know to skillfully use Stata for your data-management tasks. But if you need further assistance solving a problem, section 10.3 describes the rich array of online Stata resources available to you. I would especially recommend the Statalist listserver, which allows you to tap into the knowledge of Stata users around the world.

If you would like to contact me with comments or suggestions, I would love to hear from you. You can write me at MichaelNormanMitchell@gmail.com, or visit me on the web at http://www.MichaelNormanMitchell.com. Writing this book has been both a challenge and a pleasure. I hope that you like it!

Simi Valley, CA
April 2010

Michael N. Mitchell

1 Introduction

It has been said that data collection is like garbage collection: before you collect it you should have in mind what you are going to do with it.

—Russell Fox, Max Gorbuny, and Robert Hooke

1.1 Using this book

As stated in the title, this is a practical handbook for data management using Stata. As a practical handbook, there is no need to read the chapters in any particular order. Not only does each chapter stand alone but also most sections within each chapter stand alone as well. Each section focuses on a particular data-management task and provides examples of how to perform that particular data-management task. I imagine at least two different ways this book could be used.

You can pick a chapter, say, chapter 3 on data cleaning, and read the chapter to pick up some new tips and tricks about how to clean and prepare your data. Then the next time you need to clean data, you can use some of the tips you learned and grab the book for a quick refresher as needed.

Or, you may wish for quick help on a task you have never performed (or have not performed in a long time). For example, you may need to read a comma-separated file. You can grab the book and flip to chapter 2 on reading and writing datasets in which section 2.4 illustrates reading comma-separated and tab-separated files. Based on those examples, you can read the comma-separated file and then get back to your work.

However you read this book, each section is designed to provide you with information to solve the task at hand without getting lost in ancillary or esoteric details. If you find yourself craving more details, each section concludes with suggested references to the Stata help files for additional information. And starting with Stata 11, those help files include links to the online reference manuals. Because this book is organized by task, whereas the reference manuals are organized by command, I hope this book helps you connect data-management tasks to the corresponding reference manual entries associated with those tasks. Viewed this way, this book is not a competitor to the reference manuals but is instead a companion to them.

I encourage you to run the examples from this book for yourself. This engages you in active learning, as compared with passive learning (such as just reading the book). When you are actively engaged in typing in commands, seeing the results, and trying variations on the commands for yourself, I believe you will gain a better and deeper understanding than you would obtain from just passively reading.

To allow you to replicate the examples in this book, the datasets are available for download. You can download all the datasets used in this book into your current working directory from within Stata by typing the following commands:

```
. net from http://www.stata-press.com/data/dmus
. net get dmus1
. net get dmus2
```

After issuing these commands, you could then use a dataset, for example, `wws.dta`, just by typing the following command:

```
. use wws
```

Each section in the book is designed to be self-contained, so you can replicate the examples from the book by starting at the beginning of the section and typing the commands. At times, you might even be able to start replicating an example from the middle of a section, but that strategy might not always work. Then you will need to start from the beginning of the section to work your way through the examples. Although most sections are independent, some build on prior sections. Even in such cases, the datasets will be available so that you can execute the examples starting from the beginning of any given section.

Although the tasks illustrated in this book could be performed using the Stata point-and-click interface, this book concentrates on the use of Stata commands. However, there are two interactive/point-and-click features that are so handy, that I believe even command-oriented users (including myself) would find them useful. The Data Editor (as illustrated in section 2.10) is a very useful interactive interface for entering data into Stata. That same section illustrates the use of the Variables Manager. Although the Variables Manager is illustrated in the context of labeling variables for a newly created dataset, it is equally useful for modifying (or adding) labels for an existing dataset.

I should note that this book was written with Stata 11 in mind. Most of the examples from this book will work in versions of Stata prior to version 11. Some examples, most notably those illustrating merging datasets in chapter 6, will not work in versions of Stata prior to version 11.

This raises the issue of keeping your copy of Stata fully up to date, which is always a good practice. To verify that your copy of Stata is up to date and to obtain any free updates, type

```
. update query
```

and follow the instructions. After the update is complete, you can use the `help whatsnew` command to learn about the updates you have just received as well as prior updates documenting the evolution of Stata.

With the datasets for this book downloaded and your version of Stata fully up to date, you have what you need to dive into this book and work the examples for yourself. Before you do, however, I hope you will read the next section, which provides an overview of the book, to help you select which chapters you may want to read first.

1.2 Overview of this book

Each chapter of this book covers a different data-management topic, and each chapter pretty much stands alone. The ordering of the chapters is not like that in a traditional book, where you should read from the beginning to the end. You might get the most out of this book by reading the chapters in a different order than that in which they are presented. I would like to give you a quick overview of the book to help you get the most out of the order in which you read the chapters.

This book is composed of 11 chapters, comprising this introductory chapter (chapter 1), informational chapters 2–10, and an appendix.

The following four chapters, chapters 2–5, cover nuts-and-bolts topics that are common to every data-management project: reading and saving datasets, data cleaning, labeling datasets, and creating variables. These topics are placed at the front because I think they are the most common topics in data management; they are also placed in the front because they are the most clear-cut and concrete topics.

The next three chapters, chapters 6–8, cover tasks that occur in many (but not all) data-management projects: combining datasets, processing observations across subgroups, and changing the shape of your data.

Chapter 9 covers programming for data management. Although the topics in this chapter are common to many (if not all) data-management projects, they are a little more advanced than the topics discussed in chapters 2–5. This chapter describes how to structure your data analysis to be reproducible and describes a variety of programming shortcuts for performing repetitive tasks.

Chapter 10 contains additional resources, showing how to obtain the online resources for this book, how to find and install programs that other Stata users have written, and a list of additional recommended online resources. You might find this information more useful if you read it sooner rather than later.

Appendix A describes common elements regarding the workings of Stata. Unlike the previous chapters, these are fragments that do not pertain to a particular data-management task yet are pervasive and hence are frequently referenced throughout the book. The earlier chapters will frequently refer to the sections in the appendix, providing one explanation of these elements rather than repeating explanations each time they arise. The appendix covers topics such as comments, logical expressions, functions, `if` and `in`, missing values, and variable lists. I placed this chapter at the back to help you quickly flip to it when it is referenced. You may find it easier to read over the appendix to familiarize yourself with these elements rather than repeatedly flipping back to it.

The next section describes and explains some of the options that are used with the `list` command throughout this book.

1.3 Listing observations in this book

This book relies heavily on examples to show you how data-management commands work in Stata. I would rather show you how a command works with a simple example than explain it with lots of words. To that end, I frequently use the `list` command to illustrate the effect of commands. The default output from the `list` command is not always as clear as I might hope. Sometimes I add options to the `list` command to maximize the clarity of the output. Rather than explain the workings of these options each time they arise, I use this section to illustrate these options and explain why you might see them used throughout the book.

For the first set of examples, let's use `wws.dta`, which contains 2,246 hypothetical observations about women and their work.

```
. use wws
(Working Women Survey)
```

For files with many observations, it can be useful to list a subset of observations. I frequently use the `in` specification to show selected observations from a dataset. In the example below, observations 1–5 are listed, showing the variables `idcode`, `age`, `hours`, and `wage`.

```
. list idcode age hours wage in 1/5

     +-----------------------------------+
     | idcode   age   hours       wage   |
     |-----------------------------------|
  1. |   5159    38      38    7.15781   |
  2. |   5157    24      35   2.447664   |
  3. |   5156    26      40   3.824476   |
  4. |   5154    32      40   14.32367   |
  5. |   5153    35      35   5.517124   |
     +-----------------------------------+
```

Sometimes variable names are so long that they get abbreviated by the `list` command. This can make the listings more compact but also make the abbreviated headings harder to understand. For example, the listing below shows the variables `idcode`, `married`, `marriedyrs`, and `nevermarried` for the first five observations. Note how `marriedyrs` and `nevermarried` are abbreviated.

```
. list idcode married marriedyrs nevermarried in 1/5

     +------------------------------------------+
     | idcode   married   marrie~s   neverm~d   |
     |------------------------------------------|
  1. |   5159         0          0          0   |
  2. |   5157         1          0          0   |
  3. |   5156         1          3          0   |
  4. |   5154         1          2          0   |
  5. |   5153         0          0          1   |
     +------------------------------------------+
```

The `abbreviate()` option can be used to indicate the minimum number of characters the `list` command will use when abbreviating variables. For example, specifying `abbreviate(20)` means that none of the variables will be abbreviated to a length any shorter than 20 characters. In the book, I abbreviate this option to `abb()` (e.g., `abb(20)`, as shown below). Here this option causes all the variables to be fully spelled out.

(*Continued on next page*)

```
. list idcode married marriedyrs nevermarried in 1/5, abb(20)

     +---------------------------------------------+
     | idcode   married   marriedyrs   nevermarried |
     |---------------------------------------------|
  1. |   5159         0            0              0 |
  2. |   5157         1            0              0 |
  3. |   5156         1            3              0 |
  4. |   5154         1            2              0 |
  5. |   5153         0            0              1 |
     +---------------------------------------------+
```

When the variable listing is too wide for the page, the listing will wrap on the page. As shown below, this kind of listing is hard to follow, and so I avoid it in this book.

```
. list idcode ccity hours uniondues married marriedyrs nevermarried in 1/3,
> abb(20)

     +------------------------------------------------------------------+
  1. | idcode | ccity | hours | uniondues | married | marriedyrs |
     |   5159 |     1 |    38 |        29 |       0 |          0 |
     |------------------------------------------------------------------|
     |                     nevermarried                                  |
     |                                0                                  |
     +------------------------------------------------------------------+

     +------------------------------------------------------------------+
  2. | idcode | ccity | hours | uniondues | married | marriedyrs |
     |   5157 |     0 |    35 |         0 |       1 |          0 |
     |------------------------------------------------------------------|
     |                     nevermarried                                  |
     |                                0                                  |
     +------------------------------------------------------------------+

     +------------------------------------------------------------------+
  3. | idcode | ccity | hours | uniondues | married | marriedyrs |
     |   5156 |     0 |    40 |         0 |       1 |          3 |
     |------------------------------------------------------------------|
     |                     nevermarried                                  |
     |                                0                                  |
     +------------------------------------------------------------------+
```

Sometimes I add the `noobs` option to avoid such wrapping. The `noobs` option suppresses the display of the observation numbers, which occasionally saves just enough room to keep the listing from wrapping on the page.

The example from above is repeated below with the `noobs` option, and enough space is saved to permit the variables to be listed without wrapping.

```
. list idcode ccity hours uniondues married marriedyrs nevermarried in 1/3,
> abb(20) noobs

  +--------------------------------------------------------------------------+
  | idcode   ccity   hours   uniondues   married   marriedyrs   nevermarried |
  |--------------------------------------------------------------------------|
  |   5159       1      38          29         0            0              0 |
  |   5157       0      35           0         1            0              0 |
  |   5156       0      40           0         1            3              0 |
  +--------------------------------------------------------------------------+
```

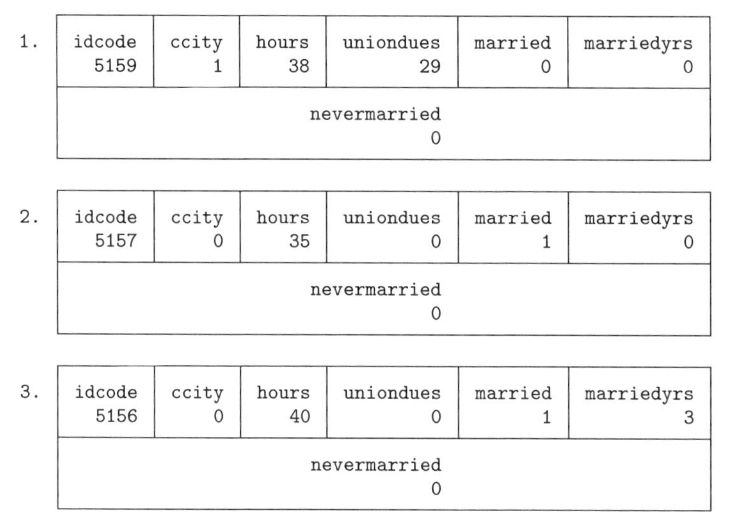

For the remaining examples, let's use `tv1.dta`, which contains 10 observations about the TV-watching habits of four different kids.

```
. use tv1
```

We can use the `list` command to see the entire dataset.

```
. list

     +----------------------------------------------+
     | kidid          dt   female   wt   tv   vac   |
     |----------------------------------------------|
  1. |     1   07jan2002        1   53    1     1   |
  2. |     1   08jan2002        1   55    3     1   |
  3. |     2   16jan2002        1   58    8     1   |
  4. |     3   18jan2002        0   60    2     0   |
  5. |     3   19jan2002        0   63    5     1   |
     |----------------------------------------------|
  6. |     3   21jan2002        0   66    1     1   |
  7. |     3   22jan2002        0   64    6     0   |
  8. |     4   10jan2002        1   62    7     0   |
  9. |     4   11jan2002        1   58    1     0   |
 10. |     4   13jan2002        1   55    4     0   |
     +----------------------------------------------+
```

Note how a separator line is displayed after every five observations. This helps make the output easier to read. Sometimes, though, I am pinched for space and suppress that separator to keep the listing on one page. The `separator(0)` option (which I abbreviate to `sep(0)`) omits the display of these separators.

```
. list, sep(0)

     +----------------------------------------------+
     | kidid          dt   female   wt   tv   vac   |
     |----------------------------------------------|
  1. |     1   07jan2002        1   53    1     1   |
  2. |     1   08jan2002        1   55    3     1   |
  3. |     2   16jan2002        1   58    8     1   |
  4. |     3   18jan2002        0   60    2     0   |
  5. |     3   19jan2002        0   63    5     1   |
  6. |     3   21jan2002        0   66    1     1   |
  7. |     3   22jan2002        0   64    6     0   |
  8. |     4   10jan2002        1   62    7     0   |
  9. |     4   11jan2002        1   58    1     0   |
 10. |     4   13jan2002        1   55    4     0   |
     +----------------------------------------------+
```

In other cases, the separators can be especially helpful in clarifying the grouping of observations. In this dataset, there are multiple observations per kid, and we can add the `sepby(kidid)` option to request that a separator be included between each level of `kidid`. This helps us clearly see the groupings of observations by kid.

(*Continued on next page*)

```
. list, sepby(kidid)

     +---------------------------------------------+
     | kidid          dt   female   wt   tv   vac |
     |---------------------------------------------|
  1. |     1   07jan2002        1   53    1     1 |
  2. |     1   08jan2002        1   55    3     1 |
     |---------------------------------------------|
  3. |     2   16jan2002        1   58    8     1 |
     |---------------------------------------------|
  4. |     3   18jan2002        0   60    2     0 |
  5. |     3   19jan2002        0   63    5     1 |
  6. |     3   21jan2002        0   66    1     1 |
  7. |     3   22jan2002        0   64    6     0 |
     |---------------------------------------------|
  8. |     4   10jan2002        1   62    7     0 |
  9. |     4   11jan2002        1   58    1     0 |
 10. |     4   13jan2002        1   55    4     0 |
     +---------------------------------------------+
```

This concludes this section describing options this book uses with the `list` command. I hope that this section helps you avoid confusion that could arise by having these options appear without any explanation of what they are or why they are being used.

2 Reading and writing datasets

Data! Data! Data! I can't make bricks without clay.

—Sherlock Holmes

2.1 Introduction

You have some raw data, and you are eager to analyze it using Stata. Before you can analyze the data in Stata, it first needs to be read into Stata. This chapter describes how you can read several common types of data files into Stata, and it shows how you can save several common types of data files. This section gives you an overview of some of the common issues you want to think about when reading and writing data files in Stata.

Changing directories

To read a data file, you first need to know the directory or folder in which it is located and how to get there.

Say that you are using Windows and you want to work with data files stored in a folder named `c:\statadata`. You could get to that directory by typing[1]

```
. cd c:/statadata
```

Say that you are using Unix (e.g., Linux, OS X, or AIX) and your data files are stored in a directory named `~/statadata`. You could go to that directory by typing

```
. cd ~/statadata
```

For further information on these kinds of navigational issues, see the *Getting Started with Stata* manual. From this point forward, I will assume that the data files of interest are in your current directory.[2]

Tip! Using the main menu to change directories

In the previous examples, the directory or folder names were short and simple, but in real life, such names are often long and typing them can be prone to error. It can be easier to point to a directory or folder than it is to type it. If you go to the **File** menu and then select **Change Working Directory...**, you can change the working directory by pointing to the directory or folder rather than having to type the full name.

---

1. Note that in Stata you can specify either `c:\statadata` or `c:/statadata`. Using the forward slash (/) is preferable because the backslash can have additional meaning in Stata.
2. Although it is possible to access files in other folders by specifying the full path of the file (e.g., directory or folder name and filename), I strongly recommend using the `cd` command to first go to the folder with your data. Then you only need to specify the file name to read and write datasets.

What kind of file are you reading?

There are several different kinds of data files that you can read into Stata, including Stata datasets, various kinds of raw data files, and SAS XPORT files. Let's consider these different kinds of data files.

As you would expect, it is simple to read Stata datasets into Stata. Section 2.2 describes how to read Stata datasets.

Raw data comes in variety of formats, including comma-separated, tab-separated, space-separated, and fixed format files. Let's look at an example of each kind of file.

Comma-separated files, sometimes referred to as CSV (comma-separated values) files, are commonly used for storing raw data. Such files often originate from spreadsheet programs and may be given a filename extension of `.csv`. Below we see an example of a comma-separated file named `dentists1.txt`. The `type` command is used to show this file.

```
. type dentists1.txt
name,years,fulltime,recom
"Y. Don Uflossmore",7.25,0,1
"Olive Tu'Drill",10.25,1,1
"Isaac O'Yerbreath",32.75,1,1
"Ruth Canaale",22,1,1
"Mike Avity",8.5,0,0
```

As implied by the name, comma-separated files use commas to separate the variables (columns) of data. Optional in such a file, this file includes the names of the variables in the first row, also separated by commas. This file contains five rows of data regarding five dentists. The four variables reflect the name of the dentist, the years she or he has been practicing, whether she or he works full time, and whether she or he recommends Quaddent gum. Note how the name of the dentist, which contains characters, is enclosed in double quotation marks. This is to avoid confusion in case the name contained commas. Section 2.4 illustrates how to read comma-separated files.

A related file is a tab-separated file. Instead of separating the variables (columns) with commas, a tab is used. The `dentists2.txt` file is an example of such a file, as shown below.

```
. type dentists2.txt
name     years   fulltime        recom
"Y. Don Uflossmore"     7.25    0       1
"Olive Tu'Drill"        10.25   1       1
"Isaac O'Yerbreath"     32.75   1       1
"Ruth Canaale"  22      1       1
"Mike Avity"    8.5     0       0
```

We do not directly see the tab characters that separate the variables, but instead we see how the presence of the tab makes the following variable line up at the next tab stop (like the tab stops in a word processor). The variables align imperfectly, in this example, largely because of the varying lengths of the names of the dentists. The first three dentists have long names and the second variable (`years`) lines up at the same tab stop. The last two dentists have short names, and the second variable lines up at an earlier tab stop. This kind of alignment of columns is commonly seen in tab-separated files. For information about how to read tab-separated files, see section 2.4.

Raw data can also be stored as a space-separated file. Such files use one (or possibly more) spaces to separate the variables (columns). The `dentists5.txt` file, shown below, is an example of such a file.

```
. type dentists5.txt
"Y. Don Uflossmore" 7.25 0 1
"Olive Tu´Drill" 10.25 1 1
"Isaac O´Yerbreath" 32.75 1 1
"Ruth Canaale" 22 1 1
"Mike Avity" 8.5 0 0
```

Note the similarity between this file and the comma-separated version. Instead of using commas to separate the variables, spaces are used. In this example, the first row does not include the variable names (as the comma-separated example did).[3] Section 2.5 illustrates how to read space-separated files.

Raw data files can also be stored as a fixed-column file. In these kinds of files, the variables are identified by their column position within the raw data file. The `dentists7.txt` file (shown below) is an example of a fixed-column file.

```
. type dentists7.txt
Y. Don Uflossmore 7.2501
Olive Tu´Drill   10.2511
Isaac O´Yerbreath32.7511
Ruth Canaale     22.0011
Mike Avity        8.5000
```

As you can see, the data are all squished together and might seem like just a jumble of numbers. To be useful, fixed-column files need to include accompanying documentation that provides the names of the variables and their column locations. For this file, the name of the dentist occupies columns 1–17, the years in practice occupies columns 18–22, whether the dentist works full time is in column 23, and whether the dentist recommends Quaddent gum is in column 24. This information about the column locations allows us to divide the information within this data file into different variables. Sections 2.6 and 2.7 illustrate how to read fixed-column files.

Perhaps you have downloaded or received a SAS XPORT file. You can read such files into Stata as described in section 2.8.

3. In my experience, comma-separated and tab-separated files commonly include the variable names in the first row while space-separated files do not.

Tip! Common errors reading files

There are several common errors that arise when reading data files. The most common errors are the "no; data in memory would be lost" error, the "you must start with an empty dataset" error, and the "no room to add more observations" error. Section 2.9 explains these errors and how to address them. You might want to jump ahead and read about these errors before you encounter them.

Sometimes you have collected data on your own and need to enter it into Stata. Section 2.10 describes how you can use the Stata Data Editor to enter data directly into Stata. And if you wish to be extra sure that such data are entered accurately, then you might want to consider double data entry as described in section 3.2.

What kind of file do you want to save?

Within Stata, you can save data in a variety of formats, including Stata datasets, various kinds of raw data files, and SAS XPORT files. Let's consider these different kinds of data files.

The most common format for saving data within Stata is a Stata dataset. Saving such files is described in section 2.3.

Section 2.11 illustrates how you can save comma-separated and tab-separated files. These kinds of files can be read by a variety of other programs, including spreadsheets.

You can save a space-separated file as illustrated in section 2.12. Such files can be useful for transferring data into software that requires data in a space-separated (sometimes called free-format) data file.

Section 2.13 shows how you can save your data as a SAS XPORT file. Such files can be used as part of submissions to the U.S. Food and Drug Administration (FDA) for new drug and new device applications (NDAs).

Tip! Reading files over the web

Most Stata commands that involve reading data will permit you to read data files over the web. For example, you can read `hsbdemo.dta` from the UCLA ATS web site by typing

```
. use http://www.ats.ucla.edu/stat/data/hsbdemo
```

Although Stata can read files stored on remote web sites, Stata cannot save files to such remote web sites.

2.2 Reading Stata datasets

This section illustrates how to read Stata datasets. For example, let's read the Stata dataset called `dentists.dta`. This dataset contains information from a survey of five dentists, including whether they recommend Quaddent gum to their patients who chew gum. We can read this dataset into Stata with the `use` command, as shown below.

```
. use dentists
. list

     +------------------------------------------------+
     |               name   years   fulltime   recom |
     |------------------------------------------------|
  1. | Y. Don Uflossmore    7.25          0       1 |
  2. |    Olive Tu´Drill   10.25          1       1 |
  3. | Isaac O´Yerbreath   32.75          1       1 |
  4. |      Ruth Canaale      22          1       1 |
  5. |        Mike Avity     8.5          0       0 |
     +------------------------------------------------+
```

As you can see, we successfully read this dataset. The `list` command shows the information from the five dentists: their names, the years they have been practicing, whether they work full time, and whether they recommend Quaddent gum. (If you get the error "no; data in memory would be lost", then you need to first use the `clear` command to clear out any data you currently have in memory.)

This same `use` command works if you had an older Stata dataset (going all the way back to version 1.0) and reads Stata datasets that were made on other computer systems. Stata figures out what kind of Stata dataset you have and reads it without the need for different commands or special options.

In addition to reading datasets from your computer, you can also read Stata datasets stored on remote web servers. For example, `dentists.dta` is located on the Stata Press web site, and you can `use` it with the following command:

```
. use http://www.stata-press.com/data/dmus/data/dentists.dta
```

Pretend with me that `dentists.dta` is an enormous dataset, and we are only interested in reading the variables `name` and `years`. We can read just these variables from the dataset as shown below. Note how the names of the variables to be read are specified after the `use` command.

```
. use name years using dentists
. list

     +---------------------------+
     |              name   years |
     |---------------------------|
  1. | Y. Don Uflossmore    7.25 |
  2. |    Olive Tu´Drill   10.25 |
  3. | Isaac O´Yerbreath   32.75 |
  4. |      Ruth Canaale      22 |
  5. |        Mike Avity     8.5 |
     +---------------------------+
```

Imagine that you only want to read a subset of observations from `dentists.dta`, just reading those dentists who have worked at least 10 years. We can do that as shown below (see section A.8 for more about the `if` qualifier).

```
. use dentists if years >= 10
. list
```

	name	years	fulltime	recom
1.	Olive Tu´Drill	10.25	1	1
2.	Isaac O´Yerbreath	32.75	1	1
3.	Ruth Canaale	22	1	1

We can even combine these to read just the variables `name` and `years` for those dentists who have worked at least 10 years, as shown below.

```
. use name years using dentists if years >= 10
. list
```

	name	years
1.	Olive Tu´Drill	10.25
2.	Isaac O´Yerbreath	32.75
3.	Ruth Canaale	22

By subsetting variables or observations, you can read Stata datasets that exceed the amount of memory you can (or want to) allocate. For example, you might have only 800 megabytes of memory free but want to read a Stata dataset that is 1,400 megabytes in size. By reading just the variables or observations you want, you might be able to read the data you want and still fit within the amount of memory you have available.

In addition to the `use` command, Stata has two other commands to help you find and use example datasets provided by Stata. The `sysuse` command allows you to find and use datasets that ship with Stata. The `sysuse dir` command lists all the example datasets that ship with Stata. The `sysuse` command reads the example dataset that you specify. `auto.dta` is one of the commonly used example datasets that ships with Stata. You can use this dataset by typing

```
. sysuse auto
(1978 Automobile Data)
```

There are many other example datasets used in the Stata manuals but not shipped with Stata. You can list these example datasets by typing `help dta contents` or selecting **File** and then **Example Datasets...** from the main menu. The `webuse` command reads the dataset you specify over the Internet. For example, I read about a competitor to `auto.dta` called `fullauto.dta` and we can use that dataset over the Internet like this:

```
. webuse fullauto
(Automobile Models)
```

For more information, see **help use**, **help sysuse**, and **help webuse**.

2.3 Saving Stata datasets

Suppose you flipped forward to one of the sections describing how to read raw datasets (e.g., section 2.4) and read the comma-separated file named `dentists1.txt`, as shown below.

```
. insheet using dentists1.txt
(4 vars, 5 obs)
```

To save this as a Stata dataset named `mydentists.dta`, you can use the **save** command, as shown below.

```
. save mydentists
file mydentists.dta saved
```

If the file `mydentists.dta` already exists, then you can add the **replace** option to indicate that it is okay to overwrite the existing file, as shown below.

```
. save mydentists, replace
file mydentists.dta saved
```

Perhaps you might not be saving the dataset for yourself but instead to give it to a friend or to share with several different people. Sometimes others might not be as quick as you to update their Stata to the latest version, so you might want to give them a dataset that will work with the prior version of Stata. You can do this using the **saveold** command. Versions 11 and 10 of Stata share the same dataset format, so when the **saveold** command is used with these versions, a file is saved that can be used with Stata version 9 and version 8 as well. This is illustrated below, saving the dataset as `dentistsold.dta`.

```
. saveold dentistsold
file dentistsold.dta saved
```

You might want to share this dataset with your best friend, but you do not know whether she uses Stata on Windows, Macintosh, Linux, Solaris, or AIX and would be embarrassed to ask. Take heart! You do not need to ask because a Stata dataset saved under one operating system can be read using Stata from any operating system.

Perhaps you like `dentists.dta` so much that you want to share it with the world via your web site http://www.iamastatgenious.net/. Suppose you upload `dentists.dta` to a folder or directory named `mydata` on your web server. Then the full path for accessing the dataset would be http://www.iamastatgenious.net/mydata/dentists.dta. You, and the whole world, could then read this dataset from that hypothetical web server from within Stata by typing

```
. use http://www.iamastatgenious.net/mydata/dentists.dta
```

Because Stata datasets are platform independent, this will work for people on all platforms. And if you use the saveold command, even those who are using previous versions of Stata could use the dataset.

> **Did you know? What's in a Stata dataset?**
>
> Have you ever wondered what is contained in a Stata dataset? Well I know you know what is in there, because you probably put it there. But I mean have you ever wondered exactly how Stata datasets are formatted? If so, see help dta, which provides some fascinating geeky details about the internal workings of Stata datasets, including how Stata is able to read datasets from different operating systems.

As shown in section 2.2, the use command allows you to specify if to read certain observations, and it allows you to specify a variable list to read certain variables. You might be tempted to try the same kind of trick on the save command, but neither of these features are supported on the save command. Instead, you can use the keep or drop command to select the variables you want to retain and use the keep if or drop if command to select the observations to retain. These commands are described in more detail in section A.9.

To illustrate this, let's first read in dentists.dta and list out the entire dataset.

```
. use dentists
. list

     +-----------------------------------------------+
     |              name   years   fulltime   recom  |
     |-----------------------------------------------|
  1. | Y. Don Uflossmore    7.25          0       1  |
  2. |    Olive Tu'Drill   10.25          1       1  |
  3. | Isaac O'Yerbreath   32.75          1       1  |
  4. |      Ruth Canaale      22          1       1  |
  5. |       Mike Avity      8.5          0       0  |
     +-----------------------------------------------+
```

Say that we want to save a dataset with just the dentists who recommend Quaddent (if recom is 1) and just the variables name and years. We can do this as illustrated below.

```
. keep if recom==1
(1 observation deleted)
. keep name years
. save dentist_subset
file dentist_subset.dta saved
```

Using the **keep if** command selected the observations we wanted to keep. (We also could have used **drop if** to select the observations to drop.) The **keep** command selected the variables we wanted to keep. (We also could have used the **drop** command to select the observations to drop.)

For more information about saving Stata datasets, see **help save**.

2.4 Reading comma-separated and tab-separated files

Raw data can be stored in several ways. If the variables are separated by commas, the file is called a comma-separated file; if the variables are separated by tabs, the file is called a tab-separated file. Such files can be read using the **insheet** command. If the data file contains the names of the variables in the first row of the data, Stata will detect and use them for naming the variables. Consider the example data file called **dentists1.txt**, which has 5 observations. This file has information about five hypothetical dentists, including whether they recommend Quaddent gum to their patients who chew gum.

```
. type dentists1.txt
name,years,fulltime,recom
"Y. Don Uflossmore",7.25,0,1
"Olive Tu´Drill",10.25,1,1
"Isaac O´Yerbreath",32.75,1,1
"Ruth Canaale",22,1,1
"Mike Avity",8.5,0,0
```

Perhaps later we will ask the fifth dentist why he did not recommend this gum, but for now let's see how we can read this data file into Stata. The first row of the data file provides the names of the variables—the dentist's name (**name**), the number of years the dentist has been practicing (**years**), whether the dentist is full time (**fulltime**), and whether the dentist recommends Quaddent (**recom**). We can read such a file with the **insheet** command, as shown below.

```
. insheet using dentists1.txt
(4 vars, 5 obs)
```

Because this is such a small file, we can verify that it was read properly by using the **list** command.

```
. list

     +-----------------------------------------------+
     |              name   years   fulltime   recom  |
     |-----------------------------------------------|
  1. | Y. Don Uflossmore    7.25          0       1  |
  2. |    Olive Tu´Drill   10.25          1       1  |
  3. | Isaac O´Yerbreath   32.75          1       1  |
  4. |      Ruth Canaale      22          1       1  |
  5. |        Mike Avity     8.5          0       0  |
     +-----------------------------------------------+
```

Another common format is a tab-separated file, where each variable is separated by a tab. The file `dentists2.txt` is a tab-separated version of the dentists file.

```
. type dentists2.txt
name     years    fulltime        recom
"Y. Don Uflossmore"     7.25    0       1
"Olive Tu'Drill"        10.25   1       1
"Isaac O'Yerbreath"     32.75   1       1
"Ruth Canaale"  22      1       1
"Mike Avity"    8.5     0       0
```

We can read such a file with the `insheet` command, but to save space, we will forgo listing the contents of the file.

```
. insheet using dentists2.txt
(4 vars, 5 obs)
```

You might have a comma-separated or tab-separated file that does not have the variable names contained in the data file. The data file `dentists3.txt` is an example of a comma-separated file that does not have the variable names in the first row of data.

```
. type dentists3.txt
"Y. Don Uflossmore",7.25,0,1
"Olive Tu'Drill",10.25,1,1
"Isaac O'Yerbreath",32.75,1,1
"Ruth Canaale",22,1,1
"Mike Avity",8.5,0,0
```

You have two choices when reading such a file: you can either let Stata assign temporary variable names for you or provide the names when you read the file. The following example shows how you can read the file and let Stata name the variables for you.

```
. insheet using dentists3.txt
(4 vars, 5 obs)
```

The `list` command shows that Stata named the variables `v1`, `v2`, `v3`, and `v4`.

```
. list

     +-----------------------------------------+
     |                v1      v2   v3   v4 |
     |-----------------------------------------|
  1. | Y. Don Uflossmore    7.25    0    1 |
  2. |    Olive Tu'Drill   10.25    1    1 |
  3. | Isaac O'Yerbreath   32.75    1    1 |
  4. |      Ruth Canaale      22    1    1 |
  5. |        Mike Avity     8.5    0    0 |
     +-----------------------------------------+
```

You can then use the `rename` command or the Variables Manager to rename the variables. See section 5.15 for more information on renaming variables in Stata or page 34 for more information on the Variables Manager.

Rather than renaming the variables after reading the data file, we can specify the desired variable names on the `insheet` command, as shown below.

```
. insheet name years fulltime recom using dentists3.txt
(4 vars, 5 obs)
```

Tip! What about files with other separators?

Stata can read files with other kinds of separators as well. The file `dentists4.txt` uses a colon (`:`) as a separator (delimiter) between the variables. You can add the `delimiter(":")` option to the `insheet` command to read the file. For example,

```
. insheet using dentists4.txt, delimiter(":")
```

See `help insheet` for more information about the `insheet` command.

2.5 Reading space-separated files

Another common format for storing raw data is a space-separated file. In such a file, variables are separated by one (or more) spaces, and if a string variable contains spaces, it is enclosed in quotes. The file `dentists5.txt` is an example of such a file with information about five dentists, including their names, the number of years they have been practicing, whether they are working full time, and whether they recommend Quaddent gum.

```
. type dentists5.txt
"Y. Don Uflossmore" 7.25 0 1
"Olive Tu´Drill" 10.25 1 1
"Isaac O´Yerbreath" 32.75 1 1
"Ruth Canaale" 22 1 1
"Mike Avity" 8.5 0 0
```

You can use the `infile` command to read this file. Because the file did not include variable names, you need to specify the variable names on the `infile` command. In addition, because the variable `name` is a string variable, you need to tell Stata that this is a string variable by prefacing `name` with `str17`, which informs Stata that this is a string variable that may be as wide as 17 characters.

```
. infile str17 name years full rec using dentists5.txt
(5 observations read)
```

Using the `list` command, we can see that the data were read properly.

```
. list
```

	name	years	full	rec
1.	Y. Don Uflossmore	7.25	0	1
2.	Olive Tu'Drill	10.25	1	1
3.	Isaac O'Yerbreath	32.75	1	1
4.	Ruth Canaale	22	1	1
5.	Mike Avity	8.5	0	0

The `infile` command does not read files with variable names in the first row. How can we read such a file? We can use the `insheet` command with the `delimiter(" ")` option to indicate that the variables are separated (delimited) by a space. We have a file called `dentists6.txt`, and it uses a space as a separator and has variable names in the first row. We can read this file using the `insheet` command like this:

```
. insheet using dentists6.txt, delimiter(" ")
(4 vars, 5 obs)
```

Sometimes you might need to read a space-separated file that has dozens or even hundreds of variables, but you are interested in only some of those variables. For example, say that you have a file called `abc.txt` that contains 26 variables named `a`, `b`, `c`, ..., `z`. Suppose you are only interested in the variables `a` and `x`. Rather than specifying all the variables on the `infile` statement, you can read `a`, then skip 22 variables (`b`–`w`), read `x` and then skip the last 2 variables `y` and `z`. This not only saves you effort (by not having to name variables you will not be keeping) but also permits you to read files that may exceed the amount of memory you have available by reading just the variables you need (see section 2.9 for more information on allocating enough memory for datasets).

```
. infile a _skip(22) x _skip(2) using abc.txt
(5 observations read)
. list
```

	a	x
1.	3	8
2.	6	5
3.	4	2
4.	5	9
5.	6	9

Sometimes you might want to read just some of the observations from a raw data file. You might be inclined to read the whole data file and then use `keep if` to drop the observations you do not want. Ordinarily, this is a good enough strategy, but you can save time and memory if you specify `if` on the `infile` command to read just the observations you want (section A.8 gives more details on `if`). For example, you can read the file `abc.txt` including just those observations where variable `a` is 5 or less.

```
. infile a _skip(22) x _skip(2) using abc.txt if (a <= 5)
(3 observations read)
. list

     +-------+
     | a   x |
     |-------|
  1. | 3   8 |
  2. | 4   2 |
  3. | 5   9 |
     +-------+
```

Tip! Reading consecutive variables

Consider a raw data file where we have an identification variable, a person's gender and age, and five measures of blood pressure and five measures of pulse. You could read this raw data file as shown below.

```
. infile id age bp1 bp2 bp3 bp4 bp5 pu1 pu2 pu3 pu4 pu5 using cardio1.txt
```

You could also use a shortcut, as shown below.

```
. infile id age bp1-bp5 pu1-pu5 using cardio1.txt
```

For more information, see `help infile`.

2.6 Reading fixed-column files

Fixed-column files can be confusing because the variables are pushed together without spaces, commas, or tabs separating them. In such files, the variables are identified by their column position(s). Such files are frugal in their use of space but are more challenging to read because you need to specify the starting and ending column position of each variable. Such information typically comes from a codebook that gives the column positions for the variables. Consider a fixed-column version of `dentists.dta` named `dentists7.txt`.

```
. type dentists7.txt
Y. Don Uflossmore 7.2501
Olive Tu'Drill   10.2511
Isaac O'Yerbreath32.7511
Ruth Canaale     22.0011
Mike Avity        8.5000
```

In this file, the name of the dentist occupies columns 1–17, the years in practice occupies columns 18–22, whether the dentist is full time is in column 23, and whether the dentist recommends Quaddent is in column 24. Knowing the column locations, you can read this file using the `infix` command like this:

```
. infix str name 1-17 years 18-22 fulltime 23 recom 24 using dentists7.txt
(5 observations read)
. list

     +------------------------------------------------+
     |              name   years   fulltime   recom   |
     |------------------------------------------------|
  1. | Y. Don Uflossmore    7.25          0       1   |
  2. |    Olive Tu'Drill   10.25          1       1   |
  3. | Isaac O'Yerbreath   32.75          1       1   |
  4. |      Ruth Canaale      22          1       1   |
  5. |        Mike Avity     8.5          0       0   |
     +------------------------------------------------+
```

You do not have to read all the variables in a fixed-column data file. In fact, when I first try to read a fixed-column data file, I start by reading just the first and last variables and check those variables before trying to read more variables. You can use the same strategy when you have many variables but want to read only a few of them. For example, you can read just the variables `name` and `fulltime`, as shown below.

```
. infix str name 1-17 fulltime 23 using dentists7.txt
(5 observations read)
. list

     +--------------------------------+
     |              name   fulltime   |
     |--------------------------------|
  1. | Y. Don Uflossmore          0   |
  2. |    Olive Tu'Drill          1   |
  3. | Isaac O'Yerbreath          1   |
  4. |      Ruth Canaale          1   |
  5. |        Mike Avity          0   |
     +--------------------------------+
```

Likewise, you do not have to read all the observations in the data file. You can specify an `in` qualifier or an `if` qualifier to read just a subset of the observations. When I read a file with many observations, I often read just the first 10 observations by adding `in 1/10` to quickly identify any simple problems before reading the entire file. If you wanted to read the first three observations from `dentists7.txt`, you could type

```
. infix years 18-22 fulltime 23 using dentists7.txt in 1/3
```

If you wanted to read just the dentists who worked full time, you could type

```
. infix years 18-22 fulltime 23 using dentists7.txt if fulltime==1
```

See section A.8 for more information about using `if` and `in`.

Stata offers another strategy for reading fixed-column files via a dictionary file. Like the `infix` command, above, a dictionary file (below) contains the variable names and column locations. This dictionary file specifically works in combination with the `infix` command, which is why it begins with `infix dictionary`.

```
. type dentists1.dct
infix dictionary {
  str name 1-17 years 18-22 fulltime 23 recom 24
}
```

Having defined this data dictionary, we can then invoke it with the `infix` command, as shown below. We could have omitted the `.dct` extension because the dictionary file is assumed to have a `.dct` extension.

```
. infix using dentists1.dct, using(dentists7.txt)
infix dictionary {
  str name 1-17 years 18-22 fulltime 23 recom 24
}
(5 observations read)

. list

              name   years   fulltime   recom

 1.   Y. Don Uflossmore    7.25          0       1
 2.     Olive Tu'Drill   10.25          1       1
 3.  Isaac O'Yerbreath   32.75          1       1
 4.       Ruth Canaale      22          1       1
 5.         Mike Avity     8.5          0       0
```

In this example, we have specified the name of the raw data file on the `infix` command with the `using(dentists7.txt)` option; however, we could have indicated the name of the raw data file within the `infix dictionary` file. Consider the dictionary file named `dentists2.dct`, shown below.

```
. type dentists2.dct
infix dictionary using dentists7.txt {
  str name 1-17 years 18-22 fulltime 23 recom 24
}
```

Note how this dictionary specifies the name of the data file. We can use the `infix` command to read this dictionary file, which, in turn, reads the `dentists7.txt` file, as shown below.

```
. infix using dentists2.dct
infix dictionary using dentists7.txt {
  str name 1-17 years 18-22 fulltime 23 recom 24
}
(5 observations read)
```

The `list` command shows that the variables have been properly read from the `dentists7.txt` file.

```
. list
```

	name	years	fulltime	recom
1.	Y. Don Uflossmore	7.25	0	1
2.	Olive Tu´Drill	10.25	1	1
3.	Isaac O´Yerbreath	32.75	1	1
4.	Ruth Canaale	22	1	1
5.	Mike Avity	8.5	0	0

Let's consider another way to read **dentists7.txt** by using the **infile** command combined with a dictionary file. The structure of an **infile dictionary** is different from an **infix dictionary**. The dictionary file named **dentists3.dct** below shows an example of how we can read the file **dentists7.txt** using an **infile dictionary**.

```
. type dentists3.dct
infile dictionary using dentists7.txt {
  str17 name      %17s  "Name of dentist"
        years     %5f   "Years in practice"
        fulltime  %1f   "Full time?"
        recom     %1f   "Recommend Quaddent?"
}
```

The dictionary starts with **infile dictionary** to specify that this dictionary goes with the **infile** command. This is followed by **using dentists7.txt**, indicating the name of the raw data file, and then an open brace to begin the process of specifying how to read each variable.

Next for each variable, we specify the variable storage type (optional for numeric variables), the variable name, the input format for reading the data, and the variable label (optional). The first variable will be stored using the **str17** type (a string variable with a width of 17). The variable will be called **name** and will be read using the format **%17s** (a string variable that is 17 characters wide). Finally, the variable will have the label **"Name of dentist"**. Specifying the storage type is optional for numeric variables and thus is skipped for the rest of the variables. The next variable is **years**, is five digits wide and hence read with the format **%5f**, and is followed by the variable label. The variable name, input format, and variable label are supplied for the third and fourth variable, followed by a close brace.

Having defined the dictionary, we can read the **dentists7.txt** data file using the **infile** command, as shown below.

```
. infile using dentists3.dct
infile dictionary using dentists7.txt {
  str17 name      %17s  "Name of dentist"
        years     %5f   "Years in practice"
        fulltime  %1f   "Full time?"
        recom     %1f   "Recommend Quaddent?"
}
(5 observations read)
```

The listing below shows that we successfully read the **dentists7.txt** data file.

```
. list

     +------------------------------------------------------+
     |                name   years   fulltime   recom       |
     |------------------------------------------------------|
  1. | Y. Don Uflossmore      7.25          0       1       |
  2. |    Olive Tu'Drill     10.25          1       1       |
  3. | Isaac O'Yerbreath     32.75          1       1       |
  4. |       Ruth Canaale       22          1       1       |
  5. |         Mike Avity      8.5          0       0       |
     +------------------------------------------------------+
```

You may ask why a data dictionary would be preferable to directly specifying the variable names and column locations on the `infix` command. Fixed-format data files can often have many variables, perhaps even several hundred. In such cases, it is much easier to specify the names and column locations using a dictionary file. Whether you use `infix dictionary` or `infile dictionary` is up to you. The `infix dictionary` command focuses on specifying the beginning and ending column locations for each variable, while the `infile dictionary` command focuses on specifying the length of each variable. The `infile dictionary` method allows you to include variable labels, while `infix dictionary` does not.

There is one additional reason you might choose to use a data dictionary for reading a fixed-column file. As illustrated in section 2.7, sometimes fixed-column files contain multiple rows of data per observation. Using a dictionary is the only way to read such raw data files. For more information about reading fixed-column files, see `help infix` and `help infile2`.

> **Warning! It's just a bunch of numbers**
>
> I once was working with a client who had a fixed-column data file. She looked at the data file and said "It's just a bunch of numbers!" and asked how to proceed. Unfortunately, she did not have a codebook, and we were unable to read her data. When you get a raw data file (especially a fixed-column data file), always ask for the codebook information that accompanies it. That way you can avoid having a data file that is "just a bunch of numbers".

2.7 Reading fixed-column files with multiple lines of raw data per observation

Sometimes fixed-column raw data files are stored using multiple lines (rows) of data per observation. This strategy was used for older data files when data were punched and stored using 80-column computer cards. If you had 140 columns of information per observation, each observation was split across two cards, the first card containing columns 1–80 and the second containing columns 81–140. Newer files use this strategy too, to avoid lines of data running off the edge of the computer screen. This section describes how you can read such raw data files using Stata.

In section 2.6, we saw how we could use the `infix` command for reading fixed-column files and how the `infix` command could be combined with a dictionary file that would specify the column locations for the variables. We will build upon that to see how we can read data files with multiple lines of data per observation. Consider the file below, named `dentists8.txt`, which contains two lines of data per dentist. The first line of data has the dentist's name in columns 1–17 and years in practice in columns 18–19. The second line of data has whether the dentist works full time in column 1 and whether the dentist recommends Quaddent in column 2. This file contains five dentists with 2 lines of data per dentist, for a total of 10 lines of data.

```
. type dentists8.txt
Y. Don Uflossmore 7.25
01
Olive Tu'Drill   10.25
11
Isaac O'Yerbreath32.75
11
Ruth Canaale     22.00
11
Mike Avity        8.50
00
```

We can read `dentists8.txt` using the dictionary file `dentists4.dct`, shown below. Note how I indicated the number of lines of raw data per observation with `2 lines`. This is followed by `1:` and then the instructions for reading the variables that appear on the first line of raw data for an observation. This is followed by `2:` and then the instructions for reading the variables that appear on the second line of raw data for an observation.

```
. type dentists4.dct
infix dictionary using dentists8.txt {
  2 lines
  1: str name 1-17 years 18-22
  2: fulltime 1 recom 2
}
```

We can then read `dentists8.txt` using `dentists4.dct` in combination with the `infix` command.

```
. infix using dentists4.dct
infix dictionary using dentists8.txt {
  2 lines
  1: str name 1-17 years 18-22
  2: fulltime 1 recom 2
}
(5 observations read)
```

(Continued on next page)

The `list` command confirms that this file has been read successfully.

```
. list

     +------------------------------------------------+
     |              name   years   fulltime    recom  |
     |------------------------------------------------|
  1. | Y. Don Uflossmore    7.25          0        1  |
  2. |    Olive Tu´Drill   10.25          1        1  |
  3. | Isaac O´Yerbreath   32.75          1        1  |
  4. |      Ruth Canaale      22          1        1  |
  5. |        Mike Avity     8.5          0        0  |
     +------------------------------------------------+
```

As illustrated in section 2.6, the `infile` command can be combined with a dictionary to read fixed-column files. The dictionary file `dentists5.dct` (below) can be used in combination with the `infile` command to read the `dentists8.txt` data file.

```
. type dentists5.dct
infile dictionary using dentists8.txt {
  _lines(2)
  _line(1)
  str17 name      %17s  "Name of dentist"
        years     %5f   "Years in practice"
  _line(2)
        fulltime %1f    "Full time?"
        recom    %1f    "Recommend Quaddent?"
}
```

The dictionary includes the `_lines(2)` specification to indicate that `dentists8.txt` has two lines of raw data per observation. Then `_line(1)` precedes the instructions for reading the first line of data, and `_line(2)` precedes the instructions for reading the second line of data. Below we use this dictionary to read `dentists8.txt`.

```
. infile using dentists5.dct
infile dictionary using dentists8.txt {
  _lines(2)
  _line(1)
  str17 name      %17s  "Name of dentist"
        years     %5f   "Years in practice"
  _line(2)
        fulltime %1f    "Full time?"
        recom    %1f    "Recommend Quaddent?"
}
(5 observations read)
```

As this section illustrated, both the `infix` and `infile` commands can be combined with a dictionary to read raw data files, which contain multiple lines of raw data per observation. For more information, see `help infix` and `help infile2`.

2.8 Reading SAS XPORT files

Stata has the ability to directly read SAS XPORT files. Say that someone gave you a copy of the dentists data file saved as a SAS XPORT file named **dentists.xpt**. You can read that file into Stata with the **fdause** command, as shown below.

```
. fdause dentists
. list

     +-----------------------------------------------+
     |              name   years   fulltime   recom  |
     |-----------------------------------------------|
  1. | Y. Don Uflossmore    7.25          0       1  |
  2. |    Olive Tu'Drill   10.25          1       1  |
  3. | Isaac O'Yerbreath   32.75          1       1  |
  4. |      Ruth Canaale      22          1       1  |
  5. |        Mike Avity     8.5          0       0  |
     +-----------------------------------------------+
```

Suppose that you were also given an XPORT version, named **formats.xpf**, of the SAS formats for this file and you placed it in the same folder as **dentlab.xpt**. When the **fdause** command reads **dentlab.xpt**, it will automatically detect **formats.xpf** without needing to specify any additional options.

```
. fdause dentlab
. list

     +------------------------------------------------------+
     |              name   years    fulltime          recom |
     |------------------------------------------------------|
  1. | Y. Don Uflossmore    7.25   part time      recommend |
  2. |    Olive Tu'Drill   10.25   full time      recommend |
  3. | Isaac O'Yerbreath   32.75   full time      recommend |
  4. |      Ruth Canaale      22   full time      recommend |
  5. |        Mike Avity     8.5   part time     do not rec |
     +------------------------------------------------------+
```

Note how the variables **fulltime** and **recom** above display the value labels. These are drawn from **formats.xpf**. The **describe** command shows that the **fulltime** variable is labeled with the value label **ftlab** and **recom** is labeled with the value label **reclab**.

```
. describe fulltime recom

              storage  display     value
variable name   type   format      label      variable label
-------------------------------------------------------------------------
fulltime        double %16.0g      ftlab
recom           double %10.0g      reclab
```

If your goal is to convert a SAS data file for use in Stata, you can use **PROC EXPORT** within SAS to create a Stata dataset. Below **PROC EXPORT** is used to convert the SAS data file named **dentists** to a Stata dataset named **c:\data\dentists.dta**.

```
PROC EXPORT DATA=dentists OUTFILE="C:\data\dentists.dta";
RUN;
```

For more information about reading SAS XPORT files in Stata, see `help fdause`.

2.9 Common errors reading files

This section describes and explains three common error messages you may see when reading data into Stata. These errors are the "no; data in memory would be lost" error, the "you must start with an empty dataset" error, and the "no room to add more observations" error.

To understand these errors better, let's first briefly explore the model that Stata uses for reading, modifying, and saving datasets. Think about how a word processor works. You read in a file (such as a letter to your mom), you make changes to the file, and then you save the file with the changes. Or, if you do not like the changes, you do not save the file and the letter to Mom saved on disk remains unchanged. Stata works using the same kind of logic. Stata datasets can be read into memory and modified, and if you like the changes, they can be saved. The dataset in memory is called the working dataset. You can use a variety of commands to analyze and modify the working dataset. But like the letter to Mom, the changes to the working dataset are temporary until saved. If you were careless, you could lose the changes you made. Fortunately, Stata helps you avoid this, as illustrated below.

The "no; data in memory would be lost" error

Stata allows you to have only one dataset in memory at a time. If you currently have unsaved changes to the working dataset, reading a new file would cause you to lose your unsaved changes. Stata wants to help you avoid losing unsaved changes and so will issue the "no; data in memory would be lost" error. For example, if you try to `use` a Stata dataset while you have unsaved changes to the working dataset, you will receive the following error:

```
. use dentists
no; data in memory would be lost
r(4);
```

This error is saying that you would lose the changes to the data in memory if the new dataset were to be read into memory, so Stata refused to read the new dataset. If you care about the data in memory, use the `save` command to save your data (see section 2.3); if you do not care about the working dataset, you can throw it away using the `clear` command.

Tip! The clear command versus the clear option

Rather than using the `clear` command, most (if not all) commands permit you to specify the `clear` option. For example, you can type

```
. use dentists, clear
```

instead of typing

```
. clear
. use dentists
```

Likewise, you can add the `clear` option to other commands like `infile`, `infix`, and `insheet`. The choice of which to use is up to you.

The "you must start with an empty dataset" error

When reading a raw dataset (using, for example, the `infile`, `infix`, or `insheet` command), there cannot be a working dataset in memory. If you have data currently in memory (saved or not), issuing one of these commands will give you the following error:

```
. insheet using dentists1.txt
you must start with an empty dataset
r(18);
```

This error is saying that you first need to clear the data currently in memory before you may issue the command. Being sure that you have saved the data in memory if you care about it, you would then issue the `clear` command. That clears any data currently in memory, permitting you to read raw data into Stata.

The "no room to add more observations" error

Although the files in this book are small, when you read your own files you might need to use the `set memory` command to allocate enough memory to read in your datasets. If you try to read a file that is larger than your current memory allocation, you will get an error message that looks like this:

(*Continued on next page*)

```
. use hypothetical_bigfile
no room to add more observations
    An attempt was made to increase the number of observations beyond what is
    currently possible.  You have the following alternatives:

     1.  Store your variables more efficiently; see help compress.  (Think of
         Stata's data area as the area of a rectangle; Stata can trade off width
         and length.)

     2.  Drop some variables or observations; see help drop.

     3.  Increase the amount of memory allocated to the data area using the set
         memory command; see help memory.
r(901);
```

You can use the `dir` command to see how big the file is.

```
. dir hypothetical_bigfile.dta
  237.4M   6/10/09 16:54  hypothetical_bigfile.dta
```

This file is 237.4 megabytes, so allocating 300 megabytes would seem to be sufficient (this permits room for additional variables to be added).

```
. set memory 300m
. use hypothetical_bigfile
```

The `set memory` command, above, changes the memory allocation only for the current Stata session. Once you close Stata and later reopen it, it will revert to its default memory allocation. You can add the `permanently` option (shown below) and Stata will use that memory allocation every time you start your copy of Stata.

```
. set memory 300m, permanently
```

Tip! Missing data in raw data files

Raw data files frequently use numeric codes for missing data. For example, −7 might be the code for "don't know"; −8, the code for "refused"; and −9, for "not applicable". In such cases, the missing values are not immediately distinguishable from nonmissing values and all Stata analysis commands would interpret these values as valid data. If you have missing data coded in this fashion (e.g., missing values specified as −7, −8, −9), see section 5.6 for information on how to convert the numeric values to missing values.

This concludes this section about common errors reading files. The next section illustrates how you can enter data directly into Stata using the Data Editor.

2.10 Entering data directly into the Stata Data Editor

In previous sections, I have assumed that your data are stored in a raw data file. But sometimes you collect your own data and need to enter it into the computer yourself. Many are tempted to use a spreadsheet program for such data entry because spreadsheets are commonly available. Further, those who are doing the data entry are often familiar with them. Nevertheless, I cannot emphasize how strongly I recommend against using spreadsheets for data entry. I have repeatedly seen problems such as rogue data in the spreadsheet, data entered in a nonrectangular form, improperly constructed variable names, no variable names, data that gets sorted on one column but not others, difficulty transferring the spreadsheet data into Stata, and so forth. To avoid such problems, I recommend entering data directly into Stata using the Data Editor. The Data Editor has the look and feel of a spreadsheet while avoiding these kinds of data-entry problems.

Before you are ready to enter data into the Data Editor, you first need to create a codebook for your data. I have created an example codebook below for a hypothetical survey of students. This survey includes variables that uniquely identify the student (`id`), the name of the student (`stuname`), their ethnicity (`race`), whether they are happy (`happy`), whether they are glad (`glad`), their date of birth (`dob`), and their hourly wage (`wage`).

```
Codebook for studentsurvey
Variable list
------------------------------------------------------------------------------
   Variable name   Label                   Var type        Coding scheme name
1. id              Unique identifier       Numeric
2. stuname         Name of student         String 30
3. race            Race of student         Numeric          racelab
4. happy           Is student happy        Numeric          yesnolab
5. glad            Is student glad         Numeric          yesnolab
6. dob             Date of birth           Date
7. wage            Hourly wage             Numeric
Coding scheme for categorical variables
------------------------------------------------------------------------------
  Name        Coding scheme
  racelab     1=White, 2=Black, 3=Hispanic, 4=Asian
  yesnolab    1=yes 0=no
```

The codebook contains a variable name and descriptive label for every variable. It also indicates a general description of the variable type, focusing on whether the variable is numeric, a date variable, or a string variable (and if a string variable, how long it can be). The `wage` variable, for example, is a numeric variable, while `dob` is a date variable. The name of the student is a string variable, and it was decided that it could be up to 30 characters long.

The final column of the variable list indicates the name of the coding scheme for categorical variables, which links to the second half of the codebook that describes the coding scheme for these variables. For example, the `race` variable is associated with the coding scheme named `racelab`,[4] which is coded as 1 = White, 2 = Black, 3 = Hispanic, and 4 = Asian. Without this coding scheme, we would never know what numeric value was assigned to each level of `race`. The variables `happy` and `glad` are both yes/no variables which share a common coding scheme named `yesnolab` in which yes is coded as 1 and no as 0.

The process of entering data into the Data Editor is a four-step process. This involves (step 1) entering the data for the first student, (step 2) labeling the variables and values, (step 3) fixing the values of date variables, and (step 4) entering the data for the rest of the observations. This process is described in more detail below. Feel free to work along, making up your own hypothetical data for the student survey data.

Before we can start, we need to clear the working dataset with the `clear` command.

Step 1: Enter the data for the first observation. Open the Stata Data Editor with the `edit` command. Now let's start entering the data for the first student. Enter the value for the `id` variable in the first column and then press the *Tab* key, which moves you to the second column. Now enter the student's name and press *Tab*, and then enter the student's race (referring to the coding scheme for `race`) and press *Tab*. Continue entering data for all the variables, except that when you encounter a date variable enter a temporary numeric value (e.g., 1). We will go back and fix these in step 3. Continue until you have entered all the variables for the first observation. After you enter the last variable, press *Tab* one last time. Figure 2.1 shows the Data Editor after I entered the first line of my hypothetical data.

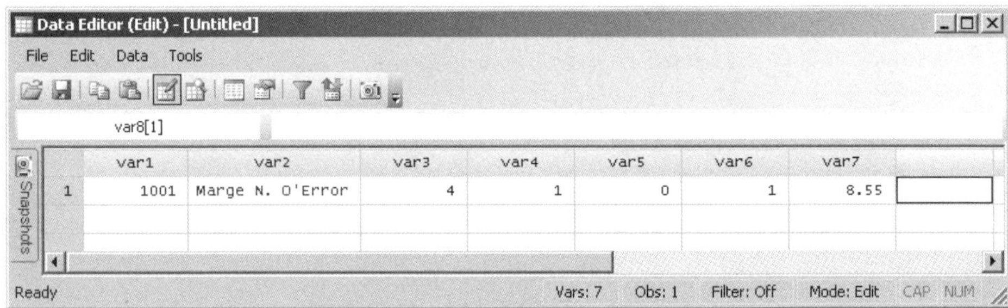

Figure 2.1. Stata Data Editor after step 1, entering data for the first observation

Step 2: Label the variables. The second step is to label the variables based on the information shown in the codebook. We will do this using the Variables Manager.[5] You

4. Some people might name the coding scheme "race", but in doing so, I have found that people then confuse the variable name `race` with the name of the coding scheme.

5. The Variables Manager is a point-and-click alternative to many of the labeling tools illustrated in chapter 4. Reading that chapter will give you an understanding of the technical aspects of the Variables Manager. Chapter 4 explains labeling using Stata commands and technical terms, while this section uses the point-and-click interface of the Variables Manager and tries to avoid any such jargon.

can open the Variables Manager window from the main menu by clicking **Tools** and then **Variables Manager** (or by clicking on the **Variables Manager** icon from the toolbar).

The first variable, `var1`, should already be selected (if not, click on it). We will use the **Variable Properties** pane (at the right) to supply the information contained in the codebook. Focusing on the first variable, change **Name** to be `id` and **Label** to be `Unique identifier`. Click on the **Apply** button and the left pane reflects these changes, as illustrated in figure 2.2.

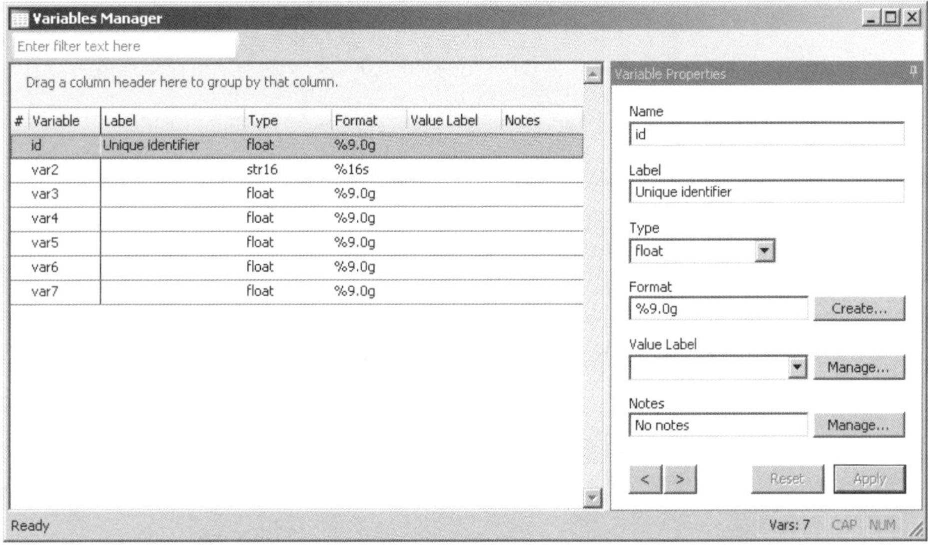

Figure 2.2. Variables Manager after labeling the first variable

You can then click on the second variable (`var2`) in the left pane and then change the Variable Properties for this variable, specifying **Name** as `stuname` and **Label** as `Name of student`. For **Type**, enter `str30` to specify that this variable is a string variable that can hold as many as 30 characters. Then change **Format** to `%30s` so that `stuname` will be displayed as a string with a width up to 30. Then click on **Apply**.

Now click on the third variable. The codebook information indicates that this variable is associated with the coding scheme `racelab`. Before doing anything (even before we specify the name or label for this variable), let's enter the information for the coding scheme `racelab`. We can do this by clicking on the **Manage...** button next to **Value Label**. Then, in the *Manage Value Labels* window, click on **Create Label**. For the **Label name**, enter `racelab`, and then enter a **Value** of `1` and a **Label** of `White`; then click on **Add**. Enter the values and labels for the three remaining race groups, clicking on **Add** after each group. At this point, the *Create Label* window will look like figure 2.3.

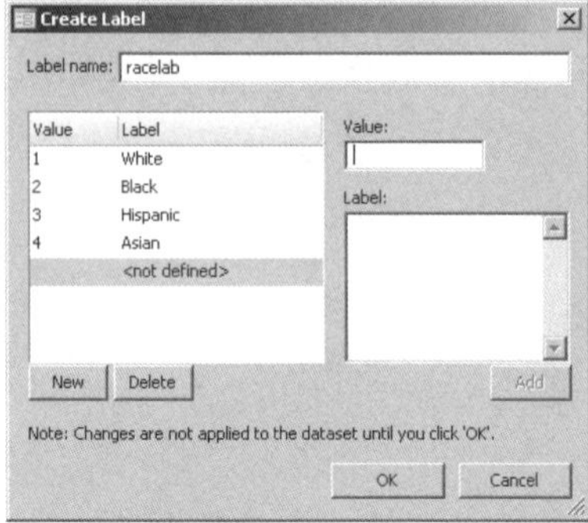

Figure 2.3. *Create Label* window showing value labels for **racelab**

You can then click on **OK** to save these changes, returning you to the *Manage Value Labels* window. While we are in the *Manage Value Labels* window, I recommend entering the coding scheme information for all other categorical variables. Referring to the codebook, we can enter the information for `yesnolab` by clicking on **Create Label**. The **Label name** is `yesnolab`, the **Value** is `1`, and the **Label** is `yes`; then click on **Add**. Then enter the **Value** of `0` and **Label** of `no`; click on **Add** and then click on **OK**. When you return to the *Manage Value Labels* window, you can click on the plus sign next to **racelab** and **yesnolab** to confirm the values and labels, as shown in figure 2.4.

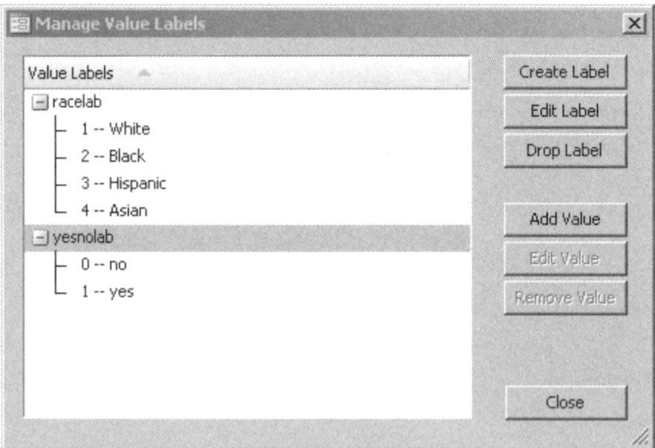

Figure 2.4. *Manage Value Labels* window showing value labels for **racelab** and **yesnolab**

In the *Manage Value Labels* window, you can now click on the **Close** button. We now have entered all the coding scheme information for `racelab` and `yesnolab`.

Now we are ready to enter the information for the variable `race` in the *Variable Properties* pane. For **Name**, enter `race` and for **Label**, enter `Race of student`. For the **Value Label**, choose **racelab** and click on **Apply**.

Labeling the variables `happy` and `glad` is much like `race`. Specify the **Name** and **Label**, and for the **Value Label**, choose `yesnolab` and click on **Apply**.

Now we have arrived at date of birth (`dob`). (Remember that we entered a temporary value of 1 for this variable and will fix it in step 3.) For **Name**, enter `dob` and for **Label**, enter `Date of birth`. To the right of **Format**, click on the **Create...** button. Under **Type of data**, choose `Daily` (because this is a date variable). The **Samples** box at the right shows examples of how this date variable can be displayed. You can choose whichever format you prefer; I will choose `April 07, 2009`. Then click on **OK** to close the *Create Format* window. Click on **Apply** to apply the changes for date of birth.

Now click on the last variable. In the *Variable Properties* pane, change **Name** to `wage` and **Label** to `Hourly wage`, and then click on **Apply**.

After I entered all the information for all the variables, my Variables Manager and Data Editor look like figure 2.5. The Data Editor shows the labeled values for `race`, `happy`, and `glad`.

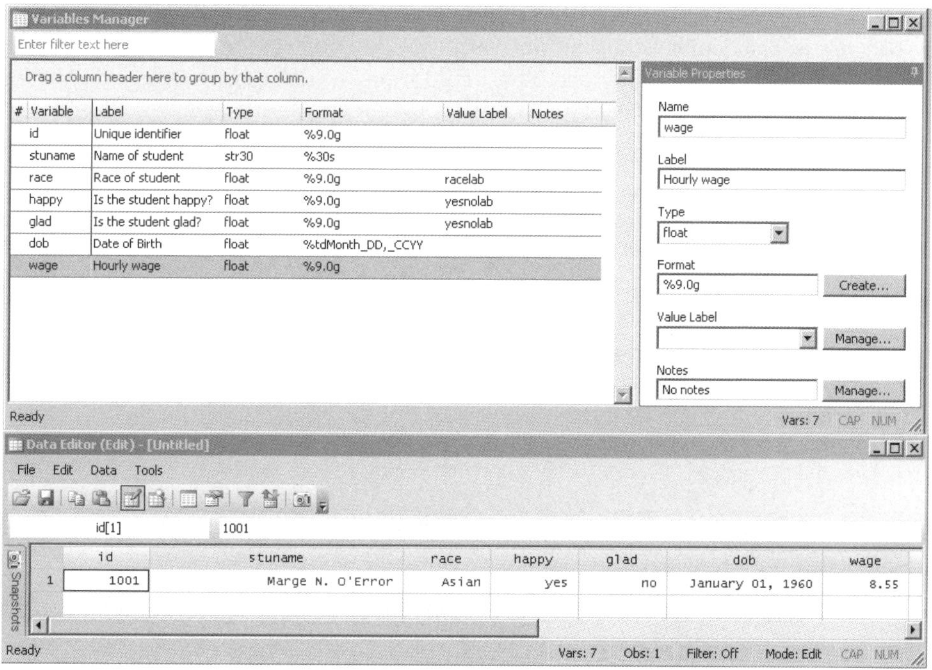

Figure 2.5. Variables Manager and Data Editor after step 2, labeling the variables

(Continued on next page)

> **Note! Red and blue values**
>
> In the Data Editor, the values for the student name are shown in red. That is to emphasize that `stuname` is a string variable. Note how the variables `race`, `happy`, and `glad` display the labeled value (e.g., Asian) in blue. The color blue signifies that the variable is numeric and the value being displayed is the labeled value. If you prefer to see the actual values, then you can go to the main menu and choose **Tools** and then **Value Labels** and then **Hide All Value Labels**. You can repeat this process to reshow the labeled values. One of the advantages of having the labeled values displayed is that it confirms the value entered for the original meaning of the variable (e.g., 4 is Asian) and gives feedback to the person entering the data if they enter an invalid value (e.g., if a value of 5 is entered for `race`, it sticks out as an unlabeled value).

Step 3: Fix date variables. In step 1, we entered a temporary value of 1 for `dob`. We did this because at that point Stata did not yet know that this was a date variable. In step 2, as part of the labeling of the variables, we informed Stata that `dob` is a date variable. Now we can properly enter the date of birth for the first observation.

In the Data Editor, click on the column for `dob`. At the right, you can select the format in which you would like to type dates into the Data Editor (see arrow in figure 2.6). The pull-down menu allows you to choose **DMY** (day month year), **MDY** (month day year), or **YMD** (year month day). I prefer and chose **MDY**. Say that this person was born on May 15, 1987. Having selected **MDY**, we can type in the date in a variety of ways, including `May 15, 1987`, `5 15 1987`, or `5/15/1987`. After entering the date of birth, my Data Editor appears like figure 2.6.

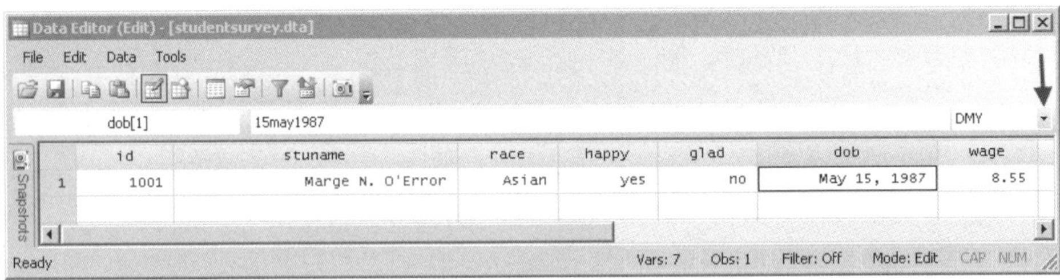

Figure 2.6. Data Editor after step 3, fixing the date variables

After investing all this effort, now is a great time to save these data. In the Data Editor, go to the main menu and click on **File** and then **Save As...**, and save the file as `studentsurvey`.

Leaving the Data Editor open, let's go to the Command window and issue the `list` and `describe` commands.

```
. list

     +----------------------------------------------------------------------------+
     |   id             stuname    race   happy   glad            dob    wage |
     |----------------------------------------------------------------------------|
  1. | 1001   Marge N. O'Error   Asian     yes     no   May 15, 1987    8.55 |
     +----------------------------------------------------------------------------+

. describe

Contains data from studentsurvey.dta
  obs:             1
 vars:             7                          15 Dec 2009 15:33
 size:            58 (99.9% of memory free)
-------------------------------------------------------------------------------
              storage  display     value
variable name   type   format      label      variable label
-------------------------------------------------------------------------------
id              float  %9.0g                  Unique identifier
stuname         str30  %30s                   Name of student
race            float  %9.0g       racelab    Race of student
happy           float  %9.0g       yesnolab   Is the student happy?
glad            float  %9.0g       yesnolab   Is the student glad?
dob             float  %td..                  Date of Birth
wage            float  %9.0g                  Hourly wage
-------------------------------------------------------------------------------
Sorted by:
```

The listing shows the labeled values for `race`, `happy`, and `glad`. The `dob` variable is displayed as a date according to the format assigned in step 2, and the value of `dob` shows the updated value we specified in step 3. The `describe` command shows the names, variable labels, and value labels specified in step 2. Now that we have successfully entered the first observation for this dataset and labeled this dataset, we are ready for the fourth step, entering the rest of the observations.

Step 4: Enter the data for rest of the observations. You can return to the Data Editor and continue entering data for the rest of the students in the survey. Note how when you enter a numeric value for `race`, `happy`, and `glad`, the number is instantly converted to its labeled value. Note how when you enter a value for `dob`, the value is instantly reformatted as a date based on the display format selected for `dob`. Once you are done entering the data for all the students, you can save the file and close the Data Editor and the Variables Manager.

You can then later retrieve the file by going to the main menu, selecting **File** and then **Open**, navigating to the folder in which you saved the file, and then choosing the file you saved. (You can, of course, also read the data with the `use` command.) You can then resume entering data using the `edit` command. Just like a spreadsheet, the data typed into the Editor is not saved until you save it. I recommend saving your data at least every 15–30 minutes so that if there is a computer glitch, you will lose a minimum amount of work.

For more information about entering data using the Stata Data Editor, see `help edit`.

2.11 Saving comma-separated and tab-separated files

Sometimes you may want to save a dataset as a comma-separated or tab-separated file. Such files can be read by a variety of other programs, including spreadsheets. The process of saving comma-separated and tab-separated files is similar, so both are illustrated in this section. Let's use a version of the dentists file named `dentlab`, which has value labels for the variables `fulltime` and `recom`.

```
. use dentlab
. list

     +-----------------------------------------------------------+
     |               name   years    fulltime              recom |
     |-----------------------------------------------------------|
  1. | Y. Don Uflossmore     7.25   part time          recommend |
  2. |    Olive Tu'Drill    10.25   full time          recommend |
  3. | Isaac O'Yerbreath    32.75   full time          recommend |
  4. |       Ruth Canaale      22   full time          recommend |
  5. |         Mike Avity     8.5   part time   do not recommend |
     +-----------------------------------------------------------+
```

The `outsheet` command is used below to write a tab-separated file called `dentists_tab.out` (the default extension is `.out`). Note that the labels for `fulltime` and `recom` are output, not their values.

```
. outsheet using dentists_tab
. type dentists_tab.out
name     years    fulltime        recom
"Y. Don Uflossmore"     7.25    "part time"     "recommend"
"Olive Tu'Drill"        10.25   "full time"     "recommend"
"Isaac O'Yerbreath"     32.75   "full time"     "recommend"
"Ruth Canaale"  22      "full time"     "recommend"
"Mike Avity"    8.5     "part time"     "do not recommend"
```

By adding the `comma` option, we can store this as a comma-separated file. We name this file `dentists_com.csv` (`.csv` for comma-separated values).

```
. outsheet using dentists_com.csv, comma
. type dentists_com.csv
name,years,fulltime,recom
"Y. Don Uflossmore",7.25,"part time","recommend"
"Olive Tu'Drill",10.25,"full time","recommend"
"Isaac O'Yerbreath",32.75,"full time","recommend"
"Ruth Canaale",22,"full time","recommend"
"Mike Avity",8.5,"part time","do not recommend"
```

To see the values of the variables, not the labels, we can add the `nolabel` option. We also add the `replace` option because we are overwriting the same file we wrote above.

```
. outsheet using dentists_com.csv, comma replace nolabel
. type dentists_com.csv
name,years,fulltime,recom
"Y. Don Uflossmore",7.25,0,1
"Olive Tu'Drill",10.25,1,1
"Isaac O'Yerbreath",32.75,1,1
"Ruth Canaale",22,1,1
"Mike Avity",8.5,0,0
```

If we want to suppress the quotes around the names of the dentists, we could add the `noquote` option. This is inadvisable if the names could have commas in them.

```
. outsheet using dentists_com.csv, comma replace nolabel noquote
. type dentists_com.csv
name,years,fulltime,recom
Y. Don Uflossmore,7.25,0,1
Olive Tu'Drill,10.25,1,1
Isaac O'Yerbreath,32.75,1,1
Ruth Canaale,22,1,1
Mike Avity,8.5,0,0
```

By default, the names of the variables are written in the first row of the raw data file. Sometimes you might want to omit the names from the raw data file. Specifying the `nonames` option omits the names from the first row of the data file.

```
. outsheet using dentists_com.csv, comma replace nolabel noquote nonames
. type dentists_com.csv
Y. Don Uflossmore,7.25,0,1
Olive Tu'Drill,10.25,1,1
Isaac O'Yerbreath,32.75,1,1
Ruth Canaale,22,1,1
Mike Avity,8.5,0,0
```

In these examples, the `replace`, `nolabel`, `noquote`, and `nonames` options were illustrated in the context of creating comma-separated files. These options work equally well when creating tab-separated files. For more information, see `help outsheet`.

2.12 Saving space-separated files

There may be times that you want to save a dataset from Stata as a space-separated file. Such files are sometimes referred to as free format files and can be read by a variety of programs. Let's see how to write a space-separated file using a version of the dentists file named `dentlab`, which has value labels for the variables `fulltime` and `recom`.

```
. use dentlab
. list

     +-------------------------------------------------------------------+
     |              name   years    fulltime                      recom |
     |-------------------------------------------------------------------|
  1. | Y. Don Uflossmore    7.25   part time                  recommend |
  2. |    Olive Tu'Drill   10.25   full time                  recommend |
  3. | Isaac O'Yerbreath   32.75   full time                  recommend |
  4. |      Ruth Canaale      22   full time                  recommend |
  5. |        Mike Avity     8.5   part time           do not recommend |
     +-------------------------------------------------------------------+
```

The outfile command shown below writes a space-separated file called dentists_space.raw (the default extension is .raw). Note how the labels for fulltime and recom are output, not their values.

```
. outfile using dentists_space
. type dentists_space.raw
"Y. Don Uflossmore"       7.25  "part time"  "recommend"
"Olive Tu'Drill"         10.25  "full time"  "recommend"
"Isaac O'Yerbreath"      32.75  "full time"  "recommend"
"Ruth Canaale"              22  "full time"  "recommend"
"Mike Avity"               8.5  "part time"  "do not recommend"
```

To display the values, not the labels, for fulltime and recom, we can add the nolabel option. We also add the replace option because we are overwriting the file from above.

```
. outfile using dentists_space, nolabel replace
. type dentists_space.raw
"Y. Don Uflossmore"       7.25           0                  1
"Olive Tu'Drill"         10.25           1                  1
"Isaac O'Yerbreath"      32.75           1                  1
"Ruth Canaale"              22           1                  1
"Mike Avity"               8.5           0                  0
```

Suppose we also have years2 (years squared) and years3 (years cubed) in the dataset. In this case, when we write the raw data file, it will exceed 80 columns, and Stata wraps the file to make sure that no lines exceed 80 columns, as shown below.

```
. outfile using dentists_space, nolabel replace
. type dentists_space.raw
"Y. Don Uflossmore"       7.25           0                  1    52.5625
 381.0781
"Olive Tu'Drill"         10.25           1                  1   105.0625
 1076.891
"Isaac O'Yerbreath"      32.75           1                  1   1072.563
 35126.42
"Ruth Canaale"              22           1                  1        484
    10648
"Mike Avity"               8.5           0                  0      72.25
  614.125
```

To avoid this wrapping, we could use the **wide** option. When using the **wide** option, one (and only one) line of raw data is written in the space-separated file for every observation in the working dataset.

```
. outfile using dentists_space, nolabel replace wide
```

Because it is hard to illustrate on the printed page, we will skip inspecting `dentists_space.raw`. But the inclusion of the **wide** option does make one line of raw data per observation. For more information on writing space-separated files, see `help outfile`.

2.13 Saving SAS XPORT files

This section shows how you can save a SAS XPORT file from within Stata. You might want to do this because you are submitting a data file to the FDA and want to provide it to them as a SAS XPORT file. You can also use this as a means of converting data from Stata to SAS (but as shown later in this section, a more direct way would be to read your Stata dataset directly into SAS). Let's illustrate how to save a SAS XPORT file using `dentists.dta`.

```
. use dentists
```

We can save this as a SAS XPORT file with the `fdasave` command.

```
. fdasave mydent
file mydent.xpt saved
```

The variables in SAS XPORT files cannot exceed 8 characters, while Stata variable names can be up to 32 characters. Suppose that the variable `fulltime` had been named `workfulltime`. Look at what happens when we try to save this as a SAS XPORT file:

```
. fdasave mydent2
the following variable(s) have names that must be changed to fit into .xpt
format:  (suggested renamings shown):

                    workfulltime -> WORKFULL

specify option rename to save .xpt file with suggested names
r(110);
```

Stata offers to rename the variable for us and shows how Stata will rename it, but we need to indicate our acceptance of this renaming by adding the `rename` option.

```
. fdasave mydent2, rename
the following variable(s) were renamed in the output file:
                    workfulltime -> WORKFULL
file mydent2.xpt saved
```

In the previous example, the dataset did not have any value labels associated with it. Consider `dentlab.dta` that has formats associated with the variables `fulltime` and `recom`.

```
. use dentlab
. list

     ┌──────────────────────────────────────────────────────────────────┐
     │               name   years    fulltime                  recom  │
     ├──────────────────────────────────────────────────────────────────┤
  1. │ Y. Don Uflossmore    7.25   part time              recommend  │
  2. │    Olive Tu'Drill   10.25   full time              recommend  │
  3. │ Isaac O'Yerbreath   32.75   full time              recommend  │
  4. │       Ruth Canaale      22   full time              recommend  │
  5. │         Mike Avity     8.5   part time       do not recommend  │
     └──────────────────────────────────────────────────────────────────┘
```

The process of saving this file is the same as saving a file that does not have formats. We use the **fdasave** command to save the data as a SAS XPORT file and a separate file containing the formats.

```
. fdasave mydentl
file mydentl.xpt saved
file formats.xpf saved
```

Now we have two files: `mydentl.xpt`, which is the dataset as a SAS XPORT format and `formats.xpf`, which contains the value labels (in SAS lingo, "formats"), also saved as a SAS XPORT file.

If your goal is to use your Stata dataset within SAS, then the most expedient way to do this is to read the Stata dataset directly within SAS using `PROC IMPORT`. The example below shows how you can use `PROC IMPORT` within SAS to read a Stata dataset named `c:\data\dentists.dta`.

```
PROC IMPORT OUT=dentists datafile="c:\data\dentists.dta";
RUN;
```

For further details, you can see your SAS documentation about `PROC IMPORT`.

As shown in this section, the **fdasave** command makes it easy to save SAS XPORT files. For more information, see `help fdasave`.

Tip! Transfers made easy

Do you frequently need to transfer data from one format to another? For example, you might need to read an SPSS data file, or an Access database, or save data as an Excel spreadsheet. The program Stat/Transfer (by Circle Systems) makes it easy to move data to and from many different statistical, database, and spreadsheet formats. You can purchase Stat/Transfer online via the Stata web site at http://www.stata.com/products/transfer.html. Before you buy, you can try a free demonstration version, available at http://www.stattransfer.com/downloads/.

3 Data cleaning

The Dirty Data Theorem states that "real world" data tends to come from bizarre and unspecifiable distributions of highly correlated variables and have unequal sample sizes, missing data points, non-independent observations, and an indeterminate number of inaccurately recorded values.

—Unknown

3.1 Introduction

Once you have read a dataset into Stata, it is tempting to immediately start analyzing the data. But the data are not ready to be analyzed until you have taken reasonable steps to clean them (you know the old saying: garbage in, garbage out). Even when you are given a dataset that is supposed to have been cleaned, it is useful to examine and check the variables. This chapter divides up the process of data cleaning into two components: checking data (searching for possible errors in the data) and correcting data (applying corrections based on confirmed errors).

I think that data checking has often been characterized as a repetitive and mindless task. It is true that some parts, like reviewing dozens of frequency tables for implausible values, can have this quality. But this is only a part of the data-checking process. Data checking is a thought-intensive process in which you imagine ways to test the integrity of your data beyond simple tabulations of frequencies. This chapter emphasizes this thought-intensive process, encouraging you to be creative in the ways that you check your data for implausible values.

I once worked on a research project involving parents and children. I was assured that the dataset was squeaky clean and ready for analysis, and everyone wanted the analyses to commence. But I wanted to take time to check the variables first. This was seen as obstructing progress on the project until I found some implausible values and implausible combinations of values in the data. Some of the parents were as young as 7 years old. There were many men who had given birth to children. There were children who were older than their mothers. Parents who were 14 years old were recorded as having graduated college, and so forth. After discovering these problems in the data, my data-cleaning efforts were recognized as a necessary step before the dataset was ready for analysis.

In looking at the types of problems that were found in this dataset, some problems concerned implausible values (e.g., parents who were 7 years old). However, many of the problems did not relate to absurd values for any particular variable but instead to absurd combinations of variables. It was not strange to have men and women in the dataset; it was not strange to have people who had given birth in the dataset; but it was strange to have men who had given birth in the dataset. Such problems were only discovered by checking variables against each other, which revealed impossible (or improbable) combinations of values.

The first data-cleaning strategy I will illustrate is double data entry (see section 3.2). This proactive method of cleaning identifies data-entry errors by entering the data twice and then comparing the two datasets. Conflicts between the two datasets indicate likely data-entry errors, which can be identified and corrected by referring to the original source of the data. If you are entering data you have collected yourself, this is an excellent way to combine data entry and data cleaning into one step.

After section 3.2, the following four sections cover four different data-checking methods. Section 3.3 covers techniques for checking individual variables for implausible values (e.g., parents who are 7 years old). The next three sections illustrate ways of checking variables against each other to discover absurd combinations of variables in your dataset. Section 3.4 covers checking categorical by categorical variables, such as gender against whether one has given birth. Section 3.5 covers checking categorical by continuous variables (e.g., checking age broken down by whether one is a college graduate). And section 3.6 covers checking continuous by continuous variables (e.g., mom's age compared to child's age).

Assuming that you have identified some problems in your data, section 3.7 shows some of the nuts and bolts of how to correct problems.

A completely different kind of problem is the presence of duplicate observations. Section 3.8 shows some of the Stata tools for identifying duplicates in your dataset and describes how to eliminate them.

The chapter concludes with section 3.9, some final thoughts on data cleaning.

I would like to mention that section 9.4 illustrates how the data-checking tasks described in this chapter can be automated. This can be an excellent method of letting the computer do the work for you and saving yourself the time of scrutinizing lengthy computer outputs searching for problems in your data.

3.2 Double data entry

An oil filter company had an advertisement in which a mechanic was rebuilding an engine and said that the rebuild could have been avoided if the engine's oil was changed at regular intervals. The mechanic said, "You can pay me now, or you can pay me later." The implication here was that you can either pay \$3 for an oil filter now, or later you can pay \$3,000 to rebuild the engine. I think this is a good analogy to the effort (price) of doing double data entry. Double data entry is like paying a small price now (expend extra effort to clean data as part of the data-entry process), rather than doing single data entry and paying a much bigger price later (check all the variables for errors and inconsistencies). If you are doing your own data entry for a questionnaire or other original data that you have collected, I highly recommend double data entry. This section describes how you can do double data entry using Stata.

As the name implies, the data are typed in twice, into two different datasets. The datasets are then compared against each other. Discrepancies between the datasets identify errors in the data entry that can be resolved by examining the original data (e.g., the original questionnaire form) to determine the correct value. The absence of discrepancies does not necessarily prove that the data are correct; it is possible that the data were entered in error in the exact same way both times. In most cases, the idea that an error occurred in the exact same way two times borders on the ridiculous, but this is not always the case. For example, suppose the data are entered based on a handwritten form and are being entered by the same person both times. A number 4 might be misread as a number 9 the first time, and then upon seeing that same written value, the same person might again be inclined to read it as a 9. This points to a couple of practices for double data entry that can reduce the chances of repeated data-entry errors.

The questionnaires should be reviewed before data entry to remove all possible ambiguities. The job of the person doing data entry is not to interpret but simply and solely to type in the data. Ambiguities in paper questionnaires can arise from poor handwriting, multiple answers being selected, stray marks, and so forth. One or more people should first review all the original forms to identify and resolve any ambiguities so there is no discretion left to the person doing the data entry. Even after this process has been completed, it still may be prudent to avoid having the same person do the double data entry because that person may have one interpretation of the data, while a second person may have a different interpretation.

The first step in the double data-entry process is to enter the data. I recommend doing so using the Stata Data Editor, as described in section 2.10. There are two exceptions (or additions) I would make to that process. First, even if you have an existing ID variable for your data, I highly recommend adding a sequential ID variable (1, 2, 3, etc.) that numbers each questionnaire form. This supplements (not replaces) any existing ID variable assigned to each questionnaire form. This sequential ID should be directly written onto the questionnaire forms before data entry begins. Second, enter the data for the first observation and label the data as described in steps 1, 2, and 3 in section 2.10. After this is completed, save two copies of the dataset. If two different people were doing the data entry, you would then give one person one of the datasets and the other person the other dataset. Each person would enter the data until completion.

Once the data entry is completed, the verification process begins by checking that each dataset has the same number of observations. If the two datasets have differing numbers of observations, the likely culprit is either an observation that was entered twice or an observation that was overlooked and not entered. Duplicates are found most easily by searching based on your ID variable. For example, if you have an ID variable named `studentid`, you can list duplicates on this variable with the command

```
. duplicates list studentid
```

If you expect to find many duplicates, you may prefer to use the `duplicates tag` command (as described in section 3.8, which goes into more detail about identifying duplicates).

Suppose you find that observation numbers 13 and 25 are duplicates of each other. You can first view the data with the Data Editor to see if there is one observation that you prefer to drop (perhaps one case was a duplicate because it was never fully entered). Say that you decide to drop observation 13. You can then type

```
. drop in 13
```

and that observation is removed from the dataset. You can repeat this process to eliminate all duplicated observations.

Finding an omitted observation is much trickier. This is why I recommended also including a sequential ID. Say that you named this variable seqid. You can identify any gaps in seqid with these commands:

```
. sort seqid
. list seqid if seqid != (seqid[_n-1] + 1) in 2/L
```

If all values are in sequence, the current value of seqid will be the same as the previous value of seqid with one added to it. This command lists the observations in which the current value of seqid is not equal to the previous value of seqid + 1 (see section 7.4 for more information on subscripting observations). Even if this command is a bit confusing, it will quickly list any observations where there are gaps in seqid. Once you identify gaps, the omitted observations can be added using the Stata Data Editor.

Once you have successfully eliminated any duplicate observations and filled in any gaps, your two datasets should have the same number of observations. Now you are ready to compare the datasets. The cf (compare files) command compares two Stata datasets observation by observation and shows any discrepancies it finds. Because the datasets are compared for each observation, the datasets should first be sorted so that the observations are in exactly the same order. Suppose your datasets are called survey1.dta and survey2.dta and that the observations are identified by studentid. I would first sort the two datasets on studentid and save them.

```
. use survey1, clear
. sort studentid
. save survey1, replace

. use survey2, clear
. sort studentid
. save survey2, replace
```

Now we are ready to compare the two datasets. I would start by making sure that the studentid variable is the same across the two datasets. We can do this with the cf (compare files) command, like this:

```
. use survey1, clear
. cf studentid using survey2, verbose
```

This first command uses survey1.dta. Then the cf command compares the values of the studentid variable in the current dataset with the values of studentid in survey2.dta. The value of studentid for the first observation from the current dataset is compared with the value of studentid for the first observation in survey2.dta. This process is repeated until all observations have been compared. Because we included the verbose option, the cf command will display a message for each observation where a discrepancy is found. This message shows the observation number with the discrepancy, followed by the value from the master dataset (e.g., survey1.dta) and the value from the using dataset (e.g., survey2.dta). You can note any discrepancies and use the Data Editor to view the datasets and resolve any discrepancies. If all values of studentid are the same, Stata will display the word "match" to indicate that all values match.

After resolving any discrepancies based on the ID variable, we are ready to examine all the variables for discrepancies using the cf command:

```
. use survey1, clear
. cf _all using survey2, all verbose
```

In contrast to the previous example, where we just compared the studentid variable, this command specifies that we want to compare all variables (indicated by _all) between survey1.dta and survey2.dta. Stata will list the name of each variable. If all the values for a variable match, it will display the word "match". Otherwise, for each discrepancy found for the variable, Stata will list the observation number along with the value from the master dataset (e.g., survey1.dta) and the value from the using dataset (e.g., survey2.dta).

You can then take this list of discrepancies and refer back to the original data forms to identify the correct values. You can select the dataset (among the two) that you feel is more accurate and apply the corrections based on the original data forms. Or if you wish to be completely fastidious, you can correct both datasets and then use the cf command to demonstrate that the two corrected datasets are completely equivalent. Either way, the list of discrepancies is your guide to making corrections to the data.

Once you have completed this process of double data entry, you can feel confident that your dataset has few, if any, data-entry errors. Of course, your dataset could still possibly have inconsistent or bizarre responses. For example, a man could have indicated that he has given birth to three children. Double data entry does not prevent people from giving bizarre or inconsistent answers, but it does help you to know that such answers are likely because of factors other than errors in data entry.

The rest of this chapter is probably most relevant for cases where double data entry was not used (but still could be useful for identifying odd responses or odd response patterns). However the data originated, the following sections discuss data cleaning (i.e., checking your data for problems and correcting problems that you identify).

3.3 Checking individual variables

This section will illustrate how you can check the values of individual variables searching for possible errors or problems in your data. This and the following sections will use a dataset called wws.dta (Working Women Survey), a purely hypothetical dataset with 2,246 observations. Let's first read in this dataset.

```
. use wws
(Working Women Survey)
```

Below we use the `describe` command to list the variables in the dataset.

```
. describe
Contains data from wws.dta
  obs:         2,246                          Working Women Survey
 vars:            30                          3 Jan 2010 00:42
 size:       172,942 (98.4% of memory free)   (_dta has notes)

              storage  display     value
variable name   type   format      label      variable label

idcode          int    %8.0g                  Unique ID
age             byte   %8.0g                  age in current year
race            byte   %8.0g                  race
married         byte   %8.0g                  married
collgrad        byte   %16.0g                 college graduate
south           byte   %8.0g                  lives in south
industry        byte   %23.0g                 industry
occupation      byte   %22.0g                 occupation
union           byte   %8.0g                  union worker
wage            float  %9.0g                  hourly wage
hours           byte   %8.0g                  usual hours worked
nevermarried    byte   %8.0g                  Woman never been married
yrschool        byte   %8.0g                  Years of school completed
metro           byte   %9.0g                  Does woman live in metro area?
ccity           byte   %8.0g                  Does woman live a city center?
currexp         float  %9.0g                  Years worked at current job
prevexp         float  %9.0g                  Years worked at previous job
everworked      float  %9.0g                  Has woman ever worked?
uniondues       float  %9.0g                  Union Dues paid last week
marriedyrs      float  %9.0g                  Years married (rounded to nearest
                                                year)
unempins        float  %9.0g                  Under/Unemployment insur.
                                                received last week
numkids         float  %9.0g                  Number of children
kidage1         float  %9.0g                  Age of first child
kidage2         float  %9.0g                  Age of second child
kidage3         float  %9.0g                  Age of third child
grade           byte   %8.0g                  current grade completed
grade4          byte   %9.0g                  4 level Current Grade Completed
wage2           float  %9.0g                  Wages, rounded to 2 digits
fwt             float  %9.0g                  Frequency weight
networth        float  %9.0g                  Net worth

Sorted by:
```

This dataset contains several demographic variables about these women and information about their work life. Let's start checking the variables, focusing on variables that are categorical. The easiest way to check categorical variables is by using the `tabulate` command (including the `missing` option to include missing values as part of the tabulation).

Below we check the variable `collgrad`, a dummy variable indicating whether the woman graduated from college. The `tabulate` command shows, as we would expect, that all values are either 0 or 1. We can also see that this variable has no missing values.

```
. tabulate collgrad, missing

    college |
   graduate |      Freq.     Percent        Cum.
------------+-----------------------------------
          0 |      1,713       76.27       76.27
          1 |        533       23.73      100.00
------------+-----------------------------------
      Total |      2,246      100.00
```

The variable **race** should range from 1 to 3, but below we see that there is one woman who is coded with a 4.

```
. tabulate race, missing

       race |      Freq.     Percent        Cum.
------------+-----------------------------------
          1 |      1,636       72.84       72.84
          2 |        583       25.96       98.80
          3 |         26        1.16       99.96
          4 |          1        0.04      100.00
------------+-----------------------------------
      Total |      2,246      100.00
```

We see that this erroneous value for **race** belongs to the woman with an **idcode** value of 543. We could then try and determine what her real value of **race** should be.

```
. list idcode race if race==4

       +---------------+
       | idcode   race |
       |---------------|
2013.  |    543      4 |
       +---------------+
```

The **summarize** command is useful for inspecting continuous variables. Below we inspect the variable **unempins**, the amount of underemployment/unemployment insurance the woman received last week. Suppose that prior knowledge tells us this variable should range from about 0 to 300 dollars. The results below are consistent with our expectations.

```
. summarize unempins

    Variable |     Obs        Mean    Std. Dev.       Min        Max
-------------+--------------------------------------------------------
    unempins |    2246    30.50401    73.16682          0        299
```

The **summarize** command (below) is used to inspect the variable **wage**, which contains the hourly wage for the previous week.

```
. summarize wage

    Variable |     Obs        Mean    Std. Dev.       Min        Max
-------------+--------------------------------------------------------
        wage |    2246    288.2885    9595.692          0     380000
```

The maximum for this was 380,000, which seems a little bit high, so we can add the **detail** option to get more information.

```
. summarize wage, detail
                         hourly wage
-------------------------------------------------------------
      Percentiles      Smallest
 1%     1.892108              0
 5%     2.801002       1.004952
10%     3.220612       1.032247       Obs                2246
25%     4.259257       1.151368       Sum of Wgt.        2246

50%     6.276297                      Mean           288.2885
                        Largest       Std. Dev.      9595.692
75%     9.661837       40.19808
90%     12.77777       40.74659       Variance       9.21e+07
95%     16.73912         250000       Skewness       35.45839
99%     38.70926         380000       Kurtosis       1297.042
```

It seems that the two largest values were entered erroneously; perhaps the respondent gave an annual wage instead of an hourly wage. Below we identify these women by showing observations with wages over 100,000. We could try to ascertain what their hourly wage should have been.

```
. list idcode wage if wage > 100000

      +-----------------+
      | idcode     wage |
      |-----------------|
 893. |   3145   380000 |
1241. |   2341   250000 |
      +-----------------+
```

Suppose that based on prior knowledge we know that the ages for this sample should range from 21 to about 50. We can use the `summarize` command to check this.

```
. summarize age
    Variable |       Obs        Mean    Std. Dev.       Min        Max
-------------+--------------------------------------------------------
         age |      2246    36.25111    5.437983         21         83
```

Seeing that the maximum age is 83, we can get more information using the `tabulate` command. But rather than tabulating all values, we create a tabulation of ages for those who are 45 and older.

```
. tabulate age if age >= 45
     age in |
    current |
       year |      Freq.     Percent        Cum.
------------+-----------------------------------
         45 |         45       90.00       90.00
         46 |          1        2.00       92.00
         47 |          1        2.00       94.00
         48 |          1        2.00       96.00
         54 |          1        2.00       98.00
         83 |          1        2.00      100.00
------------+-----------------------------------
      Total |         50      100.00
```

The ages of 54 and 83 seem suspicious. Below we list the `idcode` for these cases.

```
. list idcode age if age > 50

       +--------------+
       | idcode   age |
       |--------------|
2205.  |     80    54 |
2219.  |     51    83 |
       +--------------+
```

We could then look up the original data for these two observations to verify their values of age.

As shown in this section, the tabulate and summarize commands are useful for searching for out-of-range values in a dataset. Once an out-of-range value is found, the list command can be used to identify the actual observation with the out-of-range value so that we can further investigate the suspicious data. Section 3.7 illustrates how to correct values that are found to be in error.

The next section illustrates how to check two categorical variables against each other.

3.4 Checking categorical by categorical variables

This section shows how you can check the values of one categorical variable against another categorical variable. This draws upon a skill that you are probably familiar with and often use: creating cross-tabulations. wws.dta is again used for this section.

```
. use wws
(Working Women Survey)
```

To check categorical variables against each other, I look at my dataset and try to find implausible combinations among the categorical variables (in the hope that I do not find any). For example, consider the variables metro and ccity. The variable metro is a dummy variable that is 1 if the woman lives in a metropolitan area, while the dummy variable ccity measures whether the woman lives in a city center. If a woman lives in a city center, then she must live inside a metropolitan area. We tabulate the variables and see that this is indeed true in our data. So far, so good.

```
. tabulate metro ccity, missing

Does woman |
   live in |   Does woman live a
     metro |     city center?
     area? |         0          1 |     Total
-----------+----------------------+----------
         0 |       665          0 |       665 
         1 |       926        655 |     1,581 
-----------+----------------------+----------
     Total |     1,591        655 |     2,246 
```

Another way that we could have approached this would have been to count up the number of cases where a woman lived in a city center but not in a metropolitan area and to have verified that this count was 0. This is illustrated below. The & represents *and* and the == represents *is equal to* (see section A.6 for more details about logical expressions in Stata).

```
. count if metro == 0 & ccity == 1
  0
```

Consider the variables **married** and **nevermarried**. Although it seems obvious, if you are currently married, your value for **nevermarried** should always be 0. When we tabulate these variables, we see that there are two cases that fail this test.

```
. tabulate married nevermarried
```

married	Woman never been married 0	1	Total
0	570	234	804
1	1,440	2	1,442
Total	2,010	236	2,246

Rather than using the **tabulate** command, we can use the **count** command to count up the number of problematic cases, as shown below.

```
. count if married==1 & nevermarried==1
  2
```

Below we find the cases that fail this test by listing the cases where the person is married and has never been married. We see that women with id values of 22 and 1,758 have this problematic data pattern. We could then investigate these two cases to try to identify which variables may have been entered incorrectly.

```
. list idcode married nevermarried if married==1 & nevermarried==1, abb(20)
```

	idcode	married	nevermarried
1523.	1758	1	1
2231.	22	1	1

Let's consider one more example by checking the variable collgrad (did you graduate college?) against yrschool (how many years have you been in school?). The **table** command is used here because it produces more concise output than the **tabulate** command.

```
. table collgrad yrschool
```

college graduate	Years of school completed 8	9	10	11	12	13	14	15	16	17	18
0	69	55	84	123	943	174	180	81			
1	1					2	7	11	252	106	154

Among the college graduates, two women reported 13 years of school and seven reported 14 years of school. These women may have skipped one or two grades or graduated high school early, so these values might merit some further investigation but they are not completely implausible. However, the woman with 8 years of education who graduated college seems to be the greatest genius or has an error on one of these variables.

Cross-tabulations using the `tabulate` or the `table` command are useful for checking categorical variables against each other. The next section illustrates how to check a categorical variable against a continuous variable.

3.5 Checking categorical by continuous variables

In the previous section on checking categorical by categorical variables, cross-tabulations of the two categorical variables were used to identify suspicious combinations of values. When checking continuous variables by categorical variables, cross-tabulations are less practical because the continuous variable likely contains many values. Instead, we will focus on creating summary statistics for the continuous variable broken down by the categorical variable. Let's explore this with `wws.dta`.

```
. use wws
(Working Women Survey)
```

This dataset has a categorical (dummy) variable named `union` that is 1 if the woman belongs to a union (and 0 otherwise). There is also a variable called `uniondues`, which is the amount of union dues paid by the woman in the last week. If a woman is in a union, they may not require union dues; however, if a woman is not in a union, it would not make sense for her to be paying union dues. One way to check for problems here is by using the `summarize` command to get summary statistics on `uniondues` for women who are not in a union. For the women who are not in a union, I expect that the mean value of `uniondues` would be 0. If the value is more than 0, then it suggests that one or more nonunion women paid union dues. As the result below shows, one or more nonunion women paid dues.

```
. summarize uniondues if union==0
    Variable |       Obs        Mean    Std. Dev.       Min        Max
-------------+--------------------------------------------------------
   uniondues |      1413     .094126    1.502237          0         27
```

If we add `bysort union:` before the `summarize` command, we get summary statistics for `uniondues` by each level of `union` (see section 7.2 for more information on using the `by` prefix before commands). This is another way of showing that some nonunion women paid union dues.

```
. bysort union: summarize uniondues

-------------------------------------------------------------------------------
-> union = 0
    Variable |       Obs        Mean    Std. Dev.       Min        Max
-------------+--------------------------------------------------------
   uniondues |      1413     .094126    1.502237          0         27

-------------------------------------------------------------------------------
-> union = 1
    Variable |       Obs        Mean    Std. Dev.       Min        Max
-------------+--------------------------------------------------------
   uniondues |       461    14.65944    8.707759          0         29

-------------------------------------------------------------------------------
-> union = .
    Variable |       Obs        Mean    Std. Dev.       Min        Max
-------------+--------------------------------------------------------
   uniondues |       368    15.41304    8.815582          0         29
```

We can obtain the same output in a more concise fashion by using the `tabstat` command, as shown below.

```
. tabstat uniondues, by(union) statistics(n mean sd min max) missing
Summary for variables: uniondues
     by categories of: union (union worker)

   union |         N      mean        sd       min       max
---------+--------------------------------------------------
       0 |      1413   .094126  1.502237         0        27
       1 |       461  14.65944  8.707759         0        29
       . |       368  15.41304  8.815582         0        29
---------+--------------------------------------------------
   Total |      2242  5.603479  9.029045         0        29
------------------------------------------------------------
```

However we obtain the output, we see that there is at least one woman who was not in a union who paid some union dues. Let's use the `recode` command to create a dummy variable named `paysdues` that is 0 if a woman paid no union dues and 1 if she paid some dues (see section 5.5 for more on recoding variables).

```
. recode uniondues (0=0) (1/max=1), generate(paysdues)
(784 differences between uniondues and paysdues)
```

We can now create a table of `union` by `paysdues` to see the cross-tabulation of union membership by whether one paid union dues.

(Continued on next page)

```
. tabulate union paysdues, missing

           | RECODE of uniondues (Union Dues
     union |        paid last week)
    worker |         0          1          . |     Total
-----------+---------------------------------+----------
         0 |     1,407          6          4 |     1,417 
         1 |        17        444          0 |       461 
         . |         7        361          0 |       368 
-----------+---------------------------------+----------
     Total |     1,431        811          4 |     2,246 
```

The **tabulate** command shows that six nonunion women paid union dues. We can display those cases, as shown below.

```
. list idcode union uniondues if union==0 & (uniondues > 0) &
> ! missing(uniondues), abb(20)

      +----------------------------+
      | idcode   union   uniondues |
      |----------------------------|
   7. |   3905       0          10 |
 140. |   1411       0          27 |
 283. |   3464       0          17 |
 369. |   2541       0          27 |
 540. |    345       0          26 |
      |----------------------------|
1200. |   3848       0          26 |
      +----------------------------+
```

We included **! missing(uniondues)** as part of our **if** qualifier that excluded missing values from the display (see section A.10 for more about missing values). We could investigate further, trying to determine the appropriate values for these two variables for these six observations.

Let's turn to the variables **married** (coded 0 if not married, 1 if married) and **marriedyrs** (how many years you have been married, rounded to the nearest year). If one has been married for less than half a year, then **marriedyrs** would be coded 0. Let's use the **tabstat** command to get summary statistics for **marriedyrs** for each level of **married** and see if these results make sense.

```
. tabstat marriedyrs, by(married) statistics(n mean sd min max) missing
Summary for variables: marriedyrs
     by categories of: married (married)

 married |         N      mean        sd       min       max
---------+--------------------------------------------------
       0 |       804         0         0         0         0
       1 |      1442  5.540915  3.552138         0        11
---------+--------------------------------------------------
   Total |      2246  3.557435  3.893349         0        11
------------------------------------------------------------
```

As we would hope, the 804 women who were not married all have the appropriate value for **marriedyrs**: they are all 0. Among those who are married, some may have been married for less than six months and thus also have a value of 0. These two variables appear to be consistent with each other.

Let's check the variable **everworked** (0 if never worked, 1 if worked) against the variables **currexp** (time at current job) and **prevexp** (time at previous job). If one had never worked, the current and previous work experience should be 0. We check this below for current experience and find this to be the case.

```
. tabstat currexp, by(everworked) statistics(n mean sd min max) missing
Summary for variables: currexp
     by categories of: everworked (Has woman ever worked?)
```

everworked	N	mean	sd	min	max
0	60	0	0	0	0
1	2171	5.328881	5.042181	0	26
Total	2231	5.185567	5.048073	0	26

Also as we would expect, those who never worked have no previous work experience.

```
. tabstat prevexp, by(everworked) statistics(n mean sd min max) missing
Summary for variables: prevexp
     by categories of: everworked (Has woman ever worked?)
```

everworked	N	mean	sd	min	max
0	60	0	0	0	0
1	2171	6.248733	4.424465	0	25
Total	2231	6.080681	4.480124	0	25

Let's check the **everworked** variable against the woman's total work experience. To do this, we can create a variable called **totexp**, which is a woman's total work experience, and then check that against **everworked**. As we see below, if a woman has never worked, her total work experience is always 0, and if the woman has worked, her minimum total work experience is 1. This is exactly as we would expect.

```
. generate totexp = currexp + prevexp
(15 missing values generated)
. tabstat totexp, by(everworked) statistics(n mean sd min max) missing
Summary for variables: totexp
     by categories of: everworked (Has woman ever worked?)
```

everworked	N	mean	sd	min	max
0	60	0	0	0	0
1	2171	11.57761	4.552392	1	29
Total	2231	11.26625	4.865816	0	29

This section illustrated how we can check continuous variables against categorical variables using the `bysort` prefix with the `summarize` command or using the `tabstat` command. We can also recode the continuous variables into categorical variables and then use cross-tabulations for checking the categorical variable against the recoded version of the continuous variable. The next section illustrates how to check two continuous variables.

3.6 Checking continuous by continuous variables

This section explores how we can check one continuous variable against another continuous variable. Like the previous sections, this section uses `wws.dta`.

```
. use wws
(Working Women Survey)
```

Consider the variables `hours` (hours worked last week) and `unempins` (amount of under/unemployment insurance received last week). Suppose that only those who worked 30 or fewer hours per week would be eligible for under/unemployment insurance. If so, all values of `unempins` should be 0 when a woman works over 30 hours in a week. The `summarize` command below checks this by showing descriptive statistics for `unempins` for those who worked over 30 hours in a week and did not have a missing value for their work hours (see section A.10 for more about missing values). If all women who worked more than 30 hours did not get under/unemployment insurance, the mean and maximum for `unemins` in the output below would be 0. But as the results show, these values are not all 0, so at least one woman received under/unemployment insurance payments when working over 30 hours.

```
. summarize unempins if hours > 30 & ! missing(hours)
    Variable |       Obs        Mean    Std. Dev.       Min        Max
-------------+--------------------------------------------------------
    unempins |      1800    1.333333    16.04617          0        287
```

Although the previous `summarize` command shows that there is at least one woman who received unemployment insurance though she worked more than 30 hours, it does not show us how many women had such a pattern of data. We can use the `count` command to count up the number of women who worked over 30 hours and received under/unemployment insurance. This reveals that 19 women fit this criteria.

```
. count if (hours>30) & !missing(hours) & (unempins>0) & !missing(unempins)
  19
```

We can use the `list` command to identify the observations with these conflicting values so that we can investigate further. The output is omitted to save space.

```
. list idcode hours unempins if (hours>30) & ! missing(hours) & (unempins>0)
> & ! missing(unempins)
  (output omitted)
```

Let's say that we wanted to check the variable **age** against the amount of time married, **marriedyrs**. One way to compare these variables against each other is to create a new variable that is the age when the woman was married. This new variable can then be inspected for anomalous values. Below the **generate** command creates **agewhenmarried**.

```
. generate agewhenmarried = age - marriedyrs
```

We can then use the **tabulate** command to look for worrisome values in the new **agewhenmarried** variable. For the sake of space, we restrict this tabulation to values less than 18. We see a handful of values that might merit further investigation, such as the woman who was married when she was 13 years old.

```
. tab agewhenmarried if agewhenmarried < 18

agewhenmarr |
        ied |      Freq.     Percent        Cum.
------------+-----------------------------------
         13 |          1        2.38        2.38
         14 |          4        9.52       11.90
         15 |         11       26.19       38.10
         16 |          8       19.05       57.14
         17 |         18       42.86      100.00
------------+-----------------------------------
      Total |         42      100.00
```

We can use the same strategy to check the woman's age against her total work experience. We can create a variable, **agewhenstartwork**, that is the woman's age minus her previous plus current work experience. Like the previous example, we can then **tabulate** these values and restrict it to values less than 18 to save space. This reveals three cases where the implied age the women started working was at age 8, 9, and 12. These cases seem to merit further investigation.

```
. generate agewhenstartwork = age - (prevexp + currexp)
(15 missing values generated)

. tab agewhenstartwork if agewhenstartwork < 18

agewhenstar |
       twork |      Freq.     Percent        Cum.
------------+-----------------------------------
          8 |          1        0.50        0.50
          9 |          1        0.50        1.00
         12 |          1        0.50        1.49
         14 |         20        9.95       11.44
         15 |         44       21.89       33.33
         16 |         50       24.88       58.21
         17 |         84       41.79      100.00
------------+-----------------------------------
      Total |        201      100.00
```

The dataset has a variable, **numkids**, that contains the number of children the woman has as well as the ages of the first, second, and third child stored in **kidage1**, **kidage2**, and **kidage3**. For the women with three kids, let's compare the ages of the second and third child using the `table` command below. As we would expect, the third child is never older than the second child.

```
. table kidage2 kidage3 if numkids == 3
```

Age of second child	Age of third child 0	1	2	3	4	5	6	7
0	12							
1	10	9						
2	11	8	10					
3	10	12	6	8				
4	10	12	10	7	5			
5	12	11	9	3	6	8		
6	9	8	10	6	5	6	6	
7	7	6	7	9	4	14	12	6
8		5	11	7	6	14	6	11
9			8	13	10	7	12	9
10				15	3	10	6	12
11					9	8	3	13
12						16	9	6
13							11	5
14								8

Although not as concrete, you can also use the `count` command to verify this. Below we count the number of times the age of the third child is greater than the age of the second child when there are three children, being sure to exclude observations where `kidage3` is missing. As we would expect based on the results of the `table` command above, there are no such children.

```
. count if (kidage3 > kidage2) & (numkids == 3) & ! missing(kidage3)
  0
```

Likewise, we count the number of second children whose ages are greater than the age of the first child if the woman has two or more children, being sure to exclude observations where `kidage2` is missing. As we would hope, there are no such cases.

```
. count if (kidage2 > kidage1) & (numkids >= 2) & ! missing(kidage2)
  0
```

Another check we might perform is comparing the age of the woman with the age of her oldest child to determine the woman's age when she had her first child. We can create `agewhenfirstkid`, which is the age of the woman when she gave birth to her first child. We then tabulate `agewhenfirstkid`. This reveals either cases that need further investigation or fodder for the tabloids about the girl who gave birth at age 3.

```
. generate agewhenfirstkid = age - kidage1
(563 missing values generated)

. tabulate agewhenfirstkid if agewhenfirstkid < 18

agewhenfirs |
       tkid |      Freq.     Percent        Cum.
------------+-----------------------------------
          3 |          1        0.51        0.51
          5 |          2        1.01        1.52
          7 |          2        1.01        2.53
          8 |          5        2.53        5.05
          9 |          8        4.04        9.09
         10 |          7        3.54       12.63
         11 |         10        5.05       17.68
         12 |         10        5.05       22.73
         13 |         20       10.10       32.83
         14 |         30       15.15       47.98
         15 |         27       13.64       61.62
         16 |         39       19.70       81.31
         17 |         37       18.69      100.00
------------+-----------------------------------
      Total |        198      100.00
```

Checking continuous variables against each other can be challenging. It sometimes takes a little extra work and some creativity to come up with ways to check one continuous variable against another. But such checks can reveal inconsistencies between the variables that would not be revealed by checking each variable individually.

The next section illustrates some methods you can use for correcting problems found in your data.

3.7 Correcting errors in data

The previous sections have shown how to check for problems in your data. Now let's consider strategies you might use to correct problems. This section assumes that you entered the data yourself and that you have access to the original data, or that you have some relationship with the people who provided you with the data where they could investigate anomalies in the data. In either case, providing clear information about the problem is key. Below are some examples of problems and how you might document them.

In section 3.3, we saw that `race` was supposed to have the values 1, 2, or 3, but there was one case where `race` was 4. We not only want to document that we found a case where `race` was 4 but also note the `idcode` and a couple of other identifying demographic variables for this case. We can do this with a simple `list` command.

```
. * woman has race coded 4
. use wws, clear
(Working Women Survey)
. list idcode age yrschool race wage if race==4

       +-------------------------------------------+
       | idcode   age   yrschool   race       wage |
       |-------------------------------------------|
2013.  |    543    39          8      4   4.428341 |
       +-------------------------------------------+
```

In section 3.3, we also saw two cases where the values for hourly income seemed outrageously high. The same strategy we just employed can be used to document those possibly problematic cases.

```
. * hourly income seems too high
. list idcode age yrschool race wage if wage > 50

       +-----------------------------------------+
       | idcode   age   yrschool   race     wage |
       |-----------------------------------------|
 893.  |   3145    36         12      2   380000 |
1241.  |   2341    29         16      2   250000 |
       +-----------------------------------------+
```

In sections 3.4–3.6, we uncovered problems by checking variables against each other. In these kinds of cases, we did not find values that were intrinsically problematic, but we did find conflicts in the values among two or more variables. In these cases, documenting the problem involves noting how the values between the variables do not make sense. For example, in section 3.4 there was a woman who graduated college who had reported only eight years of school completed. This can be documented using a cross-tabulation:

```
. * some conflicts between college graduate and years of school
. table collgrad yrschool

----------------------------------------------------------------------------
college   |                      Years of school completed
graduate  |     8      9     10     11     12     13     14     15     16     17     18
----------+-----------------------------------------------------------------
        0 |    69     55     84    123    943    174    180     81
        1 |     1                                  2      7     11    252    106    154
----------------------------------------------------------------------------
```

This documentation can be supplemented with a listing showing more information about the potentially problematic cases:

```
. * college grad with 8 years of school completed, seems like a problem.
. list idcode collgrad yrschool if yrschool==8 & collgrad==1

       +------------------------------+
       | idcode   collgrad   yrschool |
       |------------------------------|
2198.  |    107          1          8 |
       +------------------------------+

. * college grad with 13, 14, 15 years of school completed, is this a problem?
. list idcode collgrad yrschool if inlist(yrschool,13,14,15) & collgrad
  (output omitted)
```

One important part about this process is distinguishing between clearly incongruent values and ones that simply merit some further investigation. I try to prioritize problems, creating terminology that distinguishes clear conflicts (e.g., the college grad with eight years of education) from observations that merely might be worth looking into. For example, a college grad with 13 years of education could be a gifted woman who skipped several years of school.

Sometimes resources for data checking are not infinite. It may be important to prioritize efforts to focus on data values that are likely to change the results of the analysis, such as the women with hourly income that exceeded $300 an hour. If there is only a finite amount of time for investigating problems, imagine the analyses you will be doing and imagine the impact various kinds of mistakes will have on the data. Try to prioritize efforts on mistakes that will be most influential on your analysis, such as values that are most extreme or conflicts that involve large numbers of cases, even if the magnitude of the error is not as large.

Once you discover corrections that need to be made to the data, it might be tempting to open up the Stata Data Editor and just start typing in corrections, but I highly recommend against this strategy for two reasons: it does not document the changes that you made to the data in a systematic way and it does not integrate into a data-checking strategy. Once you mend a problem in the data, you want to then use the same procedures that uncovered the problem to verify that you have indeed remedied the problem.

Instead of correcting problems using the Data Editor, I recommend using the `replace` command combined with an `if` qualifier that uniquely identifies the observations to be mended. For example, consider the problem with `race` described earlier in this section, where one value was coded as a 4. After investigation, we learned that the observation in error had a unique `idcode` of 543 and that the value of `race` should have been 1. You can change the value of `race` to 1 for `idcode` 543 like this:

```
. * correcting idcode 543 where race of 4 should have been 1
. replace race = 1 if idcode == 543
(1 real change made)

. tab race

       race |      Freq.     Percent        Cum.
------------+-----------------------------------
          1 |      1,637       72.89       72.89
          2 |        583       25.96       98.84
          3 |         26        1.16      100.00
------------+-----------------------------------
      Total |      2,246      100.00
```

Note that the replacement was based on `if idcode == 543` and not `if race == 4`. When corrections are identified based on an observation, then the replacements should also be based on a variable that uniquely identifies the observation (e.g., `idcode`).

It would be useful to add a note to the dataset to indicate that this value was corrected. We can do so by using the `note` command, as shown below. You can see more about adding notes in section 4.7.

```
. note race: race changed to 1 (from 4) for idcode 543
```

Likewise, we might be told that case 107 with the woman who appeared to be a college graduate with only eight years of school was not a college graduate; that was a typo. We make this correction and document it below.

```
. replace collgrad = 0 if idcode == 107
(1 real change made)
. note collgrad: collgrad changed from 1 to 0 for idcode 107
```

After applying this correction, the cross-tabulation of `collgrad` by `yrschool` looks okay.

```
. table collgrad yrschool
```

college graduate	Years of school completed 8	9	10	11	12	13	14	15	16	17	18
0	70	55	84	123	943	174	180	81			
1						2	7	11	252	106	154

In section 3.3, we saw a couple of women whose ages were higher than expected (over 50).

```
. list idcode age if age > 50
```

	idcode	age
2205.	80	54
2219.	51	83

After further inquiries, we found that the digits in these numbers were transposed. We can correct them and include notes of the corrections, as shown below.

```
. replace age = 38 if idcode == 51
(1 real change made)
. replace age = 45 if idcode == 80
(1 real change made)
. note age: the value of 83 was corrected to be 38 for idcode 51
. note age: the value of 54 was corrected to be 45 for idcode 80
```

Having corrected the values, we again list the women who are over 50 years old.

```
. list idcode age if age > 50
```

As we would hope, this output is now empty because there are no such women. We can see the notes of all the corrections that we made by using the `notes` command:

```
. notes
_dta:
  1.  This is a hypothetical dataset and should not be used for analysis
      purposes
age:
  1.  the value of 83 was corrected to be 38 for idcode 51
  2.  the value of 54 was corrected to be 45 for idcode 80
race:
  1.  race changed to 1 (from 4) for idcode 543
collgrad:
  1.  collgrad changed from 1 to 0 for idcode 107
```

After we made a correction to the data, we checked it again to ensure that the correction did the trick. In other words, data cleaning and data correcting are, ideally, an integrated process. To this end, this process is best done as part of a Stata do-file, where the commands for checking, correcting, and rechecking each variable are saved and can easily be executed. Section 9.3 provides details about how to create and use Stata do-files. Further, section 9.4 illustrates how you can automate the process of data checking.

3.8 Identifying duplicates

This section shows how you can identify duplicates in your dataset. Duplicates can arise for a variety of reasons, including the same observation being entered twice during data entry. Because finding and eliminating duplicate observations is a common problem, Stata has an entire set of commands for identifying, describing, and eliminating duplicates. This section illustrates the use of these commands, first using a tiny dataset, and then using a more realistic dataset. First, let's consider a variation of **dentists.dta** called **dentists_dups.dta**. Looking at a listing of the observations in this dataset shows that there are duplicate observations.

```
. use dentists_dups
. list

     +----------------------------------------------+
     |              name   years   fulltime   recom |
     |----------------------------------------------|
  1. |    Olive Tu´Drill   10.25          1       1 |
  2. |      Ruth Canaale      22          1       1 |
  3. |      Ruth Canaale      22          1       1 |
  4. |        Mike Avity     8.5          0       0 |
  5. |        Mary Smith       3          1       1 |
     |----------------------------------------------|
  6. |        Mike Avity     8.5          0       0 |
  7. | Y. Don Uflossmore    7.25          0       1 |
  8. |        Mike Avity     8.5          0       0 |
  9. |        Mary Smith      27          0       0 |
 10. | Isaac O´Yerbreath   32.75          1       1 |
     |----------------------------------------------|
 11. |    Olive Tu´Drill   10.25          1       1 |
     +----------------------------------------------+
```

We can use the `duplicates list` command to list the duplicates contained in this dataset.

```
. duplicates list
Duplicates in terms of all variables

  group:   obs:            name    years   fulltime   recom

      1      4      Mike Avity      8.5          0       0
      1      6      Mike Avity      8.5          0       0
      1      8      Mike Avity      8.5          0       0
      2      1   Olive Tu´Drill   10.25          1       1
      2     11   Olive Tu´Drill   10.25          1       1

      3      2    Ruth Canaale       22          1       1
      3      3    Ruth Canaale       22          1       1
```

The above command shows every observation that contains a duplicate. For example, three observations are shown for the dentist Mike Avity.

Rather than listing every duplicate, we can list one instance of each duplicate by using the `duplicates examples` command. The column labeled `#` shows the total number of duplicates (e.g., Mike Avity has three duplicate observations).

```
. duplicates examples
Duplicates in terms of all variables

  group:   #   e.g. obs             name    years   fulltime   recom

      1    3          4       Mike Avity      8.5          0       0
      2    2          1   Olive Tu´Drill    10.25          1       1
      3    2          2     Ruth Canaale       22          1       1
```

The `duplicates report` command creates a report (like the `tabulate` command) that tabulates the number of copies for each observation.

```
. duplicates report
Duplicates in terms of all variables

  copies   observations       surplus

       1              4             0
       2              4             2
       3              3             2
```

The output above shows that there are four observations in the dataset that are unique (i.e., have only one copy). There are four observations in which there are two copies of the observation. These correspond to the observations for Olive and for Ruth, each of which had two copies. The report also shows that there are three observations that have three copies; these are the three observations for Mike.

This report shows useful information about the prevalence of duplicates in the dataset, but it does not identify the duplicates. This is where the **duplicates tag** command is useful. This command creates a variable that indicates for each observation how many duplicates that observation has. We use this command to create the variable **dup**.

```
. duplicates tag, generate(dup)
Duplicates in terms of all variables
```

The listing below shows the number of duplicates (**dup**) for each observation.

```
. list, sep(0)

     +----------------------------------------------------------+
     |               name   years   fulltime   recom   dup |
     |----------------------------------------------------------|
  1. |     Olive Tu'Drill   10.25          1       1     1 |
  2. |       Ruth Canaale      22          1       1     1 |
  3. |       Ruth Canaale      22          1       1     1 |
  4. |         Mike Avity     8.5          0       0     2 |
  5. |         Mary Smith       3          1       1     0 |
  6. |         Mike Avity     8.5          0       0     2 |
  7. |  Y. Don Uflossmore    7.25          0       1     0 |
  8. |         Mike Avity     8.5          0       0     2 |
  9. |         Mary Smith      27          0       0     0 |
 10. | Isaac O'Yerbreath   32.75          1       1     0 |
 11. |     Olive Tu'Drill   10.25          1       1     1 |
     +----------------------------------------------------------+
```

To make this output easier to follow, let's sort the data by **name** and **years** and then list the observations, separating them into groups based on **name** and **years**.

```
. sort name years
. list, sepby(name years)

     +----------------------------------------------------------+
     |               name   years   fulltime   recom   dup |
     |----------------------------------------------------------|
  1. | Isaac O'Yerbreath   32.75          1       1     0 |
     |----------------------------------------------------------|
  2. |         Mary Smith       3          1       1     0 |
     |----------------------------------------------------------|
  3. |         Mary Smith      27          0       0     0 |
     |----------------------------------------------------------|
  4. |         Mike Avity     8.5          0       0     2 |
  5. |         Mike Avity     8.5          0       0     2 |
  6. |         Mike Avity     8.5          0       0     2 |
     |----------------------------------------------------------|
  7. |     Olive Tu'Drill   10.25          1       1     1 |
  8. |     Olive Tu'Drill   10.25          1       1     1 |
     |----------------------------------------------------------|
  9. |       Ruth Canaale      22          1       1     1 |
 10. |       Ruth Canaale      22          1       1     1 |
     |----------------------------------------------------------|
 11. |  Y. Don Uflossmore    7.25          0       1     0 |
     +----------------------------------------------------------+
```

Now it is easier to understand the dup variable. For the observations that were unique (such as Isaac or Y. Don), the value of dup is 0. The value of dup is 0 for Mary Smith because, even though these two dentists share the same name, they are not duplicate observations. (For example, they have a different number of years of work experience.) The observations for Olive and Ruth are identified as having a value of 1 for dup because they each have one duplicate observation. And Mike has a value of 2 for dup because he has two duplicate observations.

As you can see, duplicate observations are characterized by having a value of 1 or more for the dup variable. We can use this to list just the observations that are duplicates, as shown below.

```
. list if dup > 0

     +--------------------------------------------------------+
     |           name   years   fulltime   recom   dup |
     |--------------------------------------------------------|
  4. |     Mike Avity     8.5          0       0     2 |
  5. |     Mike Avity     8.5          0       0     2 |
  6. |     Mike Avity     8.5          0       0     2 |
  7. | Olive Tu'Drill   10.25          1       1     1 |
  8. | Olive Tu'Drill   10.25          1       1     1 |
     |--------------------------------------------------------|
  9. |   Ruth Canaale      22          1       1     1 |
 10. |   Ruth Canaale      22          1       1     1 |
     +--------------------------------------------------------+
```

If there were many variables in the dataset, you might prefer to view the duplicate observations in the Data Editor by using the browse command.[1]

```
. browse if dup > 0
```

After inspecting the observations identified as duplicates, I feel confident that these observations are genuine duplicates, and we can safely eliminate them from the dataset. We can use the duplicates drop command to eliminate duplicates from the dataset.

```
. duplicates drop
Duplicates in terms of all variables
(4 observations deleted)
```

I expected four observations to be eliminated as duplicates (one for Olive, one for Ruth, and two for Mike). Indeed, that is the number of observations deleted by the duplicates drop command. The listing below confirms that the duplicate observations have been dropped.

1. The edit command allows you to view and edit the data in the Data Editor. The browse command permits you to view (but not edit) the data, making it a safer alternative when you simply wish to view the data.

```
. list
```

	name	years	fulltime	recom	dup
1.	Isaac O´Yerbreath	32.75	1	1	0
2.	Mary Smith	3	1	1	0
3.	Mary Smith	27	0	0	0
4.	Mike Avity	8.5	0	0	2
5.	Olive Tu´Drill	10.25	1	1	1
6.	Ruth Canaale	22	1	1	1
7.	Y. Don Uflossmore	7.25	0	1	0

The previous examples using dentists_dups.dta were unrealistically small but useful for clearly seeing how these commands work. Now let's use wws.dta to explore how to identify duplicates in a more realistic example. First, let's read this dataset into memory.

```
. use wws
(Working Women Survey)
```

This dataset contains a variable uniquely identifying each observation named idcode. The first thing that I would like to do is confirm that this variable truly does uniquely identify each observation. This can be done using the isid (is this an ID?) command.

```
. isid idcode
```

Had there been duplicate values for the variable idcode, the isid command would have returned an error message. The fact that it gave no output indicates that idcode truly does uniquely identify each observation. We could also check this with the command duplicates list idcode, which displays duplicates solely based on the variable idcode. As expected, this command confirms that there are no duplicates for idcode.

```
. duplicates list idcode
Duplicates in terms of idcode
(0 observations are duplicates)
```

Now let's see if there are any duplicates in this dataset, including all the variables when checking for duplicates. Using the duplicates list command, we can see that this dataset contains no duplicates.

```
. duplicates list
Duplicates in terms of all variables
(0 observations are duplicates)
```

Let's inspect a variant of wws.dta named wws_dups.dta. As you may suspect, this dataset will give us the opportunity to discover some duplicates. In particular, I want to first search for duplicates based on idcode and then search for duplicates based on all the variables in the dataset. Below we first read this dataset into memory.

```
. use wws_dups
```

Let's first use the isid command to see if, in this dataset, the variable idcode uniquely identifies each observation. As we can see below, idcode does not uniquely identify the observations.

```
. isid idcode
variable idcode does not uniquely identify the observations
r(459);
```

Let's use the duplicates report command to determine how many duplicates we have with respect to idcode.

```
. duplicates report idcode
Duplicates in terms of idcode
```

copies	observations	surplus
1	2242	0
2	6	3

We have a total of six observations in which the idcode variable appears twice. We can use the duplicates list command to see the observations with duplicate values on idcode.

```
. duplicates list idcode, sepby(idcode)
Duplicates in terms of idcode
```

group:	obs:	idcode
1	1088	2831
1	2248	2831
2	1244	3905
2	1245	3905
3	277	4214
3	2247	4214

I do not know if these observations are duplicates of all the variables or just duplicates of idcode. Let's obtain a report showing us the number of duplicates taking all variables into consideration.

```
. duplicates report
Duplicates in terms of all variables
```

copies	observations	surplus
1	2244	0
2	4	2

The report above shows us that there are four observations that are duplicates when taking all variables into consideration. Previously, we saw that there were six observations that were duplicates just for `idcode`.

Let's use the `duplicates tag` command to identify each of these kinds of duplicates. Below the variable `iddup` is created, which identifies duplicates based solely on `idcode`. The variable `alldup` identifies observations that are duplicates when taking all the variables into consideration.

```
. duplicates tag idcode, generate(iddup)
Duplicates in terms of idcode
. duplicates tag, generate(alldup)
Duplicates in terms of all variables
```

Below we tabulate these two variables against each other. This table gives a more complete picture of what is going on. There are four observations that are duplicates for all variables, and there are two observations that are duplicates for `idcode` but not for the other variables.

```
. tabulate iddup alldup
```

	alldup		
iddup	0	1	Total
0	2,242	0	2,242
1	2	4	6
Total	2,244	4	2,248

Let's look at the two observations that are duplicates for `idcode` but not for the rest of the variables. You could do this using the `browse` command, and these observations would display in the Data Editor.

```
. browse if iddup==1 & alldup==0
```

Or, below, the `list` command is used, showing a sampling of the variables from the dataset.

```
. list idcode age race yrschool occupation wage if iddup==1 & alldup==0, abb(20)
```

	idcode	age	race	yrschool	occupation	wage
1244.	3905	36	1	14	11	4.339774
1245.	3905	41	1	10	5	7.004828

We can clearly see that these are two different women who were accidentally assigned the same value for `idcode`. We can remedy this by assigning one of the women a new and unique value for `idcode`. Let's use the `summarize` command to determine the range of values for `idcode` so that we can assign a unique value.

```
. summarize idcode

    Variable |       Obs        Mean    Std. Dev.       Min        Max
-------------+--------------------------------------------------------
      idcode |      2248    2614.776    1480.434          1       5159
```

The highest value is 5,159, so let's assign a value of 5,160 to the woman who had an `idcode` of 3,905 and who was 41 years old.

```
. replace idcode = 5160 if idcode==3905 & age==41
(1 real change made)
```

Now when we use the `duplicates report` command, we see the same number of duplicates for `idcode` and for the entire dataset. In both cases, there are four duplicate observations.

```
. duplicates report idcode

Duplicates in terms of idcode

--------------------------------------
   copies | observations       surplus
----------+---------------------------
        1 |         2244             0
        2 |            4             2
--------------------------------------

. duplicates report

Duplicates in terms of all variables

--------------------------------------
   copies | observations       surplus
----------+---------------------------
        1 |         2244             0
        2 |            4             2
--------------------------------------
```

We could further inspect these duplicate observations. Say that we do this and we determine that we are satisfied that these are genuine duplicates. We can then eliminate them using the `duplicates drop` command, as shown below.

```
. duplicates drop

Duplicates in terms of all variables

(2 observations deleted)
```

Now the `duplicates report` command confirms that there are no duplicates in this dataset.

```
. duplicates report

Duplicates in terms of all variables

--------------------------------------
   copies | observations       surplus
----------+---------------------------
        1 |         2246             0
--------------------------------------
```

This section has illustrated how you can use the suite of `duplicates` commands to create listings and reports of duplicates as well as how to identify and eliminate duplicates. You can learn more about these commands by typing `help duplicates`.

3.9 Final thoughts on data cleaning

The previous sections of this chapter have shown how to check your data for suspicious values and how to correct values that are found to be in error. After taking these steps, one might be left with the feeling that no more data cleaning needs to be done. But data cleaning is not a destination—it is a process. Every additional action you take on the dataset has the potential for introducing errors.

The process of creating and recoding variables provides opportunities for errors to sneak into your data. It is easy to make a mistake when creating or recoding a variable. Because it is easy; it is recommended that you check such variables using the same kinds of techniques illustrated in sections 3.4, 3.5, and 3.6. For example, say that you recode a continuous variable (e.g., `age`) into a categorical variable (e.g., `agecat`). You can check this recoding by using the techniques from section 3.5: check the categorical version (`agecat`) against the continuous version (`age`).

When you merge two datasets together, this might give you the chance to do additional data checking. Say that you merge two datasets, a dataset with husbands and a dataset with wives. Imagine that both datasets had a variable asking how long they have been married. You could use the techniques described in section 3.6 to check the husband's answer against the wife's answer. You could also check the age of each husband against the age of his wife with the knowledge that married couples are generally of similar age. By merging the husbands and wives datasets, more opportunities arise for data checking than you had when the datasets were separated.

Data cleaning is ideally done using a do-file, which gives you the ability to automatically repeat the data-checking and data-correcting steps. Section 9.3 describes do-files and how to use them. Further, the data-checking strategies described in this section require you to sift through a lot of output, which is not only laborious but also increases the chances that problems could be missed among the volumes of output. Section 9.4 illustrates how the process of checking can be automated to further reduce the possibility for error.

4 Labeling datasets

We must be careful not to confuse data with the abstractions we use to analyze them.

—William James

4.1 Introduction

In the previous two chapters, we have seen how to enter data into Stata (as described in chapter 2) and how to perform data checking to verify the integrity of your data (as described in chapter 3). This chapter illustrates how to label your datasets. Labeled datasets are easier for others to understand, provide better documentation for yourself, and yield output that is more readable and understandable. Plus, by labeling your datasets in such a way that others can easily understand it, you get the added benefit of making your dataset easier for you to understand at some point in the future when your memories of the data have faded.

I begin this chapter by illustrating Stata tools for describing labeled datasets (see section 4.2). The next two sections show how you can label variables (see section 4.3) and how you can label values of your variables (see section 4.4). Then section 4.5 describes some utility programs that you can use for inspecting and checking value labels in a dataset. The ability to label datasets using different languages is then covered in section 4.6. The following section illustrates how to use Stata to add comments (via the `notes` command) to your dataset (see section 4.7). Section 4.8 shows how to use the `format` command to control the display of variables. The final section shows how to order variables in your dataset (see section 4.9).

The examples in this chapter use a hypothetical survey of eight graduate students. Section 4.2 shows a fully labeled version of this dataset; the following sections begin with a completely unlabeled version to which you will add labeling information in each section until, finally, in section 4.9, you will create and save the completely labeled dataset illustrated in section 4.2.

I should note that this chapter describes how to use Stata commands for labeling datasets. If you are interested in using the point-and-click Variables Manager, you can see section 2.10, which describes how to label variables in the context of entering data using the Data Editor. Whether you are labeling a new dataset or an existing dataset, the Variables Manager works in the same way, providing a point-and-click interface for labeling your dataset.

4.2 Describing datasets

Let's have a look at an example of a well-labeled dataset. This dataset includes an overall label for the dataset, labels for the variables, labels for values of some variables, comments (notes) for some variables, and formatting to improve the display of variables. This section illustrates how such labeling improves the usability of the dataset and explores Stata tools for displaying well-documented datasets. `survey7.dta` contains the results of a hypothetical survey of eight graduate students with information about their gender, race, date of birth, and income. The survey also asks the female students if they have given birth to a child and, if so, the name, sex, and birthday of their child. Below we use the dataset and see that it has a label describing the dataset as a survey of graduate students.

```
. use survey7
(Survey of graduate students)
```

We can get even more information about this dataset using the `describe` command, as shown below.

```
. describe
Contains data from survey7.dta
  obs:             8                          Survey of graduate students
 vars:            11                          2 Feb 2010 18:48
 size:           432 (99.9% of memory free)   (_dta has notes)
-------------------------------------------------------------------------------
              storage  display     value
variable name   type   format      label      variable label
-------------------------------------------------------------------------------
id              float  %9.0g                  Unique identification variable
STUDENTVARS     float  %9.0g
gender          float  %9.0g       mf         Gender of student
race            float  %19.0g      racelab  * Race of student
bday            float  %tdNN/DD/YY            Date of birth of student
income          float  %11.1fc                Income of student
havechild       float  %18.0g      havelab  * Given birth to a child?
KIDVARS         float  %9.0g
kidname         str10  %-10s                  Name of child
ksex            float  %15.0g      mfkid    * Sex of child
kbday           float  %td..                  Date of birth of child
                                            * indicated variables have notes
-------------------------------------------------------------------------------
Sorted by:
```

The header portion of the output gives overall information about the dataset and is broken up into two columns (groups). The first (left) column tells us the name of the dataset, the number of observations and variables in the dataset, and its size. The second (right) column shows the label for the dataset, displays the last time it was saved, and mentions that the overall dataset has notes associated with it.

The body of the output shows the name of each variable, how the variable is stored (see section A.5 for more information), the format for displaying the variable (see section 4.8 for more information), the value label used for displaying the values (see section 4.4 for more information), and a variable label that describes the variable (see section 4.3 for more information). Variables with asterisks have notes associated with them (see section 4.7 for more information).

With the `short` option, we can see just the header information. This is useful if you just need to know general information about the dataset, such as the size of the dataset and the number of variables and observations it contains.

```
. describe, short
Contains data from survey7.dta
  obs:             8                          Survey of graduate students
 vars:            11                          2 Feb 2010 18:48
 size:           432 (99.9% of memory free)
Sorted by:
```

Specifying a list of variables shows just the body of the output (without the header). Below we see the information for the variables `id`, `gender`, and `race`.

```
. describe id gender race

              storage  display     value
variable name   type   format      label      variable label
---------------------------------------------------------------------
id             float  %9.0g                  Unique identification variable
gender         float  %9.0g       mf         Gender of student
race           float  %19.0g      racelab  * Race of student
```

The `codebook` command allows you to more deeply inspect the dataset, producing a kind of electronic codebook for your dataset. You can type `codebook`, and it provides such information for all the variables in the dataset.

```
. codebook
  (output omitted)
```

If you specify one or more variables, the codebook information is limited to just the variables you specify. For example, the `codebook` command below shows codebook information for the `race` variable. This output shows us that race ranges from 1 to 5, it has five unique values, and none of its values are missing. The output also shows a tabulation of the values of `race` and the labels associated with those values (i.e., value labels).

```
. codebook race

---------------------------------------------------------------------
race                                                  Race of student
---------------------------------------------------------------------

                  type:  numeric (float)
                 label:  racelab

                 range:  [1,5]                        units:  1
         unique values:  5                        missing .:  0/8

            tabulation:  Freq.   Numeric  Label
                             2         1  White
                             2         2  Asian
                             2         3  Hispanic
                             1         4  African American
                             1         5  Other
```

Adding the `notes` option to the `codebook` command shows notes associated with a variable, as shown below. The variable `havechild` has three notes (comments) attached to it.

```
. codebook havechild, notes

-------------------------------------------------------------------------------
havechild                                             Given birth to a child?
-------------------------------------------------------------------------------

                  type:  numeric (float)
                 label:  havelab

                 range:  [0,1]                        units:  1
         unique values:  2                        missing .:  0/8
       unique mv codes:  1                       missing .*:  3/8

            tabulation:  Freq.   Numeric  Label
                             1         0  Dont Have Child
                             4         1  Have Child
                             3        .n  NA

havechild:
  1.  This variable measures whether a woman has given birth to a child, not
      just whether she is a parent.
  2.  The .n (NA) missing code is used for males, because they cannot bear
      children.
  3.  The .u (Unknown) missing code for a female indicating it is unknown if
      she has a child.
```

The `mv` (missing values) option shows information about whether the missing values on a particular variable are always associated with missingness on other variables. The notation at the bottom of the output below indicates that whenever **havechild** is missing, the variable **ksex** is also always missing. Likewise, whenever **kbday** is missing, **ksex** is also missing. This is useful for understanding patterns of missing values within your dataset.

```
. codebook ksex, mv

-------------------------------------------------------------------------------
ksex                                                             Sex of child
-------------------------------------------------------------------------------

                  type:  numeric (float)
                 label:  mfkid

                 range:  [1,2]                        units:  1
         unique values:  2                        missing .:  0/8
       unique mv codes:  2                       missing .*:  5/8

            tabulation:  Freq.   Numeric  Label
                             1         1  Male
                             2         2  Female
                             4        .n  NA
                             1        .u  Unknown

        missing values:     havechild==mv --> ksex==mv
                                kbday==mv --> ksex==mv
```

So far, all the information in the variable labels and value labels has appeared in English. Stata supports labels in multiple languages. As the `label language` command shows, this dataset contains labels in two languages, `en` (English) and `de` (German).

(Continued on next page)

```
. label language
Language for variable and value labels
    Available languages:
            de
            en
    Currently set is:                 . label language en
    To select different language:     . label language <name>
    To create new language:           . label language <name>, new
    To rename current language:       . label language <name>, rename
```

After using the `label language de` command, variable labels and value labels are then displayed using German, as illustrated using the `codebook` command.

```
. label language de
. codebook ksex
-------------------------------------------------------------------------------
ksex                                                      Geschlecht des Kindes
-------------------------------------------------------------------------------
                  type:  numeric (float)
                 label:  demfkid
                 range:  [1,2]                        units:  1
         unique values:  2                        missing .:  0/8
       unique mv codes:  2                       missing .*:  5/8
            tabulation:  Freq.   Numeric  Label
                             1         1  Junge
                             2         2  Maedchen
                             4        .n  nicht anwendbar
                             1        .u  unbekannt
```

The `label language en` command returns us to English labels.

```
. label language en
```

The `lookfor` command allows us to search the current dataset for keywords. Pretend that our dataset is very large and we want to find the variable designating the birthday of the student. The `lookfor birth` command asks Stata to search the variable names and labels for any instance of the word `birth`.

```
. lookfor birth
              storage  display     value
variable name   type   format      label      variable label
-------------------------------------------------------------------------------
bday           float   %tdNN/DD/YY            Date of birth of student
havechild      float   %18.0g      havelab  * Given birth to a child?
kbday          float   %td..                  Date of birth of child
```

In this case, it found three variables, each of which included `birth` in the variable label. Had there been a variable named `birthday` or `dateofbirth`, such variables would have also been included in the list.[1]

1. Searches can also be performed using the Variables Manager by entering search text into the **Filter box**. See section 2.10 for examples of using the Variables Manager.

We can also search comments (notes) within the dataset using the `notes search` command.

```
. notes search birth
havechild:
  1.  This variable measures whether a woman has given birth to a child, not
      just whether she is a parent.
```

This command found a note associated with `havechild` that had the word `birth` in it.

Let's now list some of the variables from this dataset using the `list` command. Let's list the income and birthday for each student.

```
. list income bday

     +-------------------------+
     |      income       bday  |
     |-------------------------|
  1. |    10,500.9   01/24/61  |
  2. |    45,234.1   04/15/68  |
  3. | 1,284,354.5   05/23/71  |
  4. |   124,313.5   06/25/73  |
  5. |   120,102.3   09/22/81  |
     |-------------------------|
  6. |       545.2   10/15/73  |
  7. |   109,452.1   07/01/77  |
  8. |     4,500.9   08/03/76  |
     +-------------------------+
```

The variable `income` is displayed in an easy-to-read format, using a comma separator and rounding the income to the nearest dime. The variable `bday` is a date variable and is displayed in a format that shows the month, day, and year separated by slashes. If we `describe` these two variables, the column named "display format" shows the formatting information that was applied to each of these variables to make the values display as they do. This is described in more detail in section 4.8.

```
. describe income bday
              storage  display      value
variable name   type   format       label      variable label
--------------------------------------------------------------------------------
income          float  %11.1fc                 Income of student
bday            float  %tdNN/DD/YY             Date of birth of student
```

This section has illustrated what a labeled dataset looks like and some of the benefits of having such a labeled dataset. The rest of this chapter shows how to actually create a labeled dataset. In fact, it illustrates how this example dataset, `survey7.dta`, was created, starting with a completely unlabeled dataset, `survey1.dta`. The following sections illustrate how to label the variables, label the values, label the values with different languages, add notes to the dataset, and format the display of variables. Section 4.5 explores other labeling utilities.

4.3 Labeling variables

This section shows how you can assign labels to your variables and assign a label to the overall dataset. We will start with a completely unlabeled version of the student survey dataset named survey1.dta.

```
. use survey1
```

Using the describe command shows that this dataset has no labels, including no labels for the variables.

```
. describe
Contains data from survey1.dta
  obs:             8
 vars:             9                          1 Jan 2010 12:13
 size:           464 (99.9% of memory free)
-------------------------------------------------------------------------------
              storage  display     value
variable name   type   format      label      variable label
-------------------------------------------------------------------------------
id              float  %9.0g
gender          float  %9.0g
race            float  %9.0g
havechild       float  %9.0g
ksex            float  %9.0g
bdays           str10  %10s
income          float  %9.0g
kbdays          str10  %10s
kidname         str10  %10s
-------------------------------------------------------------------------------
Sorted by:
```

The label variable command can be used to assign labels to variables. This command can also provide more descriptive information about each variable. Below we add variable labels for the variables id and gender.

```
. label variable id "Identification variable"
. label variable gender "Gender of student"
```

The describe command shows us that these variables indeed have the labels we assigned to them.

```
. describe id gender
              storage  display     value
variable name   type   format      label      variable label
-------------------------------------------------------------------------------
id              float  %9.0g                  Identification variable
gender          float  %9.0g                  Gender of student
```

Let's apply labels to the rest of the variables, as shown below.

```
. label variable race "Race of student"
. label variable havechild "Given birth to a child?"
. label variable ksex "Sex of child"
. label variable bdays "Birthday of student"
. label variable income "Income of student"
. label variable kbdays "Birthday of child"
. label variable kidname "Name of child"
```

Now all the variables in this dataset are labeled.

```
. describe
Contains data from survey1.dta
  obs:             8
 vars:             9                          1 Jan 2010 12:13
 size:           464 (99.9% of memory free)
-------------------------------------------------------------------------------
              storage  display     value
variable name   type   format      label      variable label
-------------------------------------------------------------------------------
id              float  %9.0g                  Identification variable
gender          float  %9.0g                  Gender of student
race            float  %9.0g                  Race of student
havechild       float  %9.0g                  Given birth to a child?
ksex            float  %9.0g                  Sex of child
bdays           str10  %10s                   Birthday of student
income          float  %9.0g                  Income of student
kbdays          str10  %10s                   Birthday of child
kidname         str10  %10s                   Name of child
-------------------------------------------------------------------------------
Sorted by:
```

The `label variable` command can also be used to change a label. Below we change the label for the `id` variable and show the results.

```
. label variable id "Unique identification variable"
. describe id
              storage  display     value
variable name   type   format      label      variable label
-------------------------------------------------------------------------------
id              float  %9.0g                  Unique identification variable
```

Finally, you can assign a label for the overall dataset with the `label data` command. This label will appear whenever you `use` the dataset.

```
. label data "Survey of graduate students"
```

We now save the dataset as `survey2.dta` for use in the next section.

```
. save survey2
file survey2.dta saved
```

For more information about labeling variables, see **help label**. The next section illustrates how to create and apply value labels to label the values of variables.

4.4 Labeling values

The previous section showed how we can label variables. This section shows how we can assign labels to the values of our variables. Sometimes variables are coded with values that have no intrinsic meaning, such as 1 meaning male and 2 meaning female. Without any labels, we would not know what the meaning of a 1 or a 2 is. In fact, the variable `gender` in our dataset is coded in this way. Below we create a label named `mf` (male/female) that associates the value of 1 with male and the value of 2 with female. Once that label is created, we then associate the `gender` variable with the value label `mf`.

```
. use survey2, clear
(Survey of graduate students)
. label define mf 1 "Male" 2 "Female"
. label values gender mf
```

We could also have labeled these values using the Variables Manager; see page 34.

The `codebook` command shows us that we successfully associated the `gender` variable with the value label `mf`. We can see that 1 is associated with “Male” and 2 is associated with “Female”.

```
. codebook gender

------------------------------------------------------------------------------
gender                                                       Gender of student
------------------------------------------------------------------------------

                  type:  numeric (float)
                 label:  mf

                 range:  [1,2]                        units:  1
         unique values:  2                        missing .:  0/8

            tabulation:  Freq.   Numeric  Label
                             3         1  Male
                             5         2  Female
```

We can use the same strategy to assign labels for the variable `race`. Note how this is a two-step process. We first create the value label named `racelab` using the `label define` command, and then we use the `label values` command to say that `race` should use the value label named `racelab` to label the values.

```
. label define racelab 1 "White" 2 "Asian" 3 "Hispanic" 4 "Black"
. label values race racelab
```

We can check the results by using the `codebook` command.

```
. codebook race

race                                                       Race of student

                  type:  numeric (float)
                 label:  racelab, but 1 nonmissing value is not labeled

                 range:  [1,5]                        units:  1
         unique values:  5                        missing .:  0/8

            tabulation:  Freq.   Numeric  Label
                             2         1  White
                             2         2  Asian
                             2         3  Hispanic
                             1         4  Black
                             1         5
```

The value of 5 is not labeled for `race`. That should be labeled "Other". Using the `add` option, we add the label for this value below.

```
. label define racelab 5 "Other", add
. codebook race

race                                                       Race of student

                  type:  numeric (float)
                 label:  racelab

                 range:  [1,5]                        units:  1
         unique values:  5                        missing .:  0/8

            tabulation:  Freq.   Numeric  Label
                             2         1  White
                             2         2  Asian
                             2         3  Hispanic
                             1         4  Black
                             1         5  Other
```

Say that we would prefer to label category 4 as "African American". We can use the `modify` option to modify an existing label.

```
. label define racelab 4 "African American", modify
. codebook race

race                                                       Race of student

                  type:  numeric (float)
                 label:  racelab

                 range:  [1,5]                        units:  1
         unique values:  5                        missing .:  0/8

            tabulation:  Freq.   Numeric  Label
                             2         1  White
                             2         2  Asian
                             2         3  Hispanic
                             1         4  African American
                             1         5  Other
```

The variable `ksex` contains the sex of a woman's child. If the woman has a child, the values are coded as `1` (male), `2` (female), and `.u` (unknown). If the observation is for a man, the value is coded as `.n` (not applicable). Let's create a label named `mfkid` that reflects this coding and use this to label the values of `ksex`.

```
. label define mfkid 1 "Male" 2 "Female" .u "Unknown" .n "NA"
. label values ksex mfkid
```

We can now see the labeled version of `ksex` with the `codebook` command.

```
. codebook ksex

-------------------------------------------------------------------------------
ksex                                                              Sex of child
-------------------------------------------------------------------------------

                  type:  numeric (float)
                 label:  mfkid

                 range:  [1,2]                        units:  1
         unique values:  2                        missing .:  0/8
       unique mv codes:  2                       missing .*:  5/8

            tabulation:  Freq.   Numeric  Label
                             1         1  Male
                             2         2  Female
                             4        .n  NA
                             1        .u  Unknown
```

Let's also label the variable `havechild`. Like `mfkid`, it also has missing values of `.u` if it is unknown if a woman has a child, and it has `.n` in the case of men.

```
. label define havelab 0 "Dont Have Child" 1 "Have Child" .u "Unknown" .n "NA"
. label values havechild havelab
```

Using the `codebook` command, we can see the labeled values. Note that the value of `.u` (unknown) does not appear in the output below. This value simply never appeared among the eight observations in our dataset. If this value had appeared, it would have been properly labeled. Even if a valid value does not appear in the dataset, it is still prudent to provide the label for it.

```
. codebook havechild

-------------------------------------------------------------------------------
havechild                                                Given birth to a child?
-------------------------------------------------------------------------------

                  type:  numeric (float)
                 label:  havelab

                 range:  [0,1]                        units:  1
         unique values:  2                        missing .:  0/8
       unique mv codes:  1                       missing .*:  3/8

            tabulation:  Freq.   Numeric  Label
                             1         0  Dont Have Child
                             4         1  Have Child
                             3        .n  NA
```

Let's have a look at the output produced by the `tabulate race` command.

```
. tabulate race

 Race of student |      Freq.     Percent        Cum.
-----------------+-----------------------------------
           White |          2       25.00       25.00
           Asian |          2       25.00       50.00
        Hispanic |          2       25.00       75.00
African American |          1       12.50       87.50
           Other |          1       12.50      100.00
-----------------+-----------------------------------
           Total |          8      100.00
```

The **tabulate** command only shows the labels (but not the values) of **race**. Earlier in this section, we labeled the **race** variable using the value label **racelab**. We can display the values and labels for **racelab** using the **label list** command.

```
. label list racelab
racelab:
           1 White
           2 Asian
           3 Hispanic
           4 African American
           5 Other
```

We could manually alter these labels to insert the numeric value as a prefix in front of each label (e.g., 1. White, 2. Asian). Stata offers a convenience command called **numlabel** to insert these numeric values. The **numlabel** command below takes the value label **racelab** and adds the numeric value in front of each of the labels.

```
. numlabel racelab, add
```

Using the **label list** command shows us that each of the labels for **racelab** now includes the numeric value as well as the label.

```
. label list racelab
racelab:
           1 1. White
           2 2. Asian
           3 3. Hispanic
           4 4. African American
           5 5. Other
```

Now when we issue the **tabulate race** command, the values and labels are shown for each level of **race**.

```
. tabulate race

    Race of student |      Freq.     Percent        Cum.
--------------------+-----------------------------------
           1. White |          2       25.00       25.00
           2. Asian |          2       25.00       50.00
        3. Hispanic |          2       25.00       75.00
4. African American |          1       12.50       87.50
           5. Other |          1       12.50      100.00
--------------------+-----------------------------------
              Total |          8      100.00
```

This also applies to the `list` command. Below we see that the values and labels for `race` are displayed.

```
. list race

     +--------------------+
     |               race |
     |--------------------|
  1. |           1. White |
  2. |           2. Asian |
  3. |           1. White |
  4. |        3. Hispanic |
  5. | 4. African American |
     |--------------------|
  6. |           5. Other |
  7. |           2. Asian |
  8. |        3. Hispanic |
     +--------------------+
```

We can remove the numeric prefix from `racelab` with the `numlabel` command with the `remove` option, as shown below. Then the `label list` command shows that the numeric values have been removed from the labels defined by `racelab`.

```
. numlabel racelab, remove

. label list racelab
racelab:
           1 White
           2 Asian
           3 Hispanic
           4 African American
           5 Other
```

Now the tabulation for `race` only includes the labels.

```
. tabulate race

 Race of student |      Freq.     Percent        Cum.
-----------------+-----------------------------------
           White |          2       25.00       25.00
           Asian |          2       25.00       50.00
        Hispanic |          2       25.00       75.00
African American |          1       12.50       87.50
           Other |          1       12.50      100.00
-----------------+-----------------------------------
           Total |          8      100.00
```

We use the `mask("#=")` option below to specify a mask for combining the values and the labels for variables labeled by `mf`. Note how this impacts the tabulation of `gender`.

```
. numlabel mf, add mask("#=")

. tabulate gender

  Gender of |
    student |      Freq.     Percent        Cum.
------------+-----------------------------------
     1=Male |          3       37.50       37.50
   2=Female |          5       62.50      100.00
------------+-----------------------------------
      Total |          8      100.00
```

We can remove the mask in much the same way that we added it but by specifying the `remove` option, as shown below.

```
. numlabel mf, remove mask("#=")
. tabulate gender

  Gender of |
    student |      Freq.     Percent        Cum.
------------+-----------------------------------
       Male |          3       37.50       37.50
     Female |          5       62.50      100.00
------------+-----------------------------------
      Total |          8      100.00
```

Let's add a different mask but apply this to all the value labels in the dataset. Because no specific value label was specified in the `numlabel` command below, it applies the command to all the value labels in the current dataset.

```
. numlabel, add mask("#) ")
```

Now all the variables with value labels show the numeric value followed by a close parenthesis and then the label (e.g., `1) Male`). We can see this by tabulating all the variables that have value labels, namely, `gender`, `race`, `havechild`, and `ksex`.

```
. tab1 gender race
-> tabulation of gender

  Gender of |
    student |      Freq.     Percent        Cum.
------------+-----------------------------------
    1) Male |          3       37.50       37.50
  2) Female |          5       62.50      100.00
------------+-----------------------------------
      Total |          8      100.00
-> tabulation of race

    Race of student |      Freq.     Percent        Cum.
--------------------+-----------------------------------
           1) White |          2       25.00       25.00
           2) Asian |          2       25.00       50.00
        3) Hispanic |          2       25.00       75.00
4) African American |          1       12.50       87.50
           5) Other |          1       12.50      100.00
--------------------+-----------------------------------
              Total |          8      100.00
```

(Continued on next page)

```
. tab1 havechild ksex
-> tabulation of havechild

  Given birth to a |
            child? |      Freq.     Percent        Cum.
-------------------+-----------------------------------
0) Dont Have Child |          1       20.00       20.00
     1) Have Child |          4       80.00      100.00
-------------------+-----------------------------------
             Total |          5      100.00
-> tabulation of ksex

    Sex of |
     child |      Freq.     Percent        Cum.
-----------+-----------------------------------
   1) Male |          1       33.33       33.33
 2) Female |          2       66.67      100.00
-----------+-----------------------------------
     Total |          3      100.00
. numlabel, remove mask("#) ")
```

We now save the dataset as `survey3.dta` for use in the next section.

```
. save survey3
file survey3.dta saved
```

For more information about labeling values, see `help label` and `help numlabel`. The next section will explore utilities that you can use with value labels.

4.5 Labeling utilities

Having created some value labels, let's explore some of the utility programs that Stata has for managing them. Using `survey3.dta`, we use the `label dir` command to show a list of the value labels defined in that dataset. This shows us the four value labels we created in the previous section.

```
. use survey3, clear
(Survey of graduate students)
. label dir
havelab
mf
mfkid
racelab
```

The `label list` command can be used to inspect a value label. Below we see the labels and values for the value label `mf`.

```
. label list mf
mf:
           1 Male
           2 Female
```

We can list multiple value labels at once, as shown below.

```
. label list havelab racelab
havelab:
           0 Dont Have Child
           1 Have Child
          .n NA
          .u Unknown
racelab:
           1 White
           2 Asian
           3 Hispanic
           4 African American
           5 Other
```

If no variables are specified, then all value labels will be listed.

```
. label list
  (output omitted)
```

The `label save` command takes the value labels defined in the working dataset and writes a Stata do-file with the `label define` statements to create those value labels. This can be useful if you have a dataset with value labels that you would like to apply to a different dataset but you do not have the original `label define` commands to create the labels.

```
. label save havelab racelab using surveylabs
file surveylabs.do saved
. type surveylabs.do
label define havelab 0 `"Dont Have Child"', modify
label define havelab 1 `"Have Child"', modify
label define havelab .n `"NA"', modify
label define havelab .u `"Unknown"', modify
label define racelab 1 `"White"', modify
label define racelab 2 `"Asian"', modify
label define racelab 3 `"Hispanic"', modify
label define racelab 4 `"African American"', modify
label define racelab 5 `"Other"', modify
```

The `labelbook` command provides information about the value labels in the working dataset. The `labelbook` command below shows information about the value label `racelab`. (If we had issued the `labelbook` command alone, it would have provided information about all the labels in the working dataset.)

(*Continued on next page*)

```
. labelbook racelab

value label racelab

     values                                    labels
      range:  [1,5]                     string length:  [5,16]
          N:  5                 unique at full length:  yes
       gaps:  no                  unique at length 12:  yes
  missing .*: 0                           null string:  no
                              leading/trailing blanks:  no
                                   numeric -> numeric:  no
  definition
           1  White
           2  Asian
           3  Hispanic
           4  African American
           5  Other

   variables: race
```

Notice how three groups of information are in the output, corresponding to the headings "values", "labels", and "definition".

The values section tells us that the values range from 1 to 5 with a total of five labels that have no gaps and no missing values.

The labels section tells us that the lengths of the labels range from 8 to 19 characters wide, are all unique, and are still unique if truncated to 12 characters. In addition, none of the labels are null strings (i.e., `""`), none have blanks at the start or end of the labels, and none of the labels are just one number.

The definition section shows the definition of the label (e.g., that `1` corresponds to `White`) and the variables this value label applies to, namely, `race`. In the fourth definition (i.e., African American), you will notice that the first 12 characters of the label are underlined. Most Stata commands that display value labels only display the first 12 characters of a value label. `labelbook` underlines the first 12 characters of labels that are longer than 12 characters to help you see what will be displayed by most Stata commands.

By default, the definition section lists all values and labels, but you can use the `list()` option to restrict how many values are listed, and you can specify `list(0)` to suppress the display of this section altogether. Below we list just the values and labels sections for the variables `havelab` and `mf`.

```
. labelbook havelab mf, list(0)

value label havelab

      values                                      labels
       range:  [0,1]                     string length:  [2,15]
           N:  4                 unique at full length:  yes
        gaps:  no                  unique at length 12:  yes
  missing .*:  2                           null string:  no
                               leading/trailing blanks:  no
                                    numeric -> numeric:  no

   variables:  havechild

value label mf

      values                                      labels
       range:  [1,2]                     string length:  [4,6]
           N:  2                 unique at full length:  yes
        gaps:  no                  unique at length 12:  yes
  missing .*:  0                           null string:  no
                               leading/trailing blanks:  no
                                    numeric -> numeric:  no

   variables:  gender
```

The values and labels sections are trying to alert you to potential problems in your labels. If you have many labels, you may tire of reading this detailed output. The `problems` option can be used with the `labelbook` command to summarize the problems found with the labels. In this case, the labels were in good shape and there were no problems to report.

```
. labelbook, problems
no potential problems in dataset survey3.dta
```

Let's use a different dataset with label problems:

```
. use survey3prob, clear
(Survey of graduate students)
. labelbook, problems
   Potential problems in dataset   survey3prob.dta
               potential problem   value labels

           gaps in mapped values   racelab2
      duplicate lab. at length 12   mf2
```

The `labelbook` output is describing problems with two value labels, `racelab2` and `mf2`. Let's first ask for detailed information about the problems found with `mf2`.

(Continued on next page)

```
. labelbook mf2, detail problems

value label mf2

     values                                    labels
      range:  [1,2]                     string length:  [4,23]
          N:  4                 unique at full length:  yes
       gaps:  no                  unique at length 12:  no
   missing .*:  2                         null string:  no
                              leading/trailing blanks:  no
                                   numeric -> numeric:  no
   definition
            1   Male
            2   Female
           .n   Missing Value - Unknown
           .u   Missing Value - Refused

   variables:   gender ksex

    Potential problems in dataset    survey3prob.dta
               potential problem    value labels

     duplicate lab. at length 12    mf2
```

The problem with `mf2` is that the labels for the two missing values are the same for the first 12 characters. For example, in the `tabulate` command below, you cannot differentiate between the two types of missing values because their labels are the same for the characters that are displayed. To remedy this, we would want to choose labels where we could tell the difference between them even if the labels were shortened.

```
. tabulate gender ksex, missing

                  |                 Sex of child
Gender of student |      Male     Female  Missing V  Missing V |     Total
------------------+--------------------------------------------+----------
             Male |         0          0          3          0 |         3
           Female |         1          2          1          1 |         5
------------------+--------------------------------------------+----------
            Total |         1          2          4          1 |         8
```

The problem with `racelab2` is that it has a gap in the labels. The values 1, 2, 3, and 5 are labeled, but there is no label for the value 4. Such a gap suggests that we forgot to label one of the values.

```
. labelbook racelab2, detail problems

value label racelab2

     values                                    labels
      range:  [1,5]                     string length:  [5,8]
          N:  4                 unique at full length:  yes
       gaps:  yes                 unique at length 12:  yes
  missing .*:  0                          null string:  no
                              leading/trailing blanks:  no
                                   numeric -> numeric:  no
  definition
           1   White
           2   Asian
           3   Hispanic
           5   Other

   variables:  race

   Potential problems in dataset   survey3prob.dta

              potential problem   value labels

          gaps in mapped values   racelab2
```

Using the `codebook` command for the variable `race` (which is labeled with `racelab`) shows that the fourth value is indeed unlabeled. The label for `racelab` would need to be modified to include a label for the fourth value.

```
. codebook race

race                                                     Race of student

                  type:  numeric (float)
                 label:  racelab2, but 1 nonmissing value is not labeled

                 range:  [1,5]                        units:  1
         unique values:  5                        missing .:  0/8

            tabulation:  Freq.   Numeric  Label
                             2         1  White
                             2         2  Asian
                             2         3  Hispanic
                             1         4
                             1         5  Other
```

This concludes our exploration of labeling utilities. For more information, see `help label list` and `help labelbook`.

The next section illustrates how you can supply variable labels and value labels in multiple languages.

4.6 Labeling variables and values in different languages

Stata supports variable labels and value labels in different languages. We can use the `label language` command to see what languages the dataset currently contains.

```
. use survey3, clear
(Survey of graduate students)

. label language

Language for variable and value labels

    In this dataset, value and variable labels have been defined in only one
    language:  default

    To create new language:        . label language <name>, new
    To rename current language:    . label language <name>, rename
```

Currently, the only language defined is `default`. Let's rename the current language to be `en` for English.

```
. label language en, rename
(language default renamed en)
```

Let's now add German (`de`) as a new language. This not only creates this new language but also selects it.

```
. label language de, new
(language de now current language)
```

As the `describe` command shows, the variable labels and value labels are empty for this language (however, the variable and value labels for the language `en` still exist).

```
. describe

Contains data from survey3.dta
  obs:             8
 vars:             9                          2 Feb 2010 18:54
 size:           464 (99.9% of memory free)
------------------------------------------------------------------------------
              storage  display     value
variable name   type   format      label      variable label
------------------------------------------------------------------------------
id              float  %9.0g
gender          float  %9.0g
race            float  %19.0g
havechild       float  %18.0g
ksex            float  %11.0g
bdays           str10  %10s
income          float  %9.0g
kbdays          str10  %10s
kidname         str10  %10s
------------------------------------------------------------------------------
Sorted by:
```

Let's now add German variable labels.

```
. label variable id "Identifikationsvariable"
. label variable gender "Geschlecht"
. label variable race "Ethnische Abstammung"
. label variable havechild "Jemals ein Kind geboren?"
. label variable ksex "Geschlecht des Kindes"
. label variable bdays "Geburtstag des/der Student/-in"
. label variable income "Einkommen"
. label variable kbdays "Geburtstag des Kindes"
. label variable kidname "Name des Kindes"
```

The `describe` command shows us that these variable labels were successfully assigned.

```
. describe
Contains data from survey3.dta
  obs:             8
 vars:             9                          2 Feb 2010 18:54
 size:           464 (99.9% of memory free)
-------------------------------------------------------------------------------
              storage  display     value
variable name   type   format      label      variable label
-------------------------------------------------------------------------------
id              float  %9.0g                  Identifikationsvariable
gender          float  %9.0g                  Geschlecht
race            float  %19.0g                 Ethnische Abstammung
havechild       float  %18.0g                 Jemals ein Kind geboren?
ksex            float  %11.0g                 Geschlecht des Kindes
bdays           str10  %10s                   Geburtstag des/der Student/-in
income          float  %9.0g                  Einkommen
kbdays          str10  %10s                   Geburtstag des Kindes
kidname         str10  %10s                   Name des Kindes
-------------------------------------------------------------------------------
Sorted by:
```

Now we assign German value labels for the variables `gender`, `race`, `havechild`, and `ksex`.

```
. label define demf 1 "Mann" 2 "Frau"
. label values gender demf
. label define deracelab 1 "kaukasisch" 2 "asiatisch" 3 "lateinamerikanisch"
> 4 "afroamerikanisch" 5 "andere"
. label values race deracelab
. label define dehavelab 0 "habe kein Kind" 1 "habe ein Kind" .u "unbekannt"
> .n "nicht anwendbar"
. label values havechild dehavelab
. label define demfkid 1 "Junge" 2 "Maedchen" .u "unbekannt" .n "nicht anwendbar"
. label values ksex demfkid
```

The codebook command shows us that this was successful.

```
. codebook gender race havechild ksex

------------------------------------------------------------------------------
gender                                                              Geschlecht
------------------------------------------------------------------------------

                  type:  numeric (float)
                 label:  demf

                 range:  [1,2]                        units:  1
         unique values:  2                        missing .:  0/8

            tabulation:  Freq.   Numeric  Label
                             3         1  Mann
                             5         2  Frau

------------------------------------------------------------------------------
race                                                     Ethnische Abstammung
------------------------------------------------------------------------------

                  type:  numeric (float)
                 label:  deracelab

                 range:  [1,5]                        units:  1
         unique values:  5                        missing .:  0/8

            tabulation:  Freq.   Numeric  Label
                             2         1  kaukasisch
                             2         2  asiatisch
                             2         3  lateinamerikanisch
                             1         4  afroamerikanisch
                             1         5  andere

------------------------------------------------------------------------------
havechild                                           Jemals ein Kind geboren?
------------------------------------------------------------------------------

                  type:  numeric (float)
                 label:  dehavelab

                 range:  [0,1]                        units:  1
         unique values:  2                        missing .:  0/8
       unique mv codes:  1                       missing .*:  3/8

            tabulation:  Freq.   Numeric  Label
                             1         0  habe kein Kind
                             4         1  habe ein Kind
                             3        .n  nicht anwendbar

------------------------------------------------------------------------------
ksex                                                   Geschlecht des Kindes
------------------------------------------------------------------------------

                  type:  numeric (float)
                 label:  demfkid

                 range:  [1,2]                        units:  1
         unique values:  2                        missing .:  0/8
       unique mv codes:  2                       missing .*:  5/8

            tabulation:  Freq.   Numeric  Label
                             1         1  Junge
                             2         2  Maedchen
                             4        .n  nicht anwendbar
                             1        .u  unbekannt
```

Below we make **en** the selected language. We can see that the English language labels are still intact.

```
. label language en

. describe

Contains data from survey3.dta
  obs:             8                          Survey of graduate students
 vars:             9                          2 Feb 2010 18:54
 size:           464 (99.9% of memory free)
-------------------------------------------------------------------------------
              storage  display     value
variable name   type   format      label      variable label
-------------------------------------------------------------------------------
id              float  %9.0g                  Unique identification variable
gender          float  %9.0g       mf         Gender of student
race            float  %19.0g      racelab    Race of student
havechild       float  %18.0g      havelab    Given birth to a child?
ksex            float  %15.0g      mfkid      Sex of child
bdays           str10  %10s                   Birthday of student
income          float  %9.0g                  Income of student
kbdays          str10  %10s                   Birthday of child
kidname         str10  %10s                   Name of child
-------------------------------------------------------------------------------
Sorted by:
```

Let's make a third language named **es** for Spanish.

```
. label language es, new
(language es now current language)
```

We are now using the **es** language. The **describe** command below shows that in this new language, we have no variable labels or value labels.

```
. describe

Contains data from survey3.dta
  obs:             8
 vars:             9                          2 Feb 2010 18:54
 size:           464 (99.9% of memory free)
-------------------------------------------------------------------------------
              storage  display     value
variable name   type   format      label      variable label
-------------------------------------------------------------------------------
id              float  %9.0g
gender          float  %9.0g
race            float  %19.0g
havechild       float  %18.0g
ksex            float  %15.0g
bdays           str10  %10s
income          float  %9.0g
kbdays          str10  %10s
kidname         str10  %10s
-------------------------------------------------------------------------------
Sorted by:
```

For brevity, we will just add variable labels and value labels for `gender`.

```
. label variable gender "el genero de studenta"
. label define esmf 1 "masculino" 2 "hembra"
. label values gender esmf
```

The output of the `codebook` command shows that the new Spanish labels have been applied successfully.

```
. codebook gender
-------------------------------------------------------------------------------
gender                                                    el genero de studenta
-------------------------------------------------------------------------------

                  type:  numeric (float)
                 label:  esmf

                 range:  [1,2]                        units:  1
         unique values:  2                        missing .:  0/8

            tabulation:  Freq.   Numeric  Label
                             3         1  masculino
                             5         2  hembra
```

Let's switch back to English labels and then delete the Spanish labels from this dataset.

```
. label language en
. label language es, delete
```

Now let's save the dataset as `survey4.dta` for use in the next section. The selected language will be English, but the dataset also includes German.

```
. save survey4
file survey4.dta saved
```

For more information, see `help label language`.

The next section will illustrate how to add notes to Stata datasets to provide additional documentation.

4.7 Adding comments to your dataset using notes

This section shows how you can add notes to your dataset. `survey4.dta`, which was saved at the end of the previous section, will be used in this section.

You can add an overall note to your dataset with the `note` command.

```
. use survey4
(Survey of graduate students)
. note: This was based on the dataset called survey1.txt
```

The `notes` command displays notes that are contained in the dataset.

```
. notes
_dta:
  1.  This was based on the dataset called survey1.txt
```

You can add additional notes with the `note` command. This note also includes `TS`, which adds a time stamp.

```
. note: The missing values for havechild and childage were coded using -1 and
> -2 but were converted to .n and .u TS
```

The `notes` command now shows both of the notes that have been added to this dataset.

```
. notes
_dta:
  1.  This was based on the dataset called survey1.txt
  2.  The missing values for havechild and childage were coded using -1 and -2
      but were converted to .n and .u 2 Feb 2010 18:54
```

You can use the `note` command to add notes for specific variables as well. This is illustrated below for the variable `race`.

```
. note race: The other category includes people who specified multiple races
```

Now the `notes` command shows the notes for the overall dataset as well as the notes associated with specific variables.

```
. notes
_dta:
  1.  This was based on the dataset called survey1.txt
  2.  The missing values for havechild and childage were coded using -1 and -2
      but were converted to .n and .u 2 Feb 2010 18:54
race:
  1.  The other category includes people who specified multiple races
```

We can see just the notes for `race` via the `notes race` command.

```
. notes race
race:
  1.  The other category includes people who specified multiple races
```

We can add multiple notes for a variable. Below we add four notes for the variable `havechild` and two notes for the variable `ksex`.

(Continued on next page)

```
. note havechild: This variable measures whether a woman has given birth to a
> child, not just whether she is a parent.
. note havechild: Men cannot bear children.
. note havechild: The .n (NA) missing code is used for males, because they cannot
> bear children.
. note havechild: The .u (Unknown) missing code for a female indicating it is
> unknown if she has a child.
. note ksex: This is the sex of the woman´s child
. note ksex: .n and .u missing value codes are like for the havechild variable.
```

We can view the notes for `havechild` and `ksex` like this:

```
. notes havechild ksex
havechild:
  1.  This variable measures whether a woman has given birth to a child, not
      just whether she is a parent.
  2.  Men cannot bear children.
  3.  The .n (NA) missing code is used for males, because they cannot bear
      children.
  4.  The .u (Unknown) missing code for a female indicating it is unknown if
      she has a child.
ksex:
  1.  This is the sex of the woman´s child
  2.  .n and .u missing value codes are like for the havechild variable.
```

You can then see all the notes in the dataset with the `notes` command. This shows the notes for the overall dataset and for specific variables.

```
. notes
_dta:
  1.  This was based on the dataset called survey1.txt
  2.  The missing values for havechild and childage were coded using -1 and -2
      but were converted to .n and .u 2 Feb 2010 18:54
race:
  1.  The other category includes people who specified multiple races
havechild:
  1.  This variable measures whether a woman has given birth to a child, not
      just whether she is a parent.
  2.  Men cannot bear children.
  3.  The .n (NA) missing code is used for males, because they cannot bear
      children.
  4.  The .u (Unknown) missing code for a female indicating it is unknown if
      she has a child.
ksex:
  1.  This is the sex of the woman´s child
  2.  .n and .u missing value codes are like for the havechild variable.
```

We can view just the notes for the overall dataset with the `notes _dta` command, like this:

```
. notes _dta
_dta:
  1.  This was based on the dataset called survey1.txt
  2.  The missing values for havechild and childage were coded using -1 and -2
      but were converted to .n and .u 2 Feb 2010 18:54
```

The second note for **havechild** is not useful, so we remove it with the **notes drop** command. The following **notes** command shows that this note was indeed dropped:

```
. notes drop havechild in 2
  (1 note dropped)
. notes havechild
havechild:
  1.  This variable measures whether a woman has given birth to a child, not
      just whether she is a parent.
  3.  The .n (NA) missing code is used for males, because they cannot bear
      children.
  4.  The .u (Unknown) missing code for a female indicating it is unknown if
      she has a child.
```

The **notes renumber** command is used below to renumber the notes for **havechild**, eliminating the gap in the numbering.

```
. notes renumber havechild
. notes havechild
havechild:
  1.  This variable measures whether a woman has given birth to a child, not
      just whether she is a parent.
  2.  The .n (NA) missing code is used for males, because they cannot bear
      children.
  3.  The .u (Unknown) missing code for a female indicating it is unknown if
      she has a child.
```

The **notes search** command allows you to search the contents of the notes. We use it below to show all the notes that contain the text **.u**.

```
. notes search .u
_dta:
  2.  The missing values for havechild and childage were coded using -1 and -2
      but were converted to .n and .u 2 Feb 2010 18:54
havechild:
  3.  The .u (Unknown) missing code for a female indicating it is unknown if
      she has a child.
ksex:
  2.  .n and .u missing value codes are like for the havechild variable.
```

We now save the dataset as **survey5.dta** for use in the next section.

```
. save survey5
file survey5.dta saved
```

For more information about **notes**, see **help notes**.

The next section illustrates how to customize the display of variables.

4.8 Formatting the display of variables

Formats give you control over how variables are displayed. Let's illustrate this using `survey5.dta`, which we saved at the end of the last section. The impact of formats is most evident when using the `list` command. Below we list the variable `income` for the first five observations of this dataset.

```
. use survey5, clear
(Survey of graduate students)
. list id income in 1/5

     +---------------+
     | id     income |
     |---------------|
  1. |  1   10500.93 |
  2. |  2   45234.13 |
  3. |  3    1284355 |
  4. |  4   124313.5 |
  5. |  5   120102.3 |
     +---------------+
```

By using the `describe` command, we can see that the `income` variable is currently displayed using the `%9.0g` format. Without going into too many details, this is a general format that displays incomes using a width of nine and decides for us the best way to display the values of `income` within that width.

```
. describe income
              storage  display     value
variable name   type   format      label      variable label
-------------------------------------------------------------------
income          float  %9.0g                  Income of student
```

Other formats, such as the fixed format, give us more control over the display format. For example, below we use the `%12.2f` format, which displays incomes using a fixed format with a maximum width of 12 characters including the decimal point and 2 digits displayed after the decimal point. Note how observations 3, 4, and 5 now display `income` using two decimal places.

```
. format income %12.2f
. list id income in 1/5

     +-------------------+
     | id         income |
     |-------------------|
  1. |  1       10500.93 |
  2. |  2       45234.13 |
  3. |  3     1284354.50 |
  4. |  4      124313.45 |
  5. |  5      120102.32 |
     +-------------------+
```

In this dataset, `income` is measured to the penny, but we might be content to see it measured to the nearest whole dollar. If we format it using `%7.0f`, we can view incomes up to a million dollars (seven-digit incomes), and incomes will be rounded to the nearest dollar. Note how the first observation is rounded up to the next highest whole dollar.

```
. format income %7.0f
. list id income in 1/5

     +--------------+
     | id    income |
     |--------------|
  1. |  1     10501 |
  2. |  2     45234 |
  3. |  3   1284354 |
  4. |  4    124313 |
  5. |  5    120102 |
     +--------------+
```

We could display the income to the nearest dime by specifying a `%9.1f` format. Compared with the prior example, we need to increase the width of this format from 7 to 9 (not 7 to 8) to accommodate the decimal point.

```
. format income %9.1f
. list id income in 1/5

     +----------------+
     | id      income |
     |----------------|
  1. |  1     10500.9 |
  2. |  2     45234.1 |
  3. |  3   1284354.5 |
  4. |  4    124313.5 |
  5. |  5    120102.3 |
     +----------------+
```

For large numbers, it can help to see commas separating each group of three numbers. By adding a `c` to the format, we request that commas be displayed as well. Compared with the prior example, we expanded the overall width from 9 to 11 to accommodate the two commas that are inserted for observation 3.

```
. format income %11.1fc
. list id income in 1/5

     +------------------+
     | id        income |
     |------------------|
  1. |  1      10,500.9 |
  2. |  2      45,234.1 |
  3. |  3   1,284,354.5 |
  4. |  4     124,313.5 |
  5. |  5     120,102.3 |
     +------------------+
```

(Continued on next page)

Let's turn our attention to how to control the display of string variables, such as the variable **kidname**. As we see below, the display format for **kidname** is **%10s**, meaning that it is a string variable displayed with a width of 10.

```
. describe kidname
              storage  display     value
variable name   type   format      label      variable label
-------------------------------------------------------------------------------
kidname         str10  %10s                   Name of child
```

The listing below illustrates that this format displays the names as right-justified.

```
. list id kidname in 1/5

     +----------------+
     | id     kidname |
     |----------------|
  1. |  1             |
  2. |  2       Sally |
  3. |  3   Catherine |
  4. |  4             |
  5. |  5     Samuell |
     +----------------+
```

To specify that the variable should be shown as left-justified, you precede the width with a dash. Below we change the display format for **kidname** to have a width of 10 and to be left-justified.

```
. format kidname %-10s
. describe kidname
              storage  display     value
variable name   type   format      label      variable label
-------------------------------------------------------------------------------
kidname         str10  %-10s                  Name of child
. list id kidname in 1/5

     +----------------+
     | id   kidname   |
     |----------------|
  1. |  1             |
  2. |  2   Sally     |
  3. |  3   Catherine |
  4. |  4             |
  5. |  5   Samuell   |
     +----------------+
```

There are many options for the display of date variables. In this dataset, the variables **bdays** and **kbdays** contain the birth date of the mother and the child, but they are currently stored as string variables. First, we need to convert these variables into date variables, as shown below (see section 5.8 for more about creating date variables).

```
. generate bday = date(bdays,"MDY")

. generate kbday = date(kbdays,"MDY")
(4 missing values generated)

. list id bdays bday kbdays kbday in 1/5
```

	id	bdays	bday	kbdays	kbday
1.	1	1/24/1961	389		.
2.	2	4/15/1968	3027	4/15/1995	12888
3.	3	5/23/1971	4160	8/15/2003	15932
4.	4	6/25/1973	4924		.
5.	5	9/22/1981	7935	1/12/1999	14256

The conversion would appear faulty because the values for **bday** and **kbday** do not appear correct, but they are. Date variables are stored as—and, by default, are displayed as—the number of days since January 1, 1960. Below we request that the dates be displayed using a general date format named **%td**. Now the dates appear as we would expect.

```
. format bday kbday %td

. list id bdays bday kbdays kbday in 1/5
```

	id	bdays	bday	kbdays	kbday
1.	1	1/24/1961	24jan1961		.
2.	2	4/15/1968	15apr1968	4/15/1995	15apr1995
3.	3	5/23/1971	23may1971	8/15/2003	15aug2003
4.	4	6/25/1973	25jun1973		.
5.	5	9/22/1981	22sep1981	1/12/1999	12jan1999

Stata supports many custom ways to display dates. For example, below we specify that **bday** should be displayed with the format **%tdNN/DD/YY**. This format displays the variable as a date with the numeric month followed by a slash, then the numeric day followed by a slash, and then the two-digit year. This yields, for example, **01/24/61**.

```
. format bday %tdNN/DD/YY

. list id bdays bday in 1/5
```

	id	bdays	bday
1.	1	1/24/1961	01/24/61
2.	2	4/15/1968	04/15/68
3.	3	5/23/1971	05/23/71
4.	4	6/25/1973	06/25/73
5.	5	9/22/1981	09/22/81

Below we change the display format for kbday to %tdMonth_DD,CCYY. This format displays the name of the month followed by a space (indicated with the underscore), then the numeric day followed by a comma, and then the two-digit century (e.g., 19 or 20) followed by the two-digit year. This yields, for example, August 22,1983. For more examples, see section 5.8.

```
. format kbday %tdMonth_DD,CCYY
. list id kbdays kbday in 1/5

     +----------------------------------+
     | id      kbdays             kbday |
     |----------------------------------|
  1. |  1                             . |
  2. |  2   4/15/1995     April 15,1995 |
  3. |  3   8/15/2003    August 15,2003 |
  4. |  4                             . |
  5. |  5   1/12/1999   January 12,1999 |
     +----------------------------------+
```

The bday variable now makes the bdays variable no longer necessary, and likewise kbday makes kbdays no longer necessary. Let's label the new variables and drop the old versions.

```
. label variable bday "Date of birth of student"
. label variable kbday "Date of birth of child"
. drop bdays kbdays
```

Finally, let's save the dataset as survey6.dta.

```
. save survey6, replace
(note: file survey6.dta not found)
file survey6.dta saved
```

This concludes this section on formatting variables. For more information, see help format.

The next, and final, section will show how to order the variables in this dataset for greater clarity.

4.9 Changing the order of variables in a dataset

survey6.dta is well labeled, but the variables are unordered. Looking at the output of the describe command below, you can see that the information about the graduate student being surveyed is intermixed with information about that student's child.

```
. use survey6, clear
(Survey of graduate students)
. describe
Contains data from survey6.dta
  obs:             8                          Survey of graduate students
 vars:             9                          2 Feb 2010 18:54
 size:           368 (99.9% of memory free)   (_dta has notes)
-------------------------------------------------------------------------------
              storage  display     value
variable name   type   format      label      variable label
-------------------------------------------------------------------------------
id              float  %9.0g                  Unique identification variable
gender          float  %9.0g       mf         Gender of student
race            float  %19.0g      racelab  * Race of student
havechild       float  %18.0g      havelab  * Given birth to a child?
ksex            float  %15.0g      mfkid    * Sex of child
income          float  %11.1fc                Income of student
kidname         str10  %-10s                  Name of child
bday            float  %tdNN/DD/YY            Date of birth of student
kbday           float  %td..                  Date of birth of child
                                            * indicated variables have notes
-------------------------------------------------------------------------------
Sorted by:
```

Datasets often have natural groupings of variables. The clarity of the dataset is improved when related variables are positioned next to each other in the dataset. The `order` command below specifies the order in which we want the variables to appear in the dataset. The command indicates the variable `id` should be first, followed by `gender`, `race`, `bday`, `income`, and then `havechild`. Any remaining variables (which happen to be the child variables) will follow `havechild` in the order in which they currently appear in the dataset.

```
. order id gender race bday income havechild
. describe
Contains data from survey6.dta
  obs:             8                          Survey of graduate students
 vars:             9                          2 Feb 2010 18:54
 size:           368 (99.9% of memory free)   (_dta has notes)
-------------------------------------------------------------------------------
              storage  display     value
variable name   type   format      label      variable label
-------------------------------------------------------------------------------
id              float  %9.0g                  Unique identification variable
gender          float  %9.0g       mf         Gender of student
race            float  %19.0g      racelab  * Race of student
bday            float  %tdNN/DD/YY            Date of birth of student
income          float  %11.1fc                Income of student
havechild       float  %18.0g      havelab  * Given birth to a child?
ksex            float  %15.0g      mfkid    * Sex of child
kidname         str10  %-10s                  Name of child
kbday           float  %td..                  Date of birth of child
                                            * indicated variables have notes
-------------------------------------------------------------------------------
Sorted by:
```

This ordering is pretty good, except that it would be nice for the list of child variables to start with `kidname` instead of `ksex`. The `order` command below is used to move `kidname` before `ksex`. (We could get the same result by specifying `after(havechild)` instead of `before(ksex)`.)

```
. order kidname, before(ksex)
. describe
Contains data from survey6.dta
  obs:             8                          Survey of graduate students
 vars:             9                          2 Feb 2010 18:54
 size:           368 (99.9% of memory free)   (_dta has notes)
-------------------------------------------------------------------------------
              storage  display     value
variable name   type   format      label      variable label
-------------------------------------------------------------------------------
id              float  %9.0g                  Unique identification variable
gender          float  %9.0g       mf         Gender of student
race            float  %19.0g      racelab  * Race of student
bday            float  %tdNN/DD/YY            Date of birth of student
income          float  %11.1fc                Income of student
havechild       float  %18.0g      havelab  * Given birth to a child?
kidname         str10  %-10s                  Name of child
ksex            float  %15.0g      mfkid    * Sex of child
kbday           float  %td..                  Date of birth of child
                                            * indicated variables have notes
-------------------------------------------------------------------------------
Sorted by:
```

Now the variables are organized in a more natural fashion, and it is pretty easy to see this natural ordering. However, with datasets containing more variables, it can be harder to see the groupings of the variables. In such cases, I like to create variables that act as headers to introduce each new grouping of variables.

Below the variables `STUDENTVARS` and `KIDVARS` are created, and then the `order` command positions them at the beginning of their group of variables.

```
. generate STUDENTVARS = .
(8 missing values generated)
. generate KIDVARS = .
(8 missing values generated)
. order STUDENTVARS, before(gender)
. order KIDVARS, before(kidname)
```

Now when we look at this dataset, it is clear that the variables are grouped into variables about the students and variables about the kids. Although the variables `STUDENTVARS` and `KIDVARS` are helpful in documenting the file, they do take up extra space in the dataset. This could be an issue if the dataset has many observations.

```
. describe
Contains data from survey6.dta
  obs:             8                          Survey of graduate students
 vars:            11                          2 Feb 2010 18:54
 size:           432 (99.9% of memory free)   (_dta has notes)
--------------------------------------------------------------------------------
              storage  display     value
variable name   type   format      label      variable label
--------------------------------------------------------------------------------
id              float  %9.0g                  Unique identification variable
STUDENTVARS     float  %9.0g
gender          float  %9.0g       mf         Gender of student
race            float  %19.0g      racelab  * Race of student
bday            float  %tdNN/DD/YY            Date of birth of student
income          float  %11.1fc                Income of student
havechild       float  %18.0g      havelab  * Given birth to a child?
KIDVARS         float  %9.0g
kidname         str10  %-10s                  Name of child
ksex            float  %15.0g      mfkid    * Sex of child
kbday           float  %td..                  Date of birth of child
                                            * indicated variables have notes
--------------------------------------------------------------------------------
Sorted by:
     Note:  dataset has changed since last saved
```

We now have a nicely labeled and well-ordered dataset that looks like the one we saw in section 4.2. Let's now save this dataset as `survey7.dta`.

```
. save survey7
file survey7.dta saved
```

This section has focused on the `order` command to create a more user-friendly ordering of the variables in `survey6.dta`. More features are included in the `order` command that were not illustrated here, such as how to alphabetize variables, moving groups of variables, and moving variables to the end of the dataset. Section 5.15 discusses these and other issues related to renaming and ordering variables. For even more information, see `help order`.

5 Creating variables

Not everything that can be counted counts, and not everything that counts can be counted.

—Albert Einstein

5.1 Introduction

This chapter covers a wide variety of ways that you can create variables in Stata. I start by introducing the `generate` and `replace` commands for creating new variables and changing the contents of existing variables (see section 5.2). The next two sections describe how you can use numeric expressions and functions when creating variables (see section 5.3) and how you can use string expressions and functions when creating variables (see section 5.4). Section 5.5 illustrates tools to recode variables.

Tools for coding missing values are illustrated in section 5.6, which is followed by a discussion of dummy variables and the broader issue of factor variables (see section 5.7). Section 5.8 covers creating and using date variables, and section 5.9 covers creating and using date-and-time variables.

The next three sections illustrate the use of the `egen` command for computations across variables within each observation (section 5.10), for computations across observations (section 5.11), and for additional functions (section 5.12).

Methods for converting string variables to numeric variables are illustrated in section 5.13, and section 5.14 shows how numeric variables can be converted to string variables.

The chapter concludes with section 5.15, which illustrates how to rename and order variables.

5.2 Creating and changing variables

The two most common commands used for creating and modifying variables are the `generate` and `replace` commands. These commands are identical except that `generate` is used for creating a new variable, while `replace` is used for altering the values of an existing variable. These two commands are illustrated using `wws2.dta`, which contains demographic and labor force information regarding 2,246 women. Consider the variable `wage`, which contains the woman's hourly wages. This variable is summarized below. It has two missing values (the $N = 2244$).

```
. use wws2
(Working Women Survey w/fixes)
. summarize wage
    Variable |       Obs        Mean    Std. Dev.       Min        Max
-------------+--------------------------------------------------------
        wage |      2244    7.796781     5.82459          0   40.74659
```

Say that we want to compute a weekly wage for these women based on a 40-hour work week. We use the `generate` command to create the new variable, called `wageweek`, which contains the value of `wage` multiplied by 40.

```
. generate wageweek = wage*40
(2 missing values generated)
```

```
. summarize wageweek
    Variable |       Obs        Mean    Std. Dev.       Min        Max
-------------+--------------------------------------------------------
    wageweek |      2244    311.8712    232.9836          0   1629.864
```

This dataset also contains a variable named **hours**, which is the typical number of hours the woman works per week. Let's create **wageweek** again but use **hours** in place of 40. Because **wageweek** already exists, we must use the **replace** command to indicate that we want to replace the contents of the existing variable. Note that because **hours** has four missing observations, the **wageweek** variable now has four additional missing observations, having only 2,240 valid observations instead of 2,244.[1]

```
. replace wageweek = wage*hours
(1152 real changes made, 4 to missing)
. summarize wageweek
    Variable |       Obs        Mean    Std. Dev.       Min        Max
-------------+--------------------------------------------------------
    wageweek |      2240    300.2539    259.2544          0       1920
```

The **generate** and **replace** commands can be used together when a variable takes multiple steps to create. Consider the variables **married** (which is 1 if the woman is currently married and 0 otherwise) and **nevermarried** (which is 1 if she was never married and 0 if she is married or was previously married). We can place the women into three groups based on the cross-tabulation of these two variables.

```
. tabulate married nevermarried
           |   Woman never been
           |       married
   married |         0          1 |     Total
-----------+----------------------+----------
         0 |       570        234 |       804 
         1 |     1,440          2 |     1,442 
-----------+----------------------+----------
     Total |     2,010        236 |     2,246 
```

Say that we want to create a variable that reflects whether a woman is 1) single and has never married ($n = 234$), 2) currently married ($n = 1440$), or 3) single but previously married ($n = 570$). Those who are (nonsensically) currently married and have never been married ($n = 2$) will be assigned a value of missing. This can be done as shown below. The first **generate** command creates the variable **smd** (for single/married/divorced or widowed) and assigns a value of 1 if the woman meets the criteria for being single (and never married). The **replace** command assigns a value of 2 if the woman meets the criteria for being currently married. The second **replace** command assigns a value of 3 if the woman meets the criteria for being divorced or widowed. The third **replace** command is superfluous but clearly shows that **smd** is missing for those nonsense cases where the woman is currently married and has never been married. (For more information about the use of **if**, see section A.8.)

1. When a variable is missing as part of an arithmetic expression, then the result of the expression is missing.

```
. generate smd = 1 if (married==0) & (nevermarried==1)
(2012 missing values generated)
. replace  smd = 2 if (married==1) & (nevermarried==0)
(1440 real changes made)
. replace  smd = 3 if (married==0) & (nevermarried==0)
(570 real changes made)
. replace  smd = . if (married==1) & (nevermarried==1)
(0 real changes made)
```

We can double-check this in two ways. First, we can tabulate `smd` and see that the frequencies for `smd` match the frequencies of the two-way table we created above.

```
. tabulate smd, missing
```

smd	Freq.	Percent	Cum.
1	234	10.42	10.42
2	1,440	64.11	74.53
3	570	25.38	99.91
.	2	0.09	100.00
Total	2,246	100.00	

A more direct way to check the creation of this variable is to use the `table` command to make a three-way table of `smd` by `married` by `nevermarried`. As shown below, this also confirms that the values of `smd` properly correspond to the values of `married` and `nevermarried`.

```
. table smd married nevermarried
```

	Woman never been married and married			
	0		1	
smd	0	1	0	1
1			234	
2		1,440		
3	570			

The `generate` and `replace` commands can be combined to create a new dummy (0/1) variable based on the values of a continuous variable. For example, let's create a dummy variable called `over40hours` that will be 1 if a woman works over 40 hours and 0 if she works 40 or fewer hours. The `generate` command creates the `over40hours` variable and assigns a value of 0 when the woman works 40 or fewer hours. Then the `replace` command assigns a value of 1 when the woman works more than 40 hours.

```
. generate over40hours = 0 if (hours <= 40)
(394 missing values generated)
. replace over40hours = 1 if (hours > 40) & ! missing(hours)
(390 real changes made)
```

Note that the replace command specifies that over40hours is 1 if hours is over 40 and if hours is not missing. Without the second qualifier, people who had missing data on hours would be treated as though they had worked over 40 hours (because missing values are treated as positive infinity). See section A.10 for more on missing values.

We can double-check the creation of this dummy variable with the tabstat command, as shown below. When over40hours is 0, the value of hours ranges from 1 to 40 (as it should); when over40hours is 1, the value of hours ranges from 41 to 80.

```
. tabstat hours, by(over40hours) statistics(min max)

Summary for variables: hours
     by categories of: over40hours

over40hours |       min       max
------------+--------------------
          0 |         1        40
          1 |        41        80
------------+--------------------
      Total |         1        80
---------------------------------
```

We can combine these generate and replace commands into one generate command. This can save computation time (because Stata only needs to execute one command) and save you time (because you only need to type one command). This strategy is based on the values a logical expression assumes when it is true or false. When a logical expression is false, it takes on a value of 0; when it is true, it takes on a value of 1. From the previous example, the expression (hours > 40) would be 0 (false) when a woman works 40 or fewer hours and would be 1 (true) if a woman works over 40 hours (or had a missing value for hours).

Below we use this one-step strategy to create over40hours. Women who worked 40 or fewer hours get a 0 (because the expression is false), and women who worked more than 40 hours get a 1 (because the expression is true). Women with missing values on hours worked get a missing value because they are excluded based on the if qualifier. (See section A.6 for more details about logical expressions and examples.)

```
. generate over40hours = (hours > 40) if ! missing(hours)
(4 missing values generated)
```

The tabstat results below confirm that this variable was created correctly.

```
. tabstat hours, by(over40hours) statistics(min max)

Summary for variables: hours
     by categories of: over40hours

over40hours |       min       max
------------+--------------------
          0 |         1        40
          1 |        41        80
------------+--------------------
      Total |         1        80
---------------------------------
```

For more information, see `help generate` and see the next section, which illustrates how to use numeric expressions and functions to create variables.

5.3 Numeric expressions and functions

In the previous section, we used the `generate` and `replace` commands on simple expressions, such as creating a new variable that equaled `wage*40`. This section illustrates more complex expressions and some useful functions that can be used with the `generate` and `replace` commands.

Stata supports the standard mathematical operators of addition (+), subtraction (-), multiplication (*), division (/), and exponentiation (^) using the standard rules of the order of operators. Parentheses can be used to override the standard order of operators or to provide clarity. These operators are illustrated below to create a nonsense variable named `nonsense` using `wws2.dta`.

```
. use wws2, clear
(Working Women Survey w/fixes)
. generate nonsense = (age*2 + 10)^2 - (grade/10)
(4 missing values generated)
```

Stata also has a wide variety of mathematical functions that you can include in your `generate` and `replace` commands. The examples below illustrate the `int()` function (which removes any values after the decimal place), the `round()` function (which rounds a number to the desired number of decimal places), the `ln()` function (which yields the natural log), the `log10()` function (which computes the base-10 logarithm), and `sqrt()` (which computes the square root). The first five values are then listed to show the results of using these functions.

```
. generate intwage = int(wage)
(2 missing values generated)
. generate rndwage = round(wage,1.00)
(2 missing values generated)
. generate lnwage = ln(wage)
(3 missing values generated)
. generate logwage = log10(wage)
(3 missing values generated)
. generate sqrtwage = sqrt(wage)
(2 missing values generated)
. list wage intwage rndwage lnwage logwage sqrtwage in 1/5
```

	wage	intwage	rndwage	lnwage	logwage	sqrtwage
1.	7.15781	7	7	1.968204	.8547801	2.675408
2.	2.447664	2	2	.8951342	.3887518	1.564501
3.	3.824476	3	4	1.341422	.582572	1.955627
4.	14.32367	14	14	2.661913	1.156054	3.784662
5.	5.517124	5	6	1.707857	.7417127	2.348856

Stata has a variety of functions for creating random variables. For example, the **runiform()** (random uniform) function can be used to create a variable with a random number ranging from 0 to 1. Below the seed of the random-function generator is set to a number picked from thin air,[2] and then the **generate** command is used to make a variable, **r**, that is a random number between 0 and 1.

```
. set seed 83271
. generate r = runiform()
. summarize r
    Variable |       Obs        Mean    Std. Dev.       Min        Max
-------------+--------------------------------------------------------
           r |      2246    .4989732    .2848917    .000238    .999742
```

The **rnormal()** (random normal) function allows us to draw random values from a normal distribution with a mean of 0 and a standard deviation of 1, as illustrated below with the variable **randz**. The variable **randiq** is created, drawn from a normal distribution with a mean of 100 and a standard deviation of 15 (which is the same distribution as some IQ tests).

```
. generate randz = rnormal()
. generate randiq = rnormal(100,15)
. summarize randz randiq
    Variable |       Obs        Mean    Std. Dev.       Min        Max
-------------+--------------------------------------------------------
       randz |      2246    -.002875    1.013841  -4.121484   3.421237
      randiq |      2246    100.0628    15.03306   48.35595   160.1314
```

You can even use the **rchi2()** (random chi-squared) function to create a variable representing a random value from a chi-squared distribution. For example, below we create **randchi2**, which draws random values from a chi-squared distribution with 5 degrees of freedom.

```
. generate randchi2 = rchi2(5)
. summarize randchi2
    Variable |       Obs        Mean    Std. Dev.       Min        Max
-------------+--------------------------------------------------------
    randchi2 |      2246     5.05211     3.22275   .0928418   22.49582
```

This section has illustrated just a handful of the numeric functions that are available in Stata. For more information on functions, see section A.7.

5.4 String expressions and functions

The previous section focused on numeric expressions and functions, while this section focuses on string expressions and functions. **dentlab2.dta** will be used to illustrate

2. Setting the seed guarantees that we get the same series of random numbers every time we run the commands, making results that use random numbers reproducible.

string functions, because it contains names of dentists (shown below). We first format **name** so that it displays using left-justification (see section 4.8).

```
. use dentlab2
. format name %-17s
. list

     +-----------------------------------------------+
     | name                years    fulltime   recom |
     |-----------------------------------------------|
  1. | Y. Don Uflossmore    7.25   part time       1 |
  2. | Olive Tu'Drill      10.25   full time       1 |
  3. | isaac O'yerbreath   32.75   full time       1 |
  4. |  Ruth Canaale       22.00   full time       1 |
  5. | Mike avity           8.50   part time       0 |
     |-----------------------------------------------|
  6. | i William Crown      3.20   full time       0 |
  7. | Don b Iteme          4.10   full time       0 |
  8. | Ott W. Onthurt       1.10   full time       0 |
     +-----------------------------------------------+
```

Note how the names have some errors and inconsistencies. There is an extra space before Ruth's name. Sometimes the first letter or initial is in lowercase, and sometimes periods are omitted after initials. By cleaning up these names, we can see how to work with string expressions and functions in Stata.

We could handle the irregularities of capitalization in several ways. The **proper()** function generally does a good job of capitalizing names properly. If our goal was comparing or matching the names, we could use the **lower()** function to convert the name into all lowercase, or the **upper()** function to convert the name into all uppercase. The variables created using these functions are shown below.

```
. generate name2 = proper(name)
. generate lowname = lower(name)
. generate upname = upper(name)
. list name2 lowname upname

     +-----------------------------------------------------------+
     |             name2             lowname              upname |
     |-----------------------------------------------------------|
  1. | Y. Don Uflossmore   y. don uflossmore   Y. DON UFLOSSMORE |
  2. |     Olive Tu'Drill      olive tu'drill      OLIVE TU'DRILL |
  3. | Isaac O'Yerbreath   isaac o'yerbreath   ISAAC O'YERBREATH |
  4. |      Ruth Canaale        ruth canaale        RUTH CANAALE |
  5. |        Mike Avity          mike avity          MIKE AVITY |
     |-----------------------------------------------------------|
  6. |   I William Crown     i william crown     I WILLIAM CROWN |
  7. |       Don B Iteme         don b iteme         DON B ITEME |
  8. |    Ott W. Onthurt      ott w. onthurt      OTT W. ONTHURT |
     +-----------------------------------------------------------+
```

We can trim off the leading blanks, like the one in front of Ruth's name, using the `ltrim()` function, like this:

```
. generate name3 = ltrim(name2)
```

To see the result of the `ltrim()` function, we need to left-justify `name2` and `name3` before we list the results.

```
. format name2 name3 %-17s
. list name name2 name3

     +---------------------------------------------------------+
     | name                name2               name3              |
     |---------------------------------------------------------|
  1. | Y. Don Uflossmore   Y. Don Uflossmore   Y. Don Uflossmore  |
  2. | Olive Tu´Drill      Olive Tu´Drill      Olive Tu´Drill     |
  3. | isaac O´yerbreath   Isaac O´Yerbreath   Isaac O´Yerbreath  |
  4. |  Ruth Canaale        Ruth Canaale       Ruth Canaale       |
  5. | Mike avity          Mike Avity          Mike Avity         |
     |---------------------------------------------------------|
  6. | i William Crown     I William Crown     I William Crown    |
  7. | Don b Iteme         Don B Iteme         Don B Iteme        |
  8. | Ott W. Onthurt      Ott W. Onthurt      Ott W. Onthurt     |
     +---------------------------------------------------------+
```

Let's identify the names that start with an initial rather than with a full first name. When you look at those names, their second character is either a period or a space. We need a way to extract a piece of the name, starting with the second character and extracting that one character. The `substr()` function in the `generate` command below does exactly this, creating the variable `secondchar`. Then the value of `firstinit` gets the value of the logical expression that tests if `secondchar` is a space or a period, yielding a 1 if this expression is true and 0 if false (see section 5.2).

```
. generate secondchar = substr(name3,2,1)
. generate firstinit = (secondchar==" " | secondchar==".")
> if ! missing(secondchar)
. list name3 secondchar firstinit, abb(20)

     +------------------------------------------+
     | name3               secondchar   firstinit |
     |------------------------------------------|
  1. | Y. Don Uflossmore            .           1 |
  2. | Olive Tu´Drill               l           0 |
  3. | Isaac O´Yerbreath            s           0 |
  4. | Ruth Canaale                 u           0 |
  5. | Mike Avity                   i           0 |
     |------------------------------------------|
  6. | I William Crown                          1 |
  7. | Don B Iteme                  o           0 |
  8. | Ott W. Onthurt               t           0 |
     +------------------------------------------+
```

We might want to take the full name and break it up into first, middle, and last names. Because some of the dentists have only two names, we first need to count the number of names using the `wordcount()` function.

```
. generate namecnt = wordcount(name3)
. list name3 namecnt

     +------------------------------+
     | name3                namecnt |
     |------------------------------|
  1. | Y. Don Uflossmore          3 |
  2. | Olive Tu´Drill             2 |
  3. | Isaac O´Yerbreath          2 |
  4. | Ruth Canaale               2 |
  5. | Mike Avity                 2 |
     |------------------------------|
  6. | I William Crown            3 |
  7. | Don B Iteme                3 |
  8. | Ott W. Onthurt             3 |
     +------------------------------+
```

Now we can split name3 into first, middle, and last names using the `word()` function. The first name is the first word of the dentist's name (i.e., `word(name3,1)`). The second name is the second word if the dentist has three names (i.e., `word(name3,2) if namecnt == 3`). The last name is based on the number of names the dentist has (i.e., `word(name3,namecnt)`).

```
. generate fname = word(name3,1)
. generate mname = word(name3,2) if namecnt == 3
(4 missing values generated)
. generate lname = word(name3,namecnt)
. list name3 fname mname lname

     +----------------------------------------------------------+
     | name3                fname     mname           lname     |
     |----------------------------------------------------------|
  1. | Y. Don Uflossmore       Y.       Don     Uflossmore      |
  2. | Olive Tu´Drill       Olive                  Tu´Drill     |
  3. | Isaac O´Yerbreath    Isaac               O´Yerbreath     |
  4. | Ruth Canaale          Ruth                   Canaale     |
  5. | Mike Avity            Mike                     Avity     |
     |----------------------------------------------------------|
  6. | I William Crown          I   William           Crown     |
  7. | Don B Iteme            Don         B           Iteme     |
  8. | Ott W. Onthurt         Ott        W.         Onthurt     |
     +----------------------------------------------------------+
```

If you look at the values of `fname` and `mname` above, you can see that some of the names are composed of one initial. Sometimes the initial was entered with a period after it, and sometimes the period was ignored.

Let's make all the initials have a period after them. In the first `replace` command below, the `length()` function is used to identify observations where the first name is one character. In such instances, the `fname` variable is replaced with `fname` with a period appended to it (showing that the plus sign can be used to combine strings together). The same strategy is applied to the middle names in the next `replace` command.

```
. replace fname = fname + "." if length(fname)==1
(1 real change made)

. replace mname = mname + "." if length(mname)==1
(1 real change made)
```

Below we see that the first and middle names always have a period after them if they are one initial.

```
. list fname mname

     +-----------------+
     | fname     mname |
     |-----------------|
  1. |    Y.       Don |
  2. | Olive           |
  3. | Isaac           |
  4. |  Ruth           |
  5. |  Mike           |
     |-----------------|
  6. |    I.   William |
  7. |   Don        B. |
  8. |   Ott        W. |
     +-----------------+
```

Now that we have repaired the first and middle names, we can join the first, middle, and last names together to form a full name.

```
. generate fullname = fname + " " + lname if namecnt == 2
(4 missing values generated)

. replace fullname = fname + " " + mname + " " + lname if namecnt == 3
(4 real changes made)

. list fname mname lname fullname

     +-------------------------------------------------------+
     | fname     mname         lname                fullname |
     |-------------------------------------------------------|
  1. |    Y.       Don    Uflossmore    Y. Don Uflossmore    |
  2. | Olive                Tu´Drill       Olive Tu´Drill    |
  3. | Isaac              O´Yerbreath    Isaac O´Yerbreath   |
  4. |  Ruth                 Canaale        Ruth Canaale     |
  5. |  Mike                   Avity          Mike Avity     |
     |-------------------------------------------------------|
  6. |    I.   William         Crown     I. William Crown    |
  7. |   Don        B.         Iteme         Don B. Iteme    |
  8. |   Ott        W.       Onthurt       Ott W. Onthurt    |
     +-------------------------------------------------------+
```

For more information about string functions, see `help string functions`.

5.5 Recoding

Sometimes you want to recode the values of an existing variable to make a new variable, mapping the existing values for the existing variable to new values for the new variable. For example, consider the variable `occupation` from `wws2lab.dta`.

```
. use wws2lab
(Working Women Survey w/fixes)
. codebook occupation, tabulate(20)

occupation                                                          occupation

                  type:  numeric (byte)
                 label:  occlbl

                 range:  [1,13]                       units:  1
         unique values:  13                       missing .:  9/2246

           tabulation:  Freq.   Numeric  Label
                          319         1  Professional/technical
                          264         2  Managers/admin
                          725         3  Sales
                          101         4  Clerical/unskilled
                           53         5  Craftsmen
                          246         6  Operatives
                           28         7  Transport
                          286         8  Laborers
                            1         9  Farmers
                            9        10  Farm laborers
                           16        11  Service
                            2        12  Household workers
                          187        13  Other
                            9         .
```

Let's recode `occupation` into three categories: white collar, blue collar, and other. Say that we decide that occupations 1–3 will be white collar, 5–8 will be blue collar, and 4 and 9–13 will be other. We recode the variable below, creating a new variable called `occ3`.

```
. recode occupation (1/3=1) (5/8=2) (4 9/13=3), generate(occ3)
(1918 differences between occupation and occ3)
```

We use the `table` command to double-check that the variable `occ` was properly recoded into `occ3`.

```
. table occupation occ3
```

occupation	RECODE of occupation (occupation) 1	2	3
Professional/technical	319		
Managers/admin	264		
Sales	725		
Clerical/unskilled			101
Craftsmen		53	
Operatives		246	
Transport		28	
Laborers		286	
Farmers			1
Farm laborers			9
Service			16
Household workers			2
Other			187

This is pretty handy, but it would be nice if the values of `occ3` were labeled. Although we could use the **label define** and **label values** commands to label the values of `occ3` (as illustrated in section 4.4), the example below shows a shortcut that labels the values as part of the recoding process. Value labels are given after the new values in the **recode** command. (Continuation comments are used to make this long command more readable; see section A.4 for more information.)

```
. drop occ3
. recode occupation (1/3=1 "White Collar") ///
>                   (5/8=2 "Blue Collar") ///
>                   (4 9/13=3 "Other"), generate(occ3)
(1918 differences between occupation and occ3)
. label variable occ3 "Occupation in 3 groups"
. table occupation occ3
```

occupation	Occupation in 3 groups White Collar	Blue Collar	Other
Professional/technical	319		
Managers/admin	264		
Sales	725		
Clerical/unskilled			101
Craftsmen		53	
Operatives		246	
Transport		28	
Laborers		286	
Farmers			1
Farm laborers			9
Service			16
Household workers			2
Other			187

The `recode` command can also be useful when applied to continuous variables. Say that we wanted to recode the woman's hourly wage (`wage`) into four categories using the following rules: 0 up to 10 would be coded 1, over 10 to 20 would be coded 2, over 20 to 30 would be coded 3, and over 30 would be coded 4. We can do this as shown below. When you specify `recode` *#1*/*#2*, all values between *#1* and *#2*, including the boundaries *#1* and *#2* are included. So when we specify `recode wage (0/10=1) (10/20=2)`, 10 is included in both of these rules. In such cases, the first rule encountered takes precedence, so 10 is recoded to having a value of 1.

```
. recode wage (0/10  =1 "0   to 10") ///
>             (10/20 =2 ">10 to 20") ///
>             (20/30 =3 ">20 to 30") ///
>             (30/max=4 ">30 and up"), generate(wage4)
(2244 differences between wage and wage4)
```

We can check this using the `tabstat` command below (see section 3.5). The results confirm that `wage4` was created correctly. For example, for category 2 (over 10 up to 20), the minimum is slightly larger than 10 and the maximum is 20.

```
. tabstat wage, by(wage4) stat(min max)
Summary for variables: wage
     by categories of: wage4 (RECODE of wage (hourly wage))
```

wage4	min	max
0 to 10	0	10
>10 to 20	10.00805	20
>20 to 30	20.12883	30
>30 and up	30.19324	40.74659
Total	0	40.74659

You might want to use a rule that 0 up to (but not including) 10 would be coded 1, 10 up to (but not including) 20 would be coded 2, 20 up to (but not including) 30 would be coded 3, and 30 and over would be coded 4. By switching the order of the rules, now, for example, 10 belongs to category 2 because that rule appears first.

```
. recode wage (30/max=4 "30 and up") ///
>             (20/30 =3 "20 to <30") ///
>             (10/20 =2 "10 to <20") ///
>             (0/10  =1 "0  to <10"), generate(wage4a)
(2244 differences between wage and wage4a)
```

The results of the `tabstat` command below confirm that `wage4a` was recoded properly.

```
. tabstat wage, by(wage4a) stat(min max)

Summary for variables: wage
     by categories of: wage4a (RECODE of wage (hourly wage))

  wage4a |       min       max
---------+--------------------
0  to <10 |         0  9.999998
10 to <20 |        10  19.91143
20 to <30 |        20  29.72623
30 and up |        30  40.74659
---------+--------------------
    Total |         0  40.74659
```

The recode command is not the only way to recode variables. Stata has several functions that you can also use for recoding. The irecode() function can be used to recode a continuous variable into groups based on a series of cutpoints that you supply. For example, below, the wages are cut into four groups based on the cutpoints 10, 20, and 30. Those with wages up to 10 are coded 0, over 10 up to 20 are coded 1, over 20 up to 30 are coded 2, and over 30 are coded 3.

```
. generate mywage1 = irecode(wage,10,20,30)
(2 missing values generated)
```

The tabstat command confirms the recoding of this variable:

```
. tabstat wage, by(mywage1) stat(min max)

Summary for variables: wage
     by categories of: mywage1

 mywage1 |       min       max
---------+--------------------
       0 |         0        10
       1 |  10.00805        20
       2 |  20.12883        30
       3 |  30.19324  40.74659
---------+--------------------
   Total |         0  40.74659
```

The autocode() function recodes continuous variables into equally spaced groups. Below wage is recoded to form three equally spaced groups that span from 0 to 42. The groups are numbered according to the highest value in the group, so 14 represents 0 to 14, then 28 represents over 14 to 28, and 42 represents over 28 up to 42. The tabstat command confirms the recoding.

(Continued on next page)

```
. generate mywage2 = autocode(wage,3,0,42)
(2 missing values generated)

. tabstat wage, by(mywage2) stat(min max n)

Summary for variables: wage
     by categories of: mywage2

 mywage2 |       min       max         N
---------+------------------------------
      14 |         0   13.9694      2068
      28 |  14.00966  27.89049       127
      42 |  28.15219  40.74659        49
---------+------------------------------
   Total |         0  40.74659      2244
----------------------------------------
```

Although the `autocode()` function seeks to equalize the spacing of the groups, the `group()` option to the `egen` command seeks to equalize the number of observations in each group. Below we create `mywage3` using the `group()` option to create three equally sized groups.

```
. egen mywage3 = cut(wage), group(3)
(2 missing values generated)
```

The values of `mywage3` are numbered 0, 1, and 2. The lower and upper limits of `wage` for each group are selected to attempt to equalize the size of the groups, so the values chosen are not round numbers. The `tabstat` command below shows the lower and upper limits of wages for each of the three groups. The first group ranges from 0 to 4.904, the second group ranges from 4.911 to 8.068, and the third group ranges from 8.075 to 40.747.

```
. tabstat wage, by(mywage3) stat(min max n)

Summary for variables: wage
     by categories of: mywage3

 mywage3 |       min       max         N
---------+------------------------------
       0 |         0  4.903378       748
       1 |  4.911432  8.067631       748
       2 |  8.075683  40.74659       748
---------+------------------------------
   Total |         0  40.74659      2244
----------------------------------------
```

See `help recode`, `help irecode`, and `help autocode` for more information on recoding.

5.6 Coding missing values

As described in section A.10, Stata supports 27 different missing-value codes, including `.`, `.a`, `.b`, ..., `.z`. This section illustrates how you can assign such missing-value codes in your data. Consider this example dataset with missing values:

```
. use cardio2miss
. list
```

	id	age	pl1	pl2	pl3	pl4	pl5	bp1	bp2	bp3	bp4	bp5
1.	1	40	54	115	87	86	93	129	81	105	.b	.b
2.	2	30	92	123	88	136	125	107	87	111	58	120
3.	3	16	105	.a	97	122	128	101	57	109	68	112
4.	4	23	52	105	79	115	71	121	106	129	39	137
5.	5	18	70	116	.a	128	52	112	68	125	59	111

Note how this dataset has some missing values and uses different kinds of missing values to indicate different reasons for missing values. Here the value of `.a` is used to signify a missing value because of a recording error and `.b` is used to signify a missing value because the subject dropped out. But how did these values get assigned? Let's start with the original raw data.

```
. infile id age pl1-pl5 bp1-bp5 using cardio2miss.txt
(5 observations read)
. list
```

	id	age	pl1	pl2	pl3	pl4	pl5	bp1	bp2	bp3	bp4	bp5
1.	1	40	54	115	87	86	93	129	81	105	-2	-2
2.	2	30	92	123	88	136	125	107	87	111	58	120
3.	3	16	105	-1	97	122	128	101	57	109	68	112
4.	4	23	52	105	79	115	71	121	106	129	39	137
5.	5	18	70	116	-1	128	52	112	68	125	59	111

The value of `-1` indicates missing values because of recording errors, and `-2` indicates missing values because the subject dropped out of the study. The `recode` command can be used to convert these values into the appropriate missing-value codes, as shown below. (Note that `bp*` stands for any variable that begins with `bp`; see section A.11.)

```
. recode bp* pl* (-1=.a) (-2=.b)
  (output omitted)
. list
```

	id	age	pl1	pl2	pl3	pl4	pl5	bp1	bp2	bp3	bp4	bp5
1.	1	40	54	115	87	86	93	129	81	105	.b	.b
2.	2	30	92	123	88	136	125	107	87	111	58	120
3.	3	16	105	.a	97	122	128	101	57	109	68	112
4.	4	23	52	105	79	115	71	121	106	129	39	137
5.	5	18	70	116	.a	128	52	112	68	125	59	111

Another way to convert the values to missing-value codes would be to use the `mvdecode` command, which converts regular numbers into missing values. As the example below shows, the `mv()` option specifies that the values of `-1` should be converted to `.a` and the values of `-2` should be converted to `.b`.

```
. mvdecode bp* pl*, mv(-1=.a \ -2=.b)
        bp4: 1 missing value generated
        bp5: 1 missing value generated
        pl2: 1 missing value generated
        pl3: 1 missing value generated

. list
```

	id	age	pl1	pl2	pl3	pl4	pl5	bp1	bp2	bp3	bp4	bp5
1.	1	40	54	115	87	86	93	129	81	105	.b	.b
2.	2	30	92	123	88	136	125	107	87	111	58	120
3.	3	16	105	.a	97	122	128	101	57	109	68	112
4.	4	23	52	105	79	115	71	121	106	129	39	137
5.	5	18	70	116	.a	128	52	112	68	125	59	111

If you just wanted the values of `-1` and `-2` to be assigned to the general missing-value code (`.`), then you can do so as shown below:

```
. mvdecode bp* pl*, mv(-1 -2)
        bp4: 1 missing value generated
        bp5: 1 missing value generated
        pl2: 1 missing value generated
        pl3: 1 missing value generated

. list
```

	id	age	pl1	pl2	pl3	pl4	pl5	bp1	bp2	bp3	bp4	bp5
1.	1	40	54	115	87	86	93	129	81	105	.	.
2.	2	30	92	123	88	136	125	107	87	111	58	120
3.	3	16	105	.	97	122	128	101	57	109	68	112
4.	4	23	52	105	79	115	71	121	106	129	39	137
5.	5	18	70	116	.	128	52	112	68	125	59	111

The `mvdecode` command has a companion command called `mvencode`, which converts missing values into regular numbers. In the example below, we convert the missing values for all the blood pressure and pulse scores to be `-1`.

```
. use cardio2miss

. mvencode bp* pl*, mv(-1)
        bp4: 1 missing value recoded
        bp5: 1 missing value recoded
        pl2: 1 missing value recoded
        pl3: 1 missing value recoded

. list
```

	id	age	pl1	pl2	pl3	pl4	pl5	bp1	bp2	bp3	bp4	bp5
1.	1	40	54	115	87	86	93	129	81	105	-1	-1
2.	2	30	92	123	88	136	125	107	87	111	58	120
3.	3	16	105	-1	97	122	128	101	57	109	68	112
4.	4	23	52	105	79	115	71	121	106	129	39	137
5.	5	18	70	116	-1	128	52	112	68	125	59	111

Or as shown below, the values of `.a` are converted to `-1`, and the values of `.b` are converted to `-2`.

```
. use cardio2miss
. mvencode bp* pl*, mv(.a=-1 \ .b=-2)
        bp4: 1 missing value recoded
        bp5: 1 missing value recoded
        pl2: 1 missing value recoded
        pl3: 1 missing value recoded
. list
```

	id	age	pl1	pl2	pl3	pl4	pl5	bp1	bp2	bp3	bp4	bp5
1.	1	40	54	115	87	86	93	129	81	105	-2	-2
2.	2	30	92	123	88	136	125	107	87	111	58	120
3.	3	16	105	-1	97	122	128	101	57	109	68	112
4.	4	23	52	105	79	115	71	121	106	129	39	137
5.	5	18	70	116	-1	128	52	112	68	125	59	111

This concludes this section, which illustrated how to code missing values in Stata. For more information, see section A.10, `help mvdecode`, and `help mvencode`.

5.7 Dummy variables

Stata 11 introduced factor variables, providing built-in tools supporting categorical variables. In previous versions, you would have needed to use `xi:` combined with the `i.` prefix to create indicator (dummy) variables or you would have needed to manually create dummy variables. But starting in Stata 11, these steps are no longer needed to convert categorical variables into dummy variables. This section illustrates the creation of dummy variables using `wws2lab.dta`. Consider the variable `grade4`, which represents education level with four levels:

```
. use wws2lab
(Working Women Survey w/fixes)
. codebook grade4
-------------------------------------------------------------------------------
grade4                                    4 level Current Grade Completed
-------------------------------------------------------------------------------

                  type:  numeric (byte)
                 label:  grade4

                 range:  [1,4]                       units:  1
         unique values:  4                       missing .:  4/2246

            tabulation:  Freq.   Numeric  Label
                           332         1  Not HS
                           941         2  HS Grad
                           456         3  Some Coll
                           513         4  Coll Grad
                             4         .
```

grade4 is a categorical variable (which Stata calls a factor variable). You can perform a regression analysis predicting hourly wages, **wage**, from the levels of **grade4** like this.

```
. regress wage i.grade4
      Source |       SS       df       MS              Number of obs =    2240
-------------+------------------------------           F(  3,  2236) =   85.35
       Model |  7811.98756     3  2603.99585           Prob > F      =  0.0000
    Residual |  68221.1897  2236  30.5103711           R-squared     =  0.1027
-------------+------------------------------           Adj R-squared =  0.1015
       Total |  76033.1772  2239  33.9585428           Root MSE      =  5.5236

------------------------------------------------------------------------------
        wage |      Coef.   Std. Err.      t    P>|t|     [95% Conf. Interval]
-------------+----------------------------------------------------------------
      grade4 |
          2  |   1.490229   .3526422     4.23   0.000      .798689     2.18177
          3  |   3.769248   .3985065     9.46   0.000     2.987767    4.550729
          4  |   5.319548   .3892162    13.67   0.000     4.556285     6.08281
             |
       _cons |   5.194571    .303148    17.14   0.000      4.60009    5.789052
------------------------------------------------------------------------------
```

Stata intrinsically understands that supplying the `i.` prefix to **grade4** means to convert it into $k-1$ dummy variables (where k is the number of levels of **grade4**). By default, the first group is the omitted (base) group.

The **regress** command is not the only command that understands how to work with factor variables. In fact, most Stata commands understand how to work with factor variables, including data-management commands like `list` and **generate**. For example, below we list the first five observations for **wage**, **grade4**, and **i.grade4**. (The **nolabel** option shows the numeric values of **grade4** instead of the labeled values.)

```
. list wage grade4 i.grade4 in 1/5, nolabel

     +------------------------------------------------------------+
     |                        1b.         2.         3.        4. |
     |     wage   grade4   grade4     grade4     grade4    grade4 |
     |------------------------------------------------------------|
  1. |  7.15781        1        0          0          0         0 |
  2. | 2.447664        2        0          1          0         0 |
  3. | 3.824476        3        0          0          1         0 |
  4. | 14.32367        4        0          0          0         1 |
  5. | 5.517124        2        0          1          0         0 |
     +------------------------------------------------------------+
```

When we typed `i.`**grade4**, this was expanded into the names of four virtual dummy variables, the last three of which were used when the regression analysis was run. (The first level of **grade4** is the baseline, or omitted, category; that is why it is all 0s in the listing and is named `1b.`**grade4**.)

If, instead, we specify **ibn.grade4**, this specifies that we want no baseline group (the **bn** means baseline none).

```
. list wage grade4 ibn.grade4 in 1/5
```

	wage	grade4	1. grade4	2. grade4	3. grade4	4. grade4
1.	7.15781	Not HS	1	0	0	0
2.	2.447664	HS Grad	0	1	0	0
3.	3.824476	Some Coll	0	0	1	0
4.	14.32367	Coll Grad	0	0	0	1
5.	5.517124	HS Grad	0	1	0	0

Specifying `ibn.grade4` yields four dummy variables named `1.grade4`–`4.grade4`. The dummy variable `1.grade4` is a dummy variable that is 1 if the value of `grade4` is 1 and is 0 otherwise. Likewise, `2.grade4` is a dummy variable that is 1 if the value of `grade4` is 2 and is 0 otherwise, and so forth up to `4.grade4`.

Although #`.grade4` is not added to your dataset (typing `describe` will confirm this), you can refer to #`.grade4` just as you would any other variable in your dataset.

The `generate` commands below create our own dummy variables corresponding to the levels of `grade4`.

```
. generate noths = 1.grade4
(4 missing values generated)
. generate hs = 2.grade4
(4 missing values generated)
. generate smcl = 3.grade4
(4 missing values generated)
. generate clgr = 4.grade4
(4 missing values generated)
. list grade4 noths hs smcl clgr in 1/5, nolabel
```

	grade4	noths	hs	smcl	clgr
1.	1	1	0	0	0
2.	2	0	1	0	0
3.	3	0	0	1	0
4.	4	0	0	0	1
5.	2	0	1	0	0

The above example illustrates that the virtual variable `1.grade4` refers to the dummy variable associated with the value of 1 for `grade4` and `2.grade4` refers to the dummy variable associated with the value of 2 for `grade4` and so forth. When referring to these values individually, as we did in the `generate` command, there is no baseline or omitted value. As you can see, the value of the generated variable `noths` takes on a value of 1 if `grade4` is 1 and a value of 0 if it is not 1 (except if `grade4` is missing, and then `1.grade4` is also missing).

You can change which group is considered the base (omitted) group when using the `i.` prefix. In the previous examples, where we specified `i.grade4` with the `regress` and `list` commands, the first group was used as the omitted group; this is the default. If, instead, we specify **ib2.grade4**, the group where **grade4** equals 2 will be the omitted group, as shown below.

```
. regress wage ib2.grade4
      Source |       SS       df       MS              Number of obs =    2240
-------------+------------------------------           F(  3,  2236) =   85.35
       Model |  7811.98756     3  2603.99585           Prob > F      =  0.0000
    Residual |  68221.1897  2236  30.5103711           R-squared     =  0.1027
-------------+------------------------------           Adj R-squared =  0.1015
       Total |  76033.1772  2239  33.9585428           Root MSE      =  5.5236

------------------------------------------------------------------------------
        wage |      Coef.   Std. Err.      t    P>|t|     [95% Conf. Interval]
-------------+----------------------------------------------------------------
      grade4 |
           1 |  -1.490229   .3526422    -4.23   0.000     -2.18177   -.798689
           3 |   2.279019   .3152246     7.23   0.000     1.660855   2.897182
           4 |   3.829318   .3033948    12.62   0.000     3.234353   4.424283
             |
       _cons |     6.6848   .1801606    37.10   0.000     6.331501     7.0381
------------------------------------------------------------------------------
```

You could also specify **ib(first).grade4** to make the first group the omitted group or **ib(last).grade4** to make the last group the omitted group.

Another way to specify the omitted group is by using the **fvset** (factor-variable set) command. For example, the **fvset** command below specifies that the value of 3 will be the base (omitted) group for the variable **grade4**. You can see how this then changes the base group to 3 when we refer to **i.grade4** in the **regress** command. In fact, if you **save** the dataset, Stata will remember this setting the next time you **use** the dataset.

```
. fvset base 3 grade4
. regress wage i.grade4
      Source |       SS       df       MS              Number of obs =    2240
-------------+------------------------------           F(  3,  2236) =   85.35
       Model |  7811.98756     3  2603.99585           Prob > F      =  0.0000
    Residual |  68221.1897  2236  30.5103711           R-squared     =  0.1027
-------------+------------------------------           Adj R-squared =  0.1015
       Total |  76033.1772  2239  33.9585428           Root MSE      =  5.5236

------------------------------------------------------------------------------
        wage |      Coef.   Std. Err.      t    P>|t|     [95% Conf. Interval]
-------------+----------------------------------------------------------------
      grade4 |
           1 |  -3.769248   .3985065    -9.46   0.000    -4.550729  -2.987767
           2 |  -2.279019   .3152246    -7.23   0.000    -2.897182  -1.660855
           4 |     1.5503   .3556674     4.36   0.000     .8528268   2.247772
             |
       _cons |   8.963819   .2586672    34.65   0.000     8.456566   9.471072
------------------------------------------------------------------------------
```

For more information about using factor variables, see **help factor variables** and **help fvset**.

Tip! Interaction terms

Stata simplifies the inclusion of interaction terms in your model. For example, you can include the main effects and interactions of two categorical variables (e.g., grade4 and married) as shown below.

```
. regress wage i.grade4##i.married
```

You can include an interaction of a categorical variable (like grade4) and a continuous variable (like age) as shown below. Note that the continuous variable is prefixed with c..

```
. regress wage i.grade4##c.age
```

You can even include in the model c.age##c.age, which specifies the linear and quadratic effect of age.

```
. regress wage c.age##c.age
```

Knowing these tricks for your analysis can save you the effort of creating these variables as part of your data management.

5.8 Date variables

Stata supports both date variables (such as a birth date) as well as date-and-time variables (such as a date and time of birth). This section covers date variables, while the following section (section 5.9) covers date-and-time variables. This section covers how to read raw datasets with date information, how to create and format dates, how to perform computations with date variables, and how to perform comparisons on dates. Let's use as an example a file named momkid1.csv, which contains information about four moms, their birthdays, and the birthday of each mom's first kid.

```
. type momkid1.csv
momid,momm,momd,momy,kidbday
1,11,28,1972,1/5/1998
2,4,3,1973,4/11/2002
3,6,13,1968,5/15/1996
4,1,5,1960,1/4/2004
```

This illustrates two common formats that can be used for storing dates in raw data files. The second, third, and fourth variables in the file are the month, day, and year of the mom's birthday as three separate variables. The fifth variable contains the month, day, and year of the kid's birthday as one variable. When we read these variables into Stata using the insheet command (below), the month, day, and year of the mom's birthday are stored as three separate numeric variables, and the kid's birthday is stored as one string variable.

```
. insheet using momkid1.csv
(5 vars, 4 obs)
. list
```

	momid	momm	momd	momy	kidbday
1.	1	11	28	1972	1/5/1998
2.	2	4	3	1973	4/11/2002
3.	3	6	13	1968	5/15/1996
4.	4	1	5	1960	1/4/2004

Once we have the variables read into Stata, we can convert them into date variables. We can use the `mdy()` function to create a date variable containing the mom's birthday. The month, day, and year are then converted into the date variable `mombdate`.

```
. generate mombdate = mdy(momm,momd,momy)
```

The kid's birthday was read into the string variable `kidbday`. Below we convert this string variable into a date variable named `kidbdate` by using the `date()` function. We told the `date()` function that the date was in `"MDY"` format, meaning that first comes the month, then the day, and finally the year.

```
. generate kidbdate = date(kidbday,"MDY")
```

Let's list these variables and see the results.

```
. list momm momd momy mombdate kidbday kidbdate
```

	momm	momd	momy	mombdate	kidbday	kidbdate
1.	11	28	1972	4715	1/5/1998	13884
2.	4	3	1973	4841	4/11/2002	15441
3.	6	13	1968	3086	5/15/1996	13284
4.	1	5	1960	4	1/4/2004	16074

The `mombdate` and `kidbdate` variables seem like they are stored as some kind of strange number that does not make any sense. Looking at the fourth mom, we notice that her value for `mombdate` is 4 and her birthday is Jan 5, 1960. This helps illustrate that Stata stores each date as the number of days from Jan 1, 1960 (a completely arbitrary value). Imagine that all dates are on a number line where a date of 0 is Jan 1, 1960, 1 is Jan 2, 1960, 4 is Jan 5, 1960, and so forth. Like a number line, there can be negative values; for example, Dec 31, 1959, would be -1 and Dec 30, 1959, would be -2.

To make the dates easier to read, we can use the `format` command, which requests that `mombdate` and `kidbdate` be displayed using the `%td` format. The underlying contents of these variables remain unchanged, but they are displayed showing the two-digit day, three-letter month, and four-digit year.

```
. format mombdate kidbdate %td
. list momm momd momy mombdate kidbday kidbdate
```

	momm	momd	momy	mombdate	kidbday	kidbdate
1.	11	28	1972	28nov1972	1/5/1998	05jan1998
2.	4	3	1973	03apr1973	4/11/2002	11apr2002
3.	6	13	1968	13jun1968	5/15/1996	15may1996
4.	1	5	1960	05jan1960	1/4/2004	04jan2004

Stata supports an elaborate mixture of formatting codes that you can add to the `%td` format to customize the display of date variables. Below the moms' birthdays are displayed using the numeric month (`nn`), the day (`dd`), and two-digit year (`YY`).

```
. format mombdate %tdnn/dd/YY
. list momm momd momy mombdate
```

	momm	momd	momy	mombdate
1.	11	28	1972	11/28/72
2.	4	3	1973	4/3/73
3.	6	13	1968	6/13/68
4.	1	5	1960	1/5/60

The kids' birthdays are shown below using the name of the day of the week (`Dayname`), the name of the month (`Month`), the day of the month (`dd`), and the two-digit century combined with the two-digit year (`ccYY`). After the `%td`, a comma inserts a comma, a forward slash inserts a forward slash, and an underscore inserts a space in the display of variables.

```
. format kidbdate %tdDayname_Month_dd,_ccYY
. list kidbday kidbdate
```

	kidbday	kidbdate
1.	1/5/1998	Monday January 5, 1998
2.	4/11/2002	Thursday April 11, 2002
3.	5/15/1996	Wednesday May 15, 1996
4.	1/4/2004	Sunday January 4, 2004

No matter how you change the display format of a date, this does not change the way the dates are stored internally. This internal representation of dates facilitates calculations of the amount of time that has elapsed between two dates. For example, we can compute the mother's age (in days) when she had her first kid by subtracting `mombdate` from `kidbdate` to create a variable called `momagefb`, as shown below.

```
. generate momagefb = kidbdate - mombdate
. list mombdate kidbdate momagefb

     +---------------------------------------------------+
     | mombdate                      kidbdate   momagefb |
     |---------------------------------------------------|
  1. | 11/28/72      Monday January 5, 1998       9169 |
  2. |   4/3/73    Thursday April 11, 2002      10600 |
  3. |  6/13/68      Wednesday May 15, 1996      10198 |
  4. |   1/5/60      Sunday January 4, 2004      16070 |
     +---------------------------------------------------+
```

We normally think of ages in terms of years rather than days. We can divide the age in days by 365.25 to create `momagefbyr`, the age of the mom in years when she had her first kid.[3]

```
. generate momagefbyr = momagefb/365.25
. list momid momagefb momagefbyr, abb(20)

     +----------------------------------+
     | momid   momagefb      momagefbyr |
     |----------------------------------|
  1. |     1       9169        25.10335 |
  2. |     2      10600        29.02122 |
  3. |     3      10198         27.9206 |
  4. |     4      16070        43.99726 |
     +----------------------------------+
```

We might want to know how old the mom is as of a particular date, say, January 20, 2010. We can subtract `mombdate` from `td(20jan2010)` and divide that by 365.25 to obtain the age of the mom in years as of January 20, 2010. Note that `td(20jan2010)` is an example of the way that you can specify a particular date to Stata. We see the results of these computations below:

```
. generate momage = (td(20jan2010) - mombdate)/365.25
. list mombdate momage

     +---------------------+
     | mombdate     momage |
     |---------------------|
  1. | 11/28/72   37.14442 |
  2. |   4/3/73   36.79945 |
  3. |  6/13/68   41.60438 |
  4. |   1/5/60   50.04244 |
     +---------------------+
```

Say that we wanted to list the mothers who were born on or after January 1, 1970. We can do this by listing the cases where the mom's birth date is at least `td(01jan1970)`, as shown below. (Note the handling of the missing values; see section A.10 for more information.)

3. This is an approximation and could be slightly off depending on leap years; however, this simple approximation is likely sufficient for data analysis purposes.

```
. list momid mombdate if (mombdate >= td(01jan1970)) & ! missing(mombdate)
```

	momid	mombdate
1.	1	11/28/72
2.	2	4/3/73

You might want to extract the month, day, or year from a date. The `day()`, `month()`, and `year()` functions make this easy, as shown below.

```
. generate momday = day(mombdate)
. generate mommonth = month(mombdate)
. generate momyear = year(mombdate)
. list momid mombdate momday mommonth momyear
```

	momid	mombdate	momday	mommonth	momyear
1.	1	11/28/72	28	11	1972
2.	2	4/3/73	3	4	1973
3.	3	6/13/68	13	6	1968
4.	4	1/5/60	5	1	1960

There are many other date functions we can use with date variables. For example, the `dow()` function identifies the day of week (coded as 0 = Sunday, 1 = Monday, 2 = Tuesday, ..., 6 = Saturday). The `doy()` function returns the day of the year. The `week()` and `quarter()` functions return the week and quarter (respectively) of the year. Using these functions, we see that the first mom was born on a Tuesday that was the 333rd day of the 48th week in the 4th quarter of the year.

```
. generate momdow = dow(mombdate)
. generate momdoy = doy(mombdate)
. generate momweek = week(mombdate)
. generate momqtr = quarter(mombdate)
. list momid mombdate momdow momdoy momweek momqtr
```

	momid	mombdate	momdow	momdoy	momweek	momqtr
1.	1	11/28/72	2	333	48	4
2.	2	4/3/73	2	93	14	2
3.	3	6/13/68	4	165	24	2
4.	4	1/5/60	2	5	1	1

Let's conclude this section by considering issues that arise when dates are stored using two-digit years instead of four-digit years. Consider the file `momkid2.csv`, below. Note how the years for both the kids' and the moms' birthdays are stored using two-digit years.

```
. type momkid2.csv
momid,momm,momd,momy,kidbday
1,11,28,72,1/5/98
2,4,3,73,4/11/02
3,6,13,68,5/15/96
4,1,5,60,1/4/04
```

Let's read this file and try to convert the birthdays for the moms and kids into date variables.

```
. insheet using momkid2.csv
(5 vars, 4 obs)
. generate mombdate = mdy(momm,momd,momy)
(4 missing values generated)
. generate kidbdate = date(kidbday,"MDY")
(4 missing values generated)
```

This does not look promising. Each `generate` command gave the message `(4 missing values generated)`, suggesting that all values were missing. Nevertheless, let's apply the date format to the date variables and list the variables.

```
. format mombdate kidbdate %td
. list
```

	momid	momm	momd	momy	kidbday	mombdate	kidbdate
1.	1	11	28	72	1/5/98	.	.
2.	2	4	3	73	4/11/02	.	.
3.	3	6	13	68	5/15/96	.	.
4.	4	1	5	60	1/4/04	.	.

As we expected, all the dates are missing. Let's see why this is so by considering the birth dates for the moms. When we told Stata `mdy(momm,momd,momy)`, the values for `momy` were values like 72 or 68 or 60. Stata takes this to mean, literally, the year 72; however, Stata can only handle dates from January 1, 100, to December 31, 9999, so the year 72 is outside of the limits that Stata understands, leading to a missing value. The `mdy` function expects the year to be a full four-digit year. Because all the moms were born in the 1900s, we can simply add 1900 to all their years of birth, as shown below.

```
. generate mombdate = mdy(momm,momd,momy+1900)
. format mombdate %td
. list
```

	momid	momm	momd	momy	kidbday	kidbdate	mombdate
1.	1	11	28	72	1/5/98	.	28nov1972
2.	2	4	3	73	4/11/02	.	03apr1973
3.	3	6	13	68	5/15/96	.	13jun1968
4.	4	1	5	60	1/4/04	.	05jan1960

For the kids' birth dates, we had the same problem. We could instruct Stata to treat all the birth years as though they came from the 1900s, as shown below.

```
. generate kidbdate = date(kidbday,"MD19Y")
. format kidbdate %td
. list

     +--------------------------------------------------------------------+
     | momid   momm   momd   momy   kidbday    mombdate    kidbdate |
     |--------------------------------------------------------------------|
  1. |     1     11     28     72    1/5/98   28nov1972   05jan1998 |
  2. |     2      4      3     73   4/11/02   03apr1973   11apr1902 |
  3. |     3      6     13     68   5/15/96   13jun1968   15may1996 |
  4. |     4      1      5     60    1/4/04   05jan1960   04jan1904 |
     +--------------------------------------------------------------------+
```

This would have worked fine if all the kids were born in the 1900s (and if they had all been born in the 2000s, we could have specified "MD20Y"). What we need is a method for telling Stata when to treat the two-digit year as being from the 1900s versus being from the 2000s.

The date() function allows you to do just this by giving a cutoff year that distinguishes dates in the 1900s from dates in the 2000s. In the example below, any kid with a year of birth from 00 to 20 would be treated as from the 2000s, and any kid with a year of birth over 20 (21 to 99) would be treated as from the 1900s.

```
. generate kidbdate = date(kidbday,"MDY",2020)
. format kidbdate %td
. list

     +--------------------------------------------------------------------+
     | momid   momm   momd   momy   kidbday    mombdate    kidbdate |
     |--------------------------------------------------------------------|
  1. |     1     11     28     72    1/5/98   28nov1972   05jan1998 |
  2. |     2      4      3     73   4/11/02   03apr1973   11apr2002 |
  3. |     3      6     13     68   5/15/96   13jun1968   15may1996 |
  4. |     4      1      5     60    1/4/04   05jan1960   04jan2004 |
     +--------------------------------------------------------------------+
```

What if the kids' birth dates (which cross the boundary of 2000) were stored like the moms' birth dates: as a separate month, day, and year? Such a file is illustrated in momkid3.csv.

```
. type momkid3.csv
momid,momm,momd,momy,kidm,kidd,kidy
1,11,28,72,1,5,98
2,4,3,73,4,11,02
3,6,13,68,5,15,96
4,1,5,60,1,4,04
```

We first read in the month, day, and year of birth for both the moms and the kids.

```
. insheet using momkid3.csv
(7 vars, 4 obs)

. list

     +-------------------------------------------------------+
     | momid   momm   momd   momy   kidm   kidd   kidy |
     |-------------------------------------------------------|
  1. |     1     11     28     72      1      5     98 |
  2. |     2      4      3     73      4     11      2 |
  3. |     3      6     13     68      5     15     96 |
  4. |     4      1      5     60      1      4      4 |
     +-------------------------------------------------------+
```

Then for the kids, we use the **generate** command to create the variable **kidbdate** by adding 1900 to the year if the year of birth was over 20. We then use the **replace** command to replace the contents of **kidbdate** with 2000 added to the year if the year of birth was 20 or below.

```
. generate kidbdate = mdy(kidm,kidd,kidy+1900) if kidy > 20
(2 missing values generated)

. replace  kidbdate = mdy(kidm,kidd,kidy+2000) if kidy <= 20
(2 real changes made)
```

We can see below that the birthdays of the kids are now properly stored as date variables.

```
. format kidbdate %td

. list momid kidm kidd kidy kidbdate

     +-------------------------------------------+
     | momid   kidm   kidd   kidy    kidbdate |
     |-------------------------------------------|
  1. |     1      1      5     98   05jan1998 |
  2. |     2      4     11      2   11apr2002 |
  3. |     3      5     15     96   15may1996 |
  4. |     4      1      4      4   04jan2004 |
     +-------------------------------------------+
```

This concludes this section on dates in Stata. The following section builds upon this section, illustrating how to handle date and time values. For more information, see **help dates and times**.

5.9 Date-and-time variables

The previous section (section 5.8) illustrated how to create and work with date variables (such as date of birth). This section considers variables that are composed of both a date and a time (such as the date and time of birth). This section builds upon and is patterned after section 5.8 but instead focuses on date-and-time values. In this section, you will learn how to read raw data files with date-and-time information, how to create and format date-and-time values, how to perform computations with date-and-time variables, and how to perform comparisons on date-and-time values. We first read in a file named momkid1a.csv, which contains information about four moms with their date and time of birth and the date and time of birth of their first kid.

```
. type momkid1a.csv
id,momm,momd,momy,momh,mommin,moms,kidbday
1,11,28,1972,10,38,51,1/5/1998 15:21:05
2,4,3,1973,06,22,43,4/11/2002 10:49:12
3,6,13,1968,22,45,32,5/15/1996 01:58:29
4,1,5,1960,15,01,12,1/4/2004 23:01:19
```

This data file shows the two common formats that can be used for date-and-time values in a raw data file. The second, third, and fourth variables in the file are the month, day, and year of the mom's birthday, and the fifth, sixth, and seventh variables are the hour (using a 24-hour clock), minute, and second of the mom's birth. The eighth variable contains the kid's date and time of birth. When this file is read using the `insheet` command, the month, day, year, hour, minute, and second of the mom's birthday are stored as six separate numeric variables. The kid's birth date and time is stored as one string variable, as shown below.

```
. insheet using momkid1a.csv
(8 vars, 4 obs)
. list
```

	id	momm	momd	momy	momh	mommin	moms	kidbday
1.	1	11	28	1972	10	38	51	1/5/1998 15:21:05
2.	2	4	3	1973	6	22	43	4/11/2002 10:49:12
3.	3	6	13	1968	22	45	32	5/15/1996 01:58:29
4.	4	1	5	1960	15	1	12	1/4/2004 23:01:19

Once we have the variables read into Stata, we can convert them into date-and-time variables. Below the `mdyhms()` function is used to create the date-and-time variable named `momdt` based on the month, day, year, hour, minute, and second of birth for each mom. Because date-and-time variables can contain very large values, it is imperative that they be stored as type `double`; otherwise, precision can be lost (see section A.5 for more details about data types).

```
. generate double momdt = mdyhms(momm,momd,momy,momh,mommin,moms)
```

Let's apply the `%tc` format to `momdt` to display it as a date-and-time value and then list the observations. We can see that the values of `momdt` exactly represent the values of the date-and-time variables that were used to create it.[4]

4. For those concerned with leap seconds, Stata also includes the `%tC` format. See `help dates and times` for more details.

```
. format momdt %tc
. list id momm momd momy momh mommin moms momdt
```

	id	momm	momd	momy	momh	mommin	moms	momdt
1.	1	11	28	1972	10	38	51	28nov1972 10:38:51
2.	2	4	3	1973	6	22	43	03apr1973 06:22:43
3.	3	6	13	1968	22	45	32	13jun1968 22:45:32
4.	4	1	5	1960	15	1	12	05jan1960 15:01:12

Let's now repeat the above process but intentionally forget to create **momdt** as a double-precision variable. Note (below) how the minutes and seconds for the mom's birthday stored in **momdt** can differ from the values in **mommin** and **moms**. This is the kind of loss of precision that results from forgetting to store date-and-time values using a double-precision variable. In short, when creating date-and-time values, always create them using a double-precision data type (see section A.5 for more about data types).

```
. generate momdt = mdyhms(momm,momd,momy,momh,mommin,moms)
. format momdt %tc
. list id momm momd momy momh mommin moms momdt
```

	id	momm	momd	momy	momh	mommin	moms	momdt
1.	1	11	28	1972	10	38	51	28nov1972 10:39:01
2.	2	4	3	1973	6	22	43	03apr1973 06:22:35
3.	3	6	13	1968	22	45	32	13jun1968 22:45:34
4.	4	1	5	1960	15	1	12	05jan1960 15:01:12

The kid's birth date and time is stored in one string variable named **kidbday**. We can convert this string variable into a date-and-time variable by using the **clock()** function, as shown below. We told the **clock()** function that the date and time was in **"MDYhms"** format, meaning that elements of the date and time are arranged in the following order—month, day, year, hour, minute, second.

```
. generate double kiddt = clock(kidbday,"MDYhms")
```

Below we format the variable **kiddt** using the **%tc** format and show **kidbday** and **kiddt**. We can see that **kiddt** correctly contains the date and time of the kid's birth.

```
. format %tc kiddt
. list id kidbday kiddt
```

	id	kidbday	kiddt
1.	1	1/5/1998 15:21:05	05jan1998 15:21:05
2.	2	4/11/2002 10:49:12	11apr2002 10:49:12
3.	3	5/15/1996 01:58:29	15may1996 01:58:29
4.	4	1/4/2004 23:01:19	04jan2004 23:01:19

As with date variables, Stata supports many formatting codes that you can add to the `%tc` format to control the display of date-and-time variables. Below the mom's birthday is displayed using the numeric month (`nn`), day (`dd`), two-digit year (`YY`), and then the hour using a 24-hour clock (`HH`), minute (`MM`), and second (`SS`).

```
. format momdt %tcnn/dd/YY_HH:MM:SS
. list momm momd momy momdt

     +---------------------------------------+
     | momm   momd   momy              momdt |
     |---------------------------------------|
  1. |   11     28   1972  11/28/72 10:39:01 |
  2. |    4      3   1973    4/3/73 06:22:35 |
  3. |    6     13   1968   6/13/68 22:45:34 |
  4. |    1      5   1960    1/5/60 15:01:12 |
     +---------------------------------------+
```

The kid's birth date and time is shown below using the name of the day of the week (`Dayname`), the name of the month (`Month`), the day of the month (`dd`), the two-digit century combined with the two-digit year (`ccYY`), the hour using a 12-hour clock (`hh`), the minute (`MM`), and an indicator of AM or PM (`am`). After the `%tc`, a comma inserts a comma, a forward slash inserts a forward slash, and the underscore inserts a space in the results display.

```
. format kiddt %tcDayname_Month_dd,_ccYY_hh:MMam
. list kidbday kiddt

     +-------------------------------------------------------+
     |             kidbday                             kiddt |
     |-------------------------------------------------------|
  1. |  1/5/1998 15:21:05     Monday January 5, 1998 3:21pm |
  2. | 4/11/2002 10:49:12  Thursday April 11, 2002 10:49am |
  3. | 5/15/1996 01:58:29     Wednesday May 15, 1996 1:58am |
  4. |  1/4/2004 23:01:19    Sunday January 4, 2004 11:01pm |
     +-------------------------------------------------------+
```

No matter how you change the display format of a date-and-time value, the internal value remains the same. In the previous section, we saw that dates are stored as the number of days from January 1, 1960. Date-and-time values are stored as the number of milliseconds since midnight of January 1, 1960. Because there are 1,000 milliseconds in a second, if one of these kids was born five seconds after midnight on January 1, 1960, her value of `kiddt` would be 5,000.

When we subtract two date-and-time values, the result is the difference in the two dates and times expressed in milliseconds. We could divide the difference by `1000` to convert the results into seconds or divide the difference by `1000*60` to convert the results into minutes, or divide the difference by `1000*60*60` to get the results in hours, or divide the difference by `1000*60*60*24` to get the results in days, or divide the difference by `1000*60*60*24*365.25` to get the results in years.[5]

5. This is assuming that we disregard leap seconds.

Below we compute the mother's age (in milliseconds) when she had her first kid by subtracting `momdt` from `kiddt`, creating a variable called `momfbms`. We remember to create this variable using `double` precision.

```
. generate double momfbms = kiddt - momdt
```

Let's display the date and time of birth of the mom and her first kid, along with the age of the mom (in milliseconds) when her first kid was born. We first format the variable `momfbms` using the `%15.0f` format because these values will be very large.

```
. format momfbms %15.0f
. list momdt kiddt momfbms

     +------------------------------------------------------------------------+
     |              momdt                              kiddt          momfbms |
     |------------------------------------------------------------------------|
  1. | 11/28/72 10:39:01      Monday January 5, 1998 3:21pm     792218523368 |
  2. |   4/3/73 06:22:35   Thursday April 11, 2002 10:49am     915855996992 |
  3. |  6/13/68 22:45:34      Wednesday May 15, 1996 1:58am     881032374664 |
  4. |   1/5/60 15:01:12     Sunday January 4, 2004 11:01pm    1388476807000 |
     +------------------------------------------------------------------------+
```

We can convert the number of milliseconds (`momfbms`) into the number of days by dividing it by `1000*60*60*24`, as shown below.

```
. generate momfbdays = momfbms/(1000*60*60*24)
. list id momfbms momfbdays, abb(20)

     +--------------------------------+
     | id         momfbms   momfbdays |
     |--------------------------------|
  1. |  1    792218523368    9169.196 |
  2. |  2    915855996992    10600.19 |
  3. |  3    881032374664    10197.13 |
  4. |  4   1388476807000    16070.33 |
     +--------------------------------+
```

We might want to know how old the mom is in days as of January 20, 2010, at 6 PM. We do this by subtracting `momdt` from `tc(20jan2010 18:00:00)` and divide that by `1000*60*60*24` to obtain the age of the mom in days, as shown below. `tc(20jan2010 18:00:00)` is an example of the way that you can specify a particular date and time to Stata.

```
. generate momdays = (tc(20jan2010 18:00:00) - momdt)/(1000*60*60*24)
. list id momdt momdays

     +-----------------------------------+
     | id               momdt    momdays |
     |-----------------------------------|
  1. |  1   11/28/72 10:39:01   13567.31 |
  2. |  2     4/3/73 06:22:35   13441.48 |
  3. |  3    6/13/68 22:45:34    15195.8 |
  4. |  4     1/5/60 15:01:12   18278.13 |
     +-----------------------------------+
```

Say that we wanted to list the mothers who were born after midnight of January 1, 1970. We can do this by listing the cases where the mom's birth date is greater than `tc(01jan1970 00:00:00)`, as shown below. (Note the exclusion of missing values; see section A.10 for more information.)

```
. list id momdt if (momdt > tc(01jan1970 00:00:00)) & ! missing(momdt)

     ┌──────────────────────┐
     │ id             momdt │
     ├──────────────────────┤
  1. │  1   11/28/72 10:39:01 │
  2. │  2     4/3/73 06:22:35 │
     └──────────────────────┘
```

Below we see how you could extract the hour, minute, or second from a date-and-time variable.

```
. generate kidhh = hh(kiddt)
. generate kidmm = mm(kiddt)
. generate kidss = ss(kiddt)
. list id kiddt kidhh kidmm kidss

     ┌─────────────────────────────────────────────────────────────────┐
     │ id                           kiddt   kidhh   kidmm   kidss │
     ├─────────────────────────────────────────────────────────────────┤
  1. │  1      Monday January 5, 1998 3:21pm      15      21       5 │
  2. │  2   Thursday April 11, 2002 10:49am      10      49      12 │
  3. │  3      Wednesday May 15, 1996 1:58am       1      58      29 │
  4. │  4    Sunday January 4, 2004 11:01pm      23       1      19 │
     └─────────────────────────────────────────────────────────────────┘
```

We can convert a date-and-time variable into a date variable by using the `dofc()` (date-of-clock) function. Below we create the variable `kiddate`, which is the date portion of the date-and-time variable named `kiddt`. We then format `kiddate` as a date variable and list it.

```
. generate kiddate = dofc(kiddt)
. format kiddate %td
. list id kiddt kiddate

     ┌──────────────────────────────────────────────────────┐
     │ id                           kiddt     kiddate │
     ├──────────────────────────────────────────────────────┤
  1. │  1      Monday January 5, 1998 3:21pm   05jan1998 │
  2. │  2   Thursday April 11, 2002 10:49am   11apr2002 │
  3. │  3      Wednesday May 15, 1996 1:58am   15may1996 │
  4. │  4    Sunday January 4, 2004 11:01pm   04jan2004 │
     └──────────────────────────────────────────────────────┘
```

If you want to extract the month, day, or year from a date-and-time variable, you first need to use the `dofc()` (date-of-clock) function to convert the date-and-time variable into a date variable. Once you do that, then the `day()`, `month()`, and `year()` functions can be used, as shown below.

```
. generate kidday = day(dofc(kiddt))
. generate kidmonth = month(dofc(kiddt))
. generate kidyear = year(dofc(kiddt))
. list id kiddt kidday kidmonth kidyear
```

	id	kiddt	kidday	kidmonth	kidyear
1.	1	Monday January 5, 1998 3:21pm	5	1	1998
2.	2	Thursday April 11, 2002 10:49am	11	4	2002
3.	3	Wednesday May 15, 1996 1:58am	15	5	1996
4.	4	Sunday January 4, 2004 11:01pm	4	1	2004

After applying the `dofc()` function, we can use other date functions as well. For example, below we see that the first kid was born on a Monday (0 = Sunday, 1 = Monday, 2 = Tuesday, ..., 6 = Saturday), which was the fifth day in the first week in the first quarter of the year.

```
. generate kiddow = dow(dofc(kiddt))
. generate kiddoy = doy(dofc(kiddt))
. generate kidweek = week(dofc(kiddt))
. generate kidqtr = quarter(dofc(kiddt))
. list id kiddt kiddow kiddoy kidweek kidqtr, noobs
```

id	kiddt	kiddow	kiddoy	kidweek	kidqtr
1	Monday January 5, 1998 3:21pm	1	5	1	1
2	Thursday April 11, 2002 10:49am	4	101	15	2
3	Wednesday May 15, 1996 1:58am	3	136	20	2
4	Sunday January 4, 2004 11:01pm	0	4	1	1

For more information about date-and-time values, see `help dates and times`. For more information about issues with two-digit years, see section 5.8.

5.10 Computations across variables

There are times we wish to create new variables that are based on computations made across variables within each observation (such as obtaining the mean across variables within each observation). The `egen` command offers several functions that make such computations easy. For example, consider `cardio1miss.dta`, which has blood pressure and pulse data at five time points, as shown below.

```
. use cardio2miss
. list
```

	id	age	pl1	pl2	pl3	pl4	pl5	bp1	bp2	bp3	bp4	bp5
1.	1	40	54	115	87	86	93	129	81	105	.b	.b
2.	2	30	92	123	88	136	125	107	87	111	58	120
3.	3	16	105	.a	97	122	128	101	57	109	68	112
4.	4	23	52	105	79	115	71	121	106	129	39	137
5.	5	18	70	116	.a	128	52	112	68	125	59	111

We can use the **generate** command to get the average blood pressure across the five time points.

```
. generate avgbp = (bp1 + bp2 + bp3 + bp4 + bp5)/5
(1 missing value generated)
. list id bp1 bp2 bp3 bp4 bp5 avgbp
```

	id	bp1	bp2	bp3	bp4	bp5	avgbp
1.	1	129	81	105	.b	.b	.
2.	2	107	87	111	58	120	96.6
3.	3	101	57	109	68	112	89.4
4.	4	121	106	129	39	137	106.4
5.	5	112	68	125	59	111	95

Note how the value of **avgbp** is missing if any of the individual blood pressure values is missing (see section A.10 for more details about missing values). Instead, we can use the **egen** command with the **rowmean()** function to get the average blood pressure across the five time points.

```
. egen avgbp2 = rowmean(bp1 bp2 bp3 bp4 bp5)
. list id bp1 bp2 bp3 bp4 bp5 avgbp2
```

	id	bp1	bp2	bp3	bp4	bp5	avgbp2
1.	1	129	81	105	.b	.b	105
2.	2	107	87	111	58	120	96.6
3.	3	101	57	109	68	112	89.4
4.	4	121	106	129	39	137	106.4
5.	5	112	68	125	59	111	95

In this case, the means are computed based on the nonmissing variables. For example, in observation 1, the blood pressure information was missing for times 4 and 5, so **avgbp** is based on the three variables that had nonmissing data.

We can likewise use **egen** with the **rowmean()** function to compute the mean pulse rate across the five time points. In this example, we take advantage of the five pulse observations that are positioned next to each other, specifying **pl1-pl5** (see section A.11 for more information about referring to variable lists).

```
. egen avgpl = rowmean(pl1-pl5)
. list id pl1-pl5 avgpl
```

	id	pl1	pl2	pl3	pl4	pl5	avgpl
1.	1	54	115	87	86	93	87
2.	2	92	123	88	136	125	112.8
3.	3	105	.a	97	122	128	113
4.	4	52	105	79	115	71	84.4
5.	5	70	116	.a	128	52	91.5

The rowmin() and rowmax() functions can be used to get the minimum and maximum pulse rate among the five measures, as shown below.

```
. egen minbp = rowmin(bp1-bp5)
. egen maxbp = rowmax(bp1-bp5)
. list id bp1-bp5 minbp maxbp
```

	id	bp1	bp2	bp3	bp4	bp5	minbp	maxbp
1.	1	129	81	105	.b	.b	81	129
2.	2	107	87	111	58	120	58	120
3.	3	101	57	109	68	112	57	112
4.	4	121	106	129	39	137	39	137
5.	5	112	68	125	59	111	59	125

The rowmiss() function computes the number of missing values. The rownonmiss() function computes the number of nonmissing values among the variables specified. These functions are illustrated below.

```
. egen missbp = rowmiss(bp1-bp5)
. egen nonmissbp = rownonmiss(bp1-bp5)
. list id bp1-bp5 missbp nonmissbp, abb(20)
```

	id	bp1	bp2	bp3	bp4	bp5	missbp	nonmissbp
1.	1	129	81	105	.b	.b	2	3
2.	2	107	87	111	58	120	0	5
3.	3	101	57	109	68	112	0	5
4.	4	121	106	129	39	137	0	5
5.	5	112	68	125	59	111	0	5

The egen command supports other row computations, such as rowsd(), rowsum(), rowfirst(), and rowlast(). See help egen for more information.

5.11 Computations across observations

The previous section illustrated the use of the egen command for performing computations across variables within each observation. This section illustrates the use of the

egen command for performing computations across observations. For example, consider **gasctrysmall.dta**, which contains gas prices and inflation measures on four countries for one or more years per country.

```
. use gasctrysmall
. list, sepby(ctry)
```

	ctry	year	gas	infl
1.	1	1974	.78	1.32
2.	1	1975	.83	1.4
3.	2	1971	.69	1.15
4.	2	1971	.77	1.15
5.	2	1973	.89	1.29
6.	3	1974	.42	1.14
7.	4	1974	.82	1.12
8.	4	1975	.94	1.18

Say that we want to make a variable that has the average price of gas across all observations. We can use the **egen** command with the **mean()** function to do this, as shown below.

```
. egen avggas = mean(gas)
. list ctry year gas avggas, sepby(ctry)
```

	ctry	year	gas	avggas
1.	1	1974	.78	.7675
2.	1	1975	.83	.7675
3.	2	1971	.69	.7675
4.	2	1971	.77	.7675
5.	2	1973	.89	.7675
6.	3	1974	.42	.7675
7.	4	1974	.82	.7675
8.	4	1975	.94	.7675

Say that we instead wanted the average price of gas within each country. We can preface the **egen** command with the **bysort ctry:** prefix, and now the mean is computed separately for each country (see section A.3 for more on the **by** and **bysort** prefixes).

(Continued on next page)

```
. bysort ctry: egen avggas_ctry = mean(gas)
. list ctry year gas avggas_ctry, sepby(ctry) abb(20)

     +--------------------------------+
     | ctry   year   gas   avggas_ctry |
     |--------------------------------|
  1. |    1   1974   .78          .805 |
  2. |    1   1975   .83          .805 |
     |--------------------------------|
  3. |    2   1971   .69      .7833334 |
  4. |    2   1971   .77      .7833334 |
  5. |    2   1973   .89      .7833334 |
     |--------------------------------|
  6. |    3   1974   .42           .42 |
     |--------------------------------|
  7. |    4   1974   .82           .88 |
  8. |    4   1975   .94           .88 |
     +--------------------------------+
```

If we want to get the average gas price within each year, we could use the same strategy but with the `bysort year:` prefix:

```
. bysort year: egen avggas_year = mean(gas)
. list ctry year gas avggas_year, sepby(year) abb(20)

     +--------------------------------+
     | ctry   year   gas   avggas_year |
     |--------------------------------|
  1. |    2   1971   .69           .73 |
  2. |    2   1971   .77           .73 |
     |--------------------------------|
  3. |    2   1973   .89           .89 |
     |--------------------------------|
  4. |    4   1974   .82      .6733333 |
  5. |    1   1974   .78      .6733333 |
  6. |    3   1974   .42      .6733333 |
     |--------------------------------|
  7. |    1   1975   .83          .885 |
  8. |    4   1975   .94          .885 |
     +--------------------------------+
```

Perhaps we would like to get the minimum and maximum gas price within each country. We can do so using the `min()` and `max()` functions, as shown below.

```
. bysort ctry: egen mingas = min(gas)
. bysort ctry: egen maxgas = max(gas)
. list ctry year gas mingas maxgas, sepby(ctry)
```

	ctry	year	gas	mingas	maxgas
1.	1	1974	.78	.78	.83
2.	1	1975	.83	.78	.83
3.	2	1971	.69	.69	.89
4.	2	1973	.89	.69	.89
5.	2	1971	.77	.69	.89
6.	3	1974	.42	.42	.42
7.	4	1974	.82	.82	.94
8.	4	1975	.94	.82	.94

These are just a small sampling of the statistical functions that you can use with `egen` for collapsing across observations. Other functions that you might use include `count()`, `iqr()`, `kurt()`, `mad()`, `mdev()`, `median()`, `mode()`, `pc()`, `pctile()`, `sd()`, `skew()`, `std()`, and `sum()`. See `help egen` for more information. Also, section 7.3 gives further examples on the use of `egen` for performing computations across observations.

5.12 More examples using the egen command

The previous two sections illustrated ways that the `egen` command can be used for computations across rows (variables) and across observations. This section will illustrate additional functions supported by `egen`.

Consider `cardio1ex.dta`. It contains five measurements of pulse, systolic blood pressure, and how exhausted the subject feels rated on a scale of 1 to 4 (1 is least exhausted and 4 is most exhausted). Let's focus on the exhaustion measures.

```
. use cardio1ex
. list id ex1 ex2 ex3 ex4 ex5
```

	id	ex1	ex2	ex3	ex4	ex5
1.	1	3	1	4	2	4
2.	2	4	4	2	3	3
3.	3	4	4	2	2	3
4.	4	2	3	4	4	4
5.	5	3	4	3	4	3

One measure of fitness might be to count how many times the subject reported feeling most exhausted (a value of 4) out of the five measures. The `egen` command using the `anycount()` function does this for us.

```
. egen cntex4 = anycount(ex1 ex2 ex3 ex4 ex5), values(4)
. list id ex1 ex2 ex3 ex4 ex5 cntex4
```

	id	ex1	ex2	ex3	ex4	ex5	cntex4
1.	1	3	1	4	2	4	2
2.	2	4	4	2	3	3	2
3.	3	4	4	2	2	3	2
4.	4	2	3	4	4	4	3
5.	5	3	4	3	4	3	2

Another possible measure of fitness would be to see if the subject ever felt the least amount of exhaustion (a value of 1). Using the `anymatch()` function, we can determine which subjects ever gave the lowest exhaustion rating. For the subjects who did give the lowest exhaustion rating, the value of `exever1` is given a `1`; otherwise, it is given a `0`.

```
. egen exever1 = anymatch(ex1 ex2 ex3 ex4 ex5), values(1)
. list id ex1 ex2 ex3 ex4 ex5 exever1
```

	id	ex1	ex2	ex3	ex4	ex5	exever1
1.	1	3	1	4	2	4	1
2.	2	4	4	2	3	3	0
3.	3	4	4	2	2	3	0
4.	4	2	3	4	4	4	0
5.	5	3	4	3	4	3	0

We might be interested in assessing how consistent the exhaustion ratings are. Suppose that we focus on the last three observations and determine if the last three exhaustion measures are the same or whether there are any differences. The variable `exdiff` is a 1 if there are any differences and a 0 if the last three measures are all the same.

```
. egen exdiff = diff(ex3 ex4 ex5)
. list id ex1 ex2 ex3 ex4 ex5 exdiff
```

	id	ex1	ex2	ex3	ex4	ex5	exdiff
1.	1	3	1	4	2	4	1
2.	2	4	4	2	3	3	1
3.	3	4	4	2	2	3	1
4.	4	2	3	4	4	4	0
5.	5	3	4	3	4	3	1

These three sections have not illustrated all the features of the `egen` command. Some notable functions that were omitted include `concat()`, `ends()`, `fill()`, `ma()`, `mtr()`, `rank()`, `seq()`, and `tag()`. See `help egen` for more information.

Tip! Want more egen functions?

If you want even more `egen` functions, then you might want to check the `egenmore` package of tools, which you can find and download via the `findit egenmore` command (see section 10.2 for more about the `findit` command). This suite of user-written commands adds a wide array of extra functionality to `egen`.

5.13 Converting string variables to numeric variables

This section illustrates how you can handle variables that contain numeric data but are stored as a string. For example, consider `cardio1str.dta`. This dataset contains the person's weight, age, three systolic blood pressure measures, three resting pulse measures, income, and gender.

```
. use cardio1str
. list wt-gender
```

	wt	age	bp1	bp2	bp3	pl1	pl2	pl3	income	gender
1.	150.7	45	115	86	129	54	87	93	$25,308.92	male
2.	186.3	23	123	136	107	92	88	125	$46,213.31	male
3.	109.9	48	132	122	101	105	97	X	$65,234.11	male
4.	183.4	29	105	115	121	.	79	71	$89,234.23	male
5.	159.1	42	116	128	112	70		52	$54,989.87	female

It would seem that we would be ready to analyze this dataset, so let's summarize the weight and blood pressure measurements.

```
. summarize wt bp1 bp2 bp3
```

Variable	Obs	Mean	Std. Dev.	Min	Max
wt	0				
bp1	0				
bp2	0				
bp3	0				

These results might seem perplexing, but when you look at the `describe` command, they start to make sense.

(*Continued on next page*)

```
. describe
Contains data from cardio1str.dta
  obs:             5
 vars:            11                          22 Dec 2009 19:51
 size:           225 (99.9% of memory free)

              storage  display     value
variable name   type   format      label      variable label

id              str1   %3s                    Identification variable
wt              str5   %5s                    Weight of person
age             str2   %2s                    Age of person
bp1             str3   %3s                    Systolic BP: Trial 1
bp2             str3   %3s                    Systolic BP: Trial 2
bp3             str3   %3s                    Systolic BP: Trial 3
pl1             str3   %3s                    Pulse: Trial 1
pl2             str2   %3s                    Pulse: Trial 2
pl3             str3   %3s                    Pulse: Trial 3
income          str10  %10s                   Income
gender          str6   %6s                    Gender of person

Sorted by:
```

Even though the results of the `list` command appeared to be displaying numeric data, these numbers are stored in Stata as string variables (see section A.5 for more information about data types, including string variables). When you try to analyze a string variable (such as getting a mean), there are no valid numeric observations; hence, the `summarize` command showed no valid observations.

Let's convert these variables from string to numeric starting with the `age` variable. This variable is easy to fix because it contains only numeric values and has no missing data. Below the `destring` command is used to convert the string version of `age` into a numeric version named `agen`. Even though `age` and `agen` look the same when we list them side by side, the results from the `summarize` command reflect that `agen` is numeric and can be analyzed.

```
. destring age, generate(agen)
age has all characters numeric; agen generated as byte
. list age agen

     age   agen
 1.   45     45
 2.   23     23
 3.   48     48
 4.   29     29
 5.   42     42

. summarize age agen
    Variable |       Obs        Mean    Std. Dev.       Min        Max
-------------+--------------------------------------------------------
         age |         0
        agen |         5        37.4    10.83051         23         48
```

We can use the **order** command to position **agen** right after **age** in the variable list (see section 5.15 for more about the **order** command). This command is useful because it allows you to position related variables next to each other.

```
. order agen, after(age)
```

As a shortcut, you can use the **destring** command with multiple variables at once. But in doing so, **destring** replaces the existing string variable with its numerical equivalent. So let's start this process again by reusing **cardio1str.dta** and converting the variables **id**, **age**, **wt**, **bp1**, **bp2**, and **bp3** from string variables to numeric variables.

```
. use cardio1str, clear
. destring id age wt bp1 bp2 bp3, replace
id has all characters numeric; replaced as byte
age has all characters numeric; replaced as byte
wt has all characters numeric; replaced as double
bp1 has all characters numeric; replaced as int
bp2 has all characters numeric; replaced as int
bp3 has all characters numeric; replaced as int
```

When we use the **describe** command, we can see that these variable are now stored using numeric data types (see section A.5 for more about data types in Stata).

```
. describe id age wt bp1 bp2 bp3
              storage  display     value
variable name   type   format      label      variable label
------------------------------------------------------------------------------
id              byte   %10.0g                 Identification variable
age             byte   %10.0g                 Age of person
wt              double %10.0g                 Weight of person
bp1             int    %10.0g                 Systolic BP: Trial 1
bp2             int    %10.0g                 Systolic BP: Trial 2
bp3             int    %10.0g                 Systolic BP: Trial 3
```

Further, the **summarize** command produces valid summary statistics for all these variables.

```
. summarize id age wt bp1 bp2 bp3
    Variable |       Obs        Mean    Std. Dev.       Min        Max
-------------+--------------------------------------------------------
          id |         5           3    1.581139          1          5
         age |         5        37.4    10.83051         23         48
          wt |         5      157.88    30.87915      109.9      186.3
         bp1 |         5       118.2    10.03494        105        132
         bp2 |         5       117.4    19.17811         86        136
-------------+--------------------------------------------------------
         bp3 |         5         114    11.13553        101        129
```

So far, all the variables we have used with **destring** have had complete data with only numeric values. Let's consider the pulse variables **pl1**, **pl2**, and **pl3**. As you can see below, **pl1** has one case where a period was entered, **pl2** has one case where nothing was entered, and **pl3** has one case where an **X** was entered (to indicate missing).

```
. list pl1 pl2 pl3

     +-------------------+
     | pl1   pl2   pl3   |
     |-------------------|
  1. |  54    87    93   |
  2. |  92    88   125   |
  3. | 105    97     X   |
  4. |   .    79    71   |
  5. |  70          52   |
     +-------------------+
```

Let's try using `destring` and see how it works for converting these string variables into numeric values.

```
. destring pl1 pl2 pl3, replace
pl1 has all characters numeric; replaced as int
(1 missing value generated)
pl2 has all characters numeric; replaced as byte
(1 missing value generated)
pl3 contains nonnumeric characters; no replace
. list pl1 pl2 pl3

     +-------------------+
     | pl1   pl2   pl3   |
     |-------------------|
  1. |  54    87    93   |
  2. |  92    88   125   |
  3. | 105    97     X   |
  4. |   .    79    71   |
  5. |  70     .    52   |
     +-------------------+
```

`destring` seemed to work great for `pl1` and `pl2`, where the period and blank values were converted to missing values. But for `pl3`, a message was given that it contained nonnumeric characters. In our tiny example, we know that the nonnumeric character is an `X`, but in a more realistic dataset, we may not know this.

The following two commands can be used to reveal the nonnumeric characters contained in `pl3`.

```
. destring pl3, generate(pl3num) force
pl3 contains nonnumeric characters; pl3num generated as int
(1 missing value generated)
. list pl3 pl3num if missing(pl3num)

     +--------------+
     | pl3   pl3num |
     |--------------|
  3. |   X        . |
     +--------------+
```

The `destring` command generated a new variable named `pl3num`, and the `force` option indicated that the new variable should be created even though `pl3` contains nonnumeric values. The result is that `pl3num` contains missing values when `pl3` is nonnumeric. The `list` command takes advantage of that and lists the values of `pl3` and `pl3num` only when `pl3num` is missing, revealing the nonnumeric codes contained within `pl3`.

We can now use this information to rerun the destring command for pl3, adding the ignore(X) option. The X values are ignored as part of the conversion from string to numeric. This results in a successful conversion, and the value of X then yields a missing value.

```
. destring pl3, replace ignore(X)
pl3: characters X removed; replaced as int
(1 missing value generated)

. list pl3

     +-----+
     | pl3 |
     |-----|
  1. |  93 |
  2. | 125 |
  3. |   . |
  4. |  71 |
  5. |  52 |
     +-----+
```

The variable income poses the same kind of problem, but it includes two nonnumeric characters: the dollar sign and the comma.

```
. list id income

     +-----------------+
     | id       income |
     |-----------------|
  1. |  1   $25,308.92 |
  2. |  2   $46,213.31 |
  3. |  3   $65,234.11 |
  4. |  4   $89,234.23 |
  5. |  5   $54,989.87 |
     +-----------------+
```

In converting these dollar amounts into numeric values, we need to ignore the dollar sign and the comma. We can specify the ignore($,) option on the destring command to ignore both of these characters when converting the values of income from string to numeric, as shown below.

```
. destring income, replace ignore($,)
income: characters $ , removed; replaced as double

. list id income

     +---------------+
     | id     income |
     |---------------|
  1. |  1   25308.92 |
  2. |  2   46213.31 |
  3. |  3   65234.11 |
  4. |  4   89234.23 |
  5. |  5   54989.87 |
     +---------------+
```

Let's conclude this section by showing how to convert a genuine string variable into a numeric variable. The variable gender contains the word male or the word female.

```
. list id gender

     ┌─────────────┐
     │ id   gender │
     ├─────────────┤
  1. │  1     male │
  2. │  2     male │
  3. │  3     male │
  4. │  4     male │
  5. │  5   female │
     └─────────────┘
```

We would need to convert this variable into a numeric variable if we wanted to include such a variable in an analysis (such as using it as a predictor in a regression model). We can use the `encode` command for this, creating the numeric variable named `ngender`, as shown below.

```
. encode gender, generate(ngender)
```

When we list the original `gender` variable and the numeric `ngender` version, it looks like these are the same:

```
. list gender ngender

     ┌──────────────────┐
     │ gender   ngender │
     ├──────────────────┤
  1. │   male      male │
  2. │   male      male │
  3. │   male      male │
  4. │   male      male │
  5. │ female    female │
     └──────────────────┘
```

The displayed values of `ngender` are the labeled values. Using the `codebook` command, we can see that `ngender` is a numeric variable that has value labels indicating which values correspond to male and female.

```
. codebook ngender

-------------------------------------------------------------------------------
ngender                                                        Gender of person
-------------------------------------------------------------------------------

                  type:  numeric (long)
                 label:  ngender

                 range:  [1,2]                        units:  1
         unique values:  2                        missing .:  0/5

            tabulation:  Freq.   Numeric  Label
                             1         1  female
                             4         2  male
```

In this example, `ngender` is coded using 1 for female and 2 for male. These are based on the alphabetic order of the character values. When sorted alphabetically, female is first and male is second.

The `encode` command allows us to specify the values that should be assigned for each string variable. Suppose that we wanted to create a dummy variable named `female` that would be coded 0 if gender is male, and 1 if gender is female. We can do this as shown below.

```
. label define femlab 0 "male" 1 "female"
. encode gender, generate(female) label(femlab)
```

The label define femlab command works in conjunction with the label(femlab) option on the encode command to specify that male should be coded as 0 and female should be coded as 1. We can see the result of this by using the codebook command.

```
. codebook female

female                                                   Gender of person

                 type:  numeric (long)
                label:  femlab

                range:  [0,1]                        units:  1
        unique values:  2                        missing .:  0/5

           tabulation:  Freq.   Numeric  Label
                            4         0  male
                            1         1  female
```

We now have two different numeric versions of gender: ngender and female. Let's position these variables after the original string variable, gender. Then, if we drop gender, these new variables will be positioned in the spot that was occupied by gender.

```
. order ngender female, after(gender)
```

This section has illustrated different ways that you can convert string variables into numeric variables in Stata. For more information, see help destring and help encode. Also see section A.5, which goes into more details about the different variable data types used in Stata.

5.14 Converting numeric variables to string variables

This section illustrates how to convert numeric variables to string variables. Even though we generally want to store numeric values as numeric variables, there are some exceptions where such values are better stored as a string. For example, social security numbers are sometimes stored as numbers, but we want to store them as strings. Another example is a zip code. I often match-merge files based on zip code, but I frequently find that one file has the zip code stored as a numeric value while the other has it stored as a string value. I then need to convert the zip codes to be of the same type (e.g., see page 215), preferably converting the numeric zip codes to string values. For example, a zip code of 00034 when stored as a number is stored as 34. We have such a problem with the zip code in cardio3.dta, as shown below.

```
. use cardio3, clear
. list zipcode

     +---------+
     | zipcode |
     |---------|
  1. |      34 |
  2. |   90095 |
  3. |   43409 |
  4. |   23219 |
  5. |   66214 |
     +---------+
```

We can convert the zip code from a numeric value into a string value by using the `string()` function, as shown below. The format `%05.0f` means that the resulting value will have a fixed width of 5 with zero decimal places. Further, when values are less than five digits wide, leading 0s will be added (e.g., converting `34` to `00034`).

```
. gen zipcodes = string(zipcode,"%05.0f")
. list zipcode zipcodes

     +--------------------+
     | zipcode   zipcodes |
     |--------------------|
  1. |      34      00034 |
  2. |   90095      90095 |
  3. |   43409      43409 |
  4. |   23219      23219 |
  5. |   66214      66214 |
     +--------------------+
```

The `order` command can be used to position `zipcodes` after `zipcode` in the variable list, as shown below (see section 5.15 for more about the `order` command). This new ordering helps by logically grouping related variables together.

```
. order zipcodes, after(zipcode)
```

The `tostring` command can also be used to convert numeric variables into string variables. One of the advantages of the `tostring` command is that it permits you to convert multiple variables at one time. For example, suppose that we wanted to convert the three blood pressure measurements—`bp1`, `bp2`, and `bp3`—from numeric to string. Using the `tostring` command, we can do so in one step.

```
. tostring bp1 bp2 bp3, gen(bp1s bp2s bp3s)
bp1s generated as str3
bp2s generated as str3
bp3s generated as str3

. list bp1 bp2 bp3 bp1s bp2s bp3s

     +-------------------------------------+
     | bp1   bp2   bp3   bp1s   bp2s   bp3s |
     |-------------------------------------|
  1. | 115    86   129    115     86    129 |
  2. | 123   136   107    123    136    107 |
  3. | 124   122   101    124    122    101 |
  4. | 105   115   121    105    115    121 |
  5. | 116   128   112    116    128    112 |
     +-------------------------------------+
```

We could likewise convert the three pulse measurements into string values all at once. Rather than creating a new variable, let's instead replace the existing variables. Further, let's use the `%03.0f` format, meaning that if the pulse is two digits, a leading 0 will be added.

```
. tostring pl1 pl2 pl3, replace format(%03.0f)
pl1 was int now str3
pl2 was int now str3
pl3 was int now str3

. list pl1 pl2 pl3

     +-----------------+
     | pl1   pl2   pl3 |
     |-----------------|
  1. | 054   087   093 |
  2. | 092   088   125 |
  3. | 105   097   128 |
  4. | 052   079   071 |
  5. | 070   064   052 |
     +-----------------+
```

The `decode` command is another alternative for converting numeric values to string values. This is useful if, during the conversion from numeric to string values, you want the value labels to be used when making the new string value. Consider the variable **`famhist`**, which is a 0 if one has no family history of heart disease (labeled `No HD`) and is a 1 if one does have a family history of heart disease (labeled as **`Yes HD`**).

```
. codebook famhist

-------------------------------------------------------------------------------
famhist                                       Family history of heart disease
-------------------------------------------------------------------------------

                  type:  numeric (long)
                 label:  famhistl

                 range:  [0,1]                        units:  1
         unique values:  2                        missing .:  0/5

            tabulation:  Freq.   Numeric  Label
                             2         0  No HD
                             3         1  Yes HD
```

The decode command below is used to create the variable famhists, which will contain the labeled value of famhist (i.e, either No HD or Yes HD).

```
. decode famhist, generate(famhists)
. list famhist famhists, nolabel

     +--------------------+
     | famhist   famhists |
     |--------------------|
  1. |       0      No HD |
  2. |       1     Yes HD |
  3. |       0      No HD |
  4. |       1     Yes HD |
  5. |       1     Yes HD |
     +--------------------+
```

That covers the basics of how to convert numeric variables to string variables. For more information, see help string, help tostring, and help decode. Also see section A.5 for more details about the different variable data types used in Stata.

5.15 Renaming and ordering variables

This section shows how you can rename and change the order of the variables in your dataset, using cardio2.dta as an example. The variables in this dataset are shown below using the describe command.

```
. use cardio2
. describe
Contains data from cardio2.dta
  obs:             5
 vars:            12                          22 Dec 2009 19:51
 size:           120 (99.9% of memory free)
-------------------------------------------------------------------------------
              storage  display     value
variable name   type   format      label      variable label
-------------------------------------------------------------------------------
id              byte   %3.0f                  Identification variable
age             byte   %3.0f                  Age of person
pl1             int    %3.0f                  Pulse: Trial 1
bp1             int    %3.0f                  Systolic BP: Trial 1
pl2             byte   %3.0f                  Pulse: Trial 2
bp2             int    %3.0f                  Systolic BP: Trial 2
pl3             int    %3.0f                  Pulse: Trial 3
bp3             int    %3.0f                  Systolic BP: Trial 3
pl4             int    %3.0f                  Pulse: Trial 4
bp4             int    %3.0f                  Systolic BP: Trial 4
pl5             byte   %3.0f                  Pulse: Trial 5
bp5             int    %3.0f                  Systolic BP: Trial 5
-------------------------------------------------------------------------------
Sorted by:
```

We can use the rename command (or the Variables Manager as discussed on page 34) to rename the variable age to be age_yrs.

```
. rename age age_yrs
```

Say that we want to rename the variables pl1 to pl5 to be pulse1 to pulse5. We could issue five rename commands, or we could use a foreach loop, as shown below (see section 9.9 for more information on looping over numbers).

```
. foreach t of numlist 1/5 {
  2.   rename pl`t´ pulse`t´
  3. }
```

The describe command shows that this worked as planned:

```
. describe
Contains data from cardio2.dta
  obs:             5
 vars:            12                          22 Dec 2009 19:51
 size:           120 (99.9% of memory free)
-------------------------------------------------------------------------------
              storage  display     value
variable name   type   format      label      variable label
-------------------------------------------------------------------------------
id              byte   %3.0f                  Identification variable
age_yrs         byte   %3.0f                  Age of person
pulse1          int    %3.0f                  Pulse: Trial 1
bp1             int    %3.0f                  Systolic BP: Trial 1
pulse2          byte   %3.0f                  Pulse: Trial 2
bp2             int    %3.0f                  Systolic BP: Trial 2
pulse3          int    %3.0f                  Pulse: Trial 3
bp3             int    %3.0f                  Systolic BP: Trial 3
pulse4          int    %3.0f                  Pulse: Trial 4
bp4             int    %3.0f                  Systolic BP: Trial 4
pulse5          byte   %3.0f                  Pulse: Trial 5
bp5             int    %3.0f                  Systolic BP: Trial 5
-------------------------------------------------------------------------------
Sorted by:
     Note:  dataset has changed since last saved
```

Although the foreach command was fairly convenient, Stata offers an even more powerful and convenient command for this kind of task called renpfix (rename prefix). This is illustrated below, renaming the variables bp1–bp5 to become bpress1–bpress5. We use describe to confirm that the renaming took place.

(Continued on next page)

```
. renpfix bp bpress
. describe
Contains data from cardio2.dta
  obs:             5
 vars:            12                          22 Dec 2009 19:51
 size:           120 (99.9% of memory free)
-------------------------------------------------------------------------------
              storage  display     value
variable name   type   format      label      variable label
-------------------------------------------------------------------------------
id              byte   %3.0f                  Identification variable
age_yrs         byte   %3.0f                  Age of person
pulse1          int    %3.0f                  Pulse: Trial 1
bpress1         int    %3.0f                  Systolic BP: Trial 1
pulse2          byte   %3.0f                  Pulse: Trial 2
bpress2         int    %3.0f                  Systolic BP: Trial 2
pulse3          int    %3.0f                  Pulse: Trial 3
bpress3         int    %3.0f                  Systolic BP: Trial 3
pulse4          int    %3.0f                  Pulse: Trial 4
bpress4         int    %3.0f                  Systolic BP: Trial 4
pulse5          byte   %3.0f                  Pulse: Trial 5
bpress5         int    %3.0f                  Systolic BP: Trial 5
-------------------------------------------------------------------------------
Sorted by:
     Note:  dataset has changed since last saved
```

Note that `renpfix` would rename any other variable that started with `bp` to start with `bpress`, so if you had a variable called `bpollen`, it would have been renamed to `bpressollen` because `bp` would have been replaced with `bpress`.

Say that you wanted to rename the variables `pulse1`–`pulse5` to become `time1pulse`–`time5pulse`. Because this is not just a simple change of prefix, this is a case where we would need to use a loop like a `foreach` loop, as shown below.

```
. foreach t of numlist 1/5 {
  2.   rename pulse`t´ time`t´pulse
  3. }
. describe
Contains data from cardio2.dta
  obs:             5
 vars:            12                          22 Dec 2009 19:51
 size:           120 (99.9% of memory free)
───────────────────────────────────────────────────────────────────────────────
              storage  display     value
variable name   type   format      label      variable label
───────────────────────────────────────────────────────────────────────────────
id              byte   %3.0f                  Identification variable
age_yrs         byte   %3.0f                  Age of person
time1pulse      int    %3.0f                  Pulse: Trial 1
bpress1         int    %3.0f                  Systolic BP: Trial 1
time2pulse      byte   %3.0f                  Pulse: Trial 2
bpress2         int    %3.0f                  Systolic BP: Trial 2
time3pulse      int    %3.0f                  Pulse: Trial 3
bpress3         int    %3.0f                  Systolic BP: Trial 3
time4pulse      int    %3.0f                  Pulse: Trial 4
bpress4         int    %3.0f                  Systolic BP: Trial 4
time5pulse      byte   %3.0f                  Pulse: Trial 5
bpress5         int    %3.0f                  Systolic BP: Trial 5
───────────────────────────────────────────────────────────────────────────────
Sorted by:
     Note:  dataset has changed since last saved
```

Tip! Renaming variables

The user-written program called **renvars** contains a variety of useful features beyond the **rename** and **renpfix** commands. It permits you to rename groups of variables at once, convert variables to uppercase or lowercase, and add, change, or remove prefixes/suffixes to variable names. You can locate and download this program by typing **findit renvars** (see section 10.2 for more about using the **findit** command).

Let's now consider the order of the variables in this dataset. Suppose that we wanted the variables id and age_yrs to appear at the end of the dataset. The order command shown below specifies that the variables id and age_yrs should be moved to the end of the dataset.

```
. order id age_yrs, last
```

(Continued on next page)

Below we can see that id and age_yrs were moved to the end of the dataset.

```
. describe

Contains data from cardio2.dta
  obs:             5
 vars:            12                          22 Dec 2009 19:51
 size:           120 (99.9% of memory free)

              storage  display     value
variable name   type   format      label      variable label

time1pulse      int    %3.0f                  Pulse: Trial 1
bpress1         int    %3.0f                  Systolic BP: Trial 1
time2pulse      byte   %3.0f                  Pulse: Trial 2
bpress2         int    %3.0f                  Systolic BP: Trial 2
time3pulse      int    %3.0f                  Pulse: Trial 3
bpress3         int    %3.0f                  Systolic BP: Trial 3
time4pulse      int    %3.0f                  Pulse: Trial 4
bpress4         int    %3.0f                  Systolic BP: Trial 4
time5pulse      byte   %3.0f                  Pulse: Trial 5
bpress5         int    %3.0f                  Systolic BP: Trial 5
id              byte   %3.0f                  Identification variable
age_yrs         byte   %3.0f                  Age of person

Sorted by:
     Note:  dataset has changed since last saved
```

We could move id to the front of dataset by using the order command below. The default is to move the variable to the start (first) position in the dataset.[6]

```
. order id

. describe

Contains data from cardio2.dta
  obs:             5
 vars:            12                          22 Dec 2009 19:51
 size:           120 (99.9% of memory free)

              storage  display     value
variable name   type   format      label      variable label

id              byte   %3.0f                  Identification variable
time1pulse      int    %3.0f                  Pulse: Trial 1
bpress1         int    %3.0f                  Systolic BP: Trial 1
time2pulse      byte   %3.0f                  Pulse: Trial 2
bpress2         int    %3.0f                  Systolic BP: Trial 2
time3pulse      int    %3.0f                  Pulse: Trial 3
bpress3         int    %3.0f                  Systolic BP: Trial 3
time4pulse      int    %3.0f                  Pulse: Trial 4
bpress4         int    %3.0f                  Systolic BP: Trial 4
time5pulse      byte   %3.0f                  Pulse: Trial 5
bpress5         int    %3.0f                  Systolic BP: Trial 5
age_yrs         byte   %3.0f                  Age of person

Sorted by:
     Note:  dataset has changed since last saved
```

6. To be explicit, we could have added the first option to indicate that id should be moved to the first position in the dataset.

We can move age to be located after id by using the order command with the after() option. (The order command also supports a before() option to move one or more variables before a particular variable.)

```
. order age_yrs, after(id)
. describe
Contains data from cardio2.dta
  obs:             5
 vars:            12                          22 Dec 2009 19:51
 size:           120 (99.9% of memory free)
-------------------------------------------------------------------------------
              storage  display     value
variable name   type   format      label      variable label
-------------------------------------------------------------------------------
id              byte   %3.0f                  Identification variable
age_yrs         byte   %3.0f                  Age of person
time1pulse      int    %3.0f                  Pulse: Trial 1
bpress1         int    %3.0f                  Systolic BP: Trial 1
time2pulse      byte   %3.0f                  Pulse: Trial 2
bpress2         int    %3.0f                  Systolic BP: Trial 2
time3pulse      int    %3.0f                  Pulse: Trial 3
bpress3         int    %3.0f                  Systolic BP: Trial 3
time4pulse      int    %3.0f                  Pulse: Trial 4
bpress4         int    %3.0f                  Systolic BP: Trial 4
time5pulse      byte   %3.0f                  Pulse: Trial 5
bpress5         int    %3.0f                  Systolic BP: Trial 5
-------------------------------------------------------------------------------
Sorted by:
     Note:  dataset has changed since last saved
```

Tip! Ordering newly created variables

When you create a new variable, it is positioned at the end of the dataset, which is not necessarily the most logical position for the variable. You can use the order command to place a newly created variable in a more logical position in the dataset. For example, if you have a variable age and create a recoded version called agerecoded, then you can place agerecoded after age with the following order command. Then, if you drop the original age variable, agerecoded will be in the right position in the dataset.

```
. order agerecoded, after(age)
```

The blood pressure and pulse variables are not ordered optimally. For example, if bpress1 to bpress5 were positioned consecutively, we could refer them as bpress1 - bpress5. The order command below specifies that we want to reorder the variables that start with time and bpress, positioning them after age_yrs, and organize these variables alphabetically.

```
. order time* bpress*, after(age_yrs) alphabetic
```

The describe command shows that the blood pressure and pulse variables are now alphabetized and positioned after age_yrs.

```
. describe

Contains data from cardio2.dta
  obs:             5
 vars:            12                          22 Dec 2009 19:51
 size:           120 (99.9% of memory free)

              storage  display     value
variable name   type   format      label      variable label

id              byte   %3.0f                  Identification variable
age_yrs         byte   %3.0f                  Age of person
bpress1         int    %3.0f                  Systolic BP: Trial 1
bpress2         int    %3.0f                  Systolic BP: Trial 2
bpress3         int    %3.0f                  Systolic BP: Trial 3
bpress4         int    %3.0f                  Systolic BP: Trial 4
bpress5         int    %3.0f                  Systolic BP: Trial 5
time1pulse      int    %3.0f                  Pulse: Trial 1
time2pulse      byte   %3.0f                  Pulse: Trial 2
time3pulse      int    %3.0f                  Pulse: Trial 3
time4pulse      int    %3.0f                  Pulse: Trial 4
time5pulse      byte   %3.0f                  Pulse: Trial 5

Sorted by:
     Note:  dataset has changed since last saved
```

If you wanted all the variables alphabetized, then you could use order command with the _all keyword to indicate that all variables should be alphabetized (see section A.11).

```
. order _all, alphabetic
```

Tip! Alphabetic order versus sequential order

Suppose that we had a survey with 99 questions, with variable names q1, q2, q3, ..., q10, q11, ..., q99. Using the order command with the alphabetic option would place q1 before q10 and q2 before q20, and so forth. Using the sequential option would instead order the variables from q1 to q9 and then from q10 to q99.

This concludes this section on renaming and ordering variables. For more information, see help rename, help renpfix, and help order.

6 Combining datasets

Statistics are like a drunk with a lamppost: used more for support than illumination.

—Sir Winston Churchill

6.1 Introduction

This chapter describes how to combine datasets using Stata. It also covers problems that can arise when combining datasets, how you can detect them, and how to resolve them. This chapter covers four general methods of combining datasets: appending, merging, joining, and crossing. Section 6.2 covers the basics of how to append datasets, and section 6.3 illustrates problems that can arise when appending datasets. The next four sections cover four different kinds of merging—one-to-one match-merging (section 6.4), one-to-many match-merging (section 6.5), merging multiple datasets (section 6.6), and update merges (see section 6.7). Then section 6.8 discusses options that are common to each of these merging situations, and section 6.9 illustrates problems that can arise when merging datasets. The concluding sections cover joining datasets (section 6.10) and crossing datasets (section 6.11).

I should note that a new syntax was introduced in Stata 11 for the `merge` command. This new syntax introduces several new safeguards and features. This chapter exclusively illustrates this new syntax for the `merge` command, and thus these examples will not work in versions of Stata prior to version 11. Although not presented here, the syntax for the `merge` command from earlier versions of Stata continues to work using Stata 11.

6.2 Appending: Appending datasets

Consider `moms.dta` and `dad.dta`, presented below. Each dataset has four observations, the first about four moms and the second about four dads. Each dataset contains a family ID, the age of the person, his or her race, and whether he or she is a high school graduate.

```
. use moms

. list

     +--------------------------+
     | famid   age   race   hs |
     |--------------------------|
  1. |     3    24      2    1 |
  2. |     2    28      1    1 |
  3. |     4    21      1    0 |
  4. |     1    33      2    1 |
     +--------------------------+

. use dads

. list

     +--------------------------+
     | famid   age   race   hs |
     |--------------------------|
  1. |     1    21      1    0 |
  2. |     4    25      2    1 |
  3. |     2    25      1    1 |
  4. |     3    31      2    1 |
     +--------------------------+
```

Suppose that we wanted to stack these datasets on top of each other so that we would have a total of eight observations in the combined dataset. The append command is used for combining datasets like this, as illustrated below. First, we clear any data from memory. Then, after the append command, we list all the datasets we want to append together. Although we specified only two datasets, we could have specified more than two datasets on the append command.

```
. clear
. append using moms dads
```

The list command below shows us that these two files were appended successfully.

```
. list

     +------------------------+
     | famid   age   race   hs |
     |------------------------|
  1. |     3    24      2    1 |
  2. |     2    28      1    1 |
  3. |     4    21      1    0 |
  4. |     1    33      2    1 |
  5. |     1    21      1    0 |
     |------------------------|
  6. |     4    25      2    1 |
  7. |     2    25      1    1 |
  8. |     3    31      2    1 |
     +------------------------+
```

Suppose that you already had moms.dta loaded in memory, as shown below.

```
. use moms
```

At this point, you can append dads.dta like this:

```
. append using dads
. list

     +------------------------+
     | famid   age   race   hs |
     |------------------------|
  1. |     3    24      2    1 |
  2. |     2    28      1    1 |
  3. |     4    21      1    0 |
  4. |     1    33      2    1 |
  5. |     1    21      1    0 |
     |------------------------|
  6. |     4    25      2    1 |
  7. |     2    25      1    1 |
  8. |     3    31      2    1 |
     +------------------------+
```

(Continued on next page)

Tip! Appending jargon

In the last example, we call moms.dta the *master* dataset because it is the dataset in memory when the append is initiated. dads.dta is called the *using* dataset because it is specified after the using keyword.

However we append these datasets, the combined file does not identify the source of the data. We cannot tell whether an observation originated from moms.dta or from dads.dta. To solve this, we can add the generate() option, which will create a new variable that tells us from which dataset each observation came. You can name this variable anything you like; I called it datasrc.

```
. clear
. append using moms dads, generate(datasrc)
. list, sepby(datasrc)

     +------------------------------------+
     | datasrc   famid   age   race   hs  |
     |------------------------------------|
  1. |       1       3    24      2    1  |
  2. |       1       2    28      1    1  |
  3. |       1       4    21      1    0  |
  4. |       1       1    33      2    1  |
     |------------------------------------|
  5. |       2       1    21      1    0  |
  6. |       2       4    25      2    1  |
  7. |       2       2    25      1    1  |
  8. |       2       3    31      2    1  |
     +------------------------------------+
```

Looking back at the original data, we can see that when datasrc is 1, the data originate from moms.dta. When datasrc is 2, the data originate from dads.dta. If we had a third dataset on the append command, datasrc would have been 3 for the observations from that dataset.

Contrast this with the strategy where we first use the moms.dta dataset and then append the dataset dads.dta, as shown below.

```
. use moms
. append using dads, generate(datasrc)
. list, sepby(datasrc)

     +--------------------------------------+
     | famid   age   race   hs    datasrc   |
     |--------------------------------------|
  1. |     3    24      2    1          0   |
  2. |     2    28      1    1          0   |
  3. |     4    21      1    0          0   |
  4. |     1    33      2    1          0   |
     |--------------------------------------|
  5. |     1    21      1    0          1   |
  6. |     4    25      2    1          1   |
  7. |     2    25      1    1          1   |
  8. |     3    31      2    1          1   |
     +--------------------------------------+
```

Here a 0 means that the data came from the master dataset (i.e., `moms.dta`), and having a 1 means that the data came from the first using dataset (i.e., `dads.dta`). Had a second dataset been added after `dads` on the `append` command, the value for `datasrc` for those observations would have been 2.

The `label define` and `label values` commands below are used to label the values of `datasrc` (as described in section 4.4). Although I think labeling values is useful, it is optional.

```
. label define source 0 "From moms.dta" 1 "From dads.dta"
. label values datasrc source
. list, sepby(datasrc)

     +---------------------------------------------+
     | famid   age   race   hs           datasrc   |
     |---------------------------------------------|
  1. |     3    24      2    1     From moms.dta   |
  2. |     2    28      1    1     From moms.dta   |
  3. |     4    21      1    0     From moms.dta   |
  4. |     1    33      2    1     From moms.dta   |
     |---------------------------------------------|
  5. |     1    21      1    0     From dads.dta   |
  6. |     4    25      2    1     From dads.dta   |
  7. |     2    25      1    1     From dads.dta   |
  8. |     3    31      2    1     From dads.dta   |
     +---------------------------------------------+
```

As mentioned earlier, you can append multiple datasets at one time. For example, we have three datasets that contain book review information from three different reviewers: Clarence, Isaac, and Sally. The datasets are listed below using the `dir` command.

```
. dir br*.dta
   0.8k   2/02/10 18:48  br_clarence.dta
   0.8k   2/02/10 18:48  br_isaac.dta
   0.8k   2/02/10 18:48  br_sally.dta
```

The datasets all have the same variables in them. Below we can see the dataset containing the reviews from Clarence. This includes a variable identifying the book number (booknum), the name of the book (book), and the rating of the book (rating).

```
. use br_clarence
. list
```

	booknum	book	rating
1.	1	A Fistful of Significance	5
2.	2	For Whom the Null Hypothesis is Rejected	10
3.	3	Journey to the Center of the Normal Curve	6

Let's use the append command to combine all three datasets together. In doing so, we will use the generate() option to create a variable named rev that indicates the source of the data (i.e., the reviewer).

```
. clear
. append using br_clarence br_isaac br_sally, generate(rev)
. list, sepby(rev)
```

	rev	booknum	book	rating
1.	1	1	A Fistful of Significance	5
2.	1	2	For Whom the Null Hypothesis is Rejected	10
3.	1	3	Journey to the Center of the Normal Curve	6
4.	2	1	The Dreaded Type I Error	6
5.	2	2	How to Find Power	9
6.	2	3	The Outliers	8
7.	3	1	Random Effects for Fun and Profit	6
8.	3	2	A Tale of t-tests	9
9.	3	3	Days of Correlation and Regression	8

The value of rev is 1, 2, or 3 for the observations that came from br_clarence, br_isaac, or br_sally, respectively.

This covers the basics of using the append command. The next section covers some of the problems that can arise when appending datasets.

6.3 Appending: Problems

The last section showed how easy it is to append datasets, but it ignored some of the problems that can arise when appending datasets. This section describes five problems that can arise when appending datasets: differing variable names across datasets, conflicting variable labels, conflicting value labels, inconsistent variable coding, and mixing variable types across datasets. These are discussed one at a time below.

Differing variable names across datasets

Consider moms1.dta and dads1.dta, shown below. Even though the two datasets contain variables measuring the same idea (age, race, and whether one graduated high school), they are named differently in the two datasets.

```
. use moms1
. list
```

	famid	mage	mrace	mhs
1.	1	33	2	1
2.	2	28	1	1
3.	3	24	2	1
4.	4	21	1	0

```
. use dads1
. list
```

	famid	dage	drace	dhs
1.	1	21	1	0
2.	2	25	1	1
3.	3	31	2	1
4.	4	25	2	1

Because the variables with the moms' information are named differently from the variables with the dads' information, Stata cannot know how to put similar variables together when appending the datasets.

If we append these two datasets, the resulting dataset contains different variables for the moms and for the dads, as shown below.

```
. use moms1
. append using dads1
. list
```

	famid	mage	mrace	mhs	dage	drace	dhs
1.	1	33	2	1	.	.	.
2.	2	28	1	1	.	.	.
3.	3	24	2	1	.	.	.
4.	4	21	1	0	.	.	.
5.	1	.	.	.	21	1	0
6.	2	.	.	.	25	1	1
7.	3	.	.	.	31	2	1
8.	4	.	.	.	25	2	1

Tip! Good for merging

If you look ahead to the section on merging datasets (section 6.4), you will see that moms1.dta and dads1.dta may not be useful for appending but are ideal for merging. For datasets you intend to combine, the best naming scheme for the variables depends on whether you intend to append or merge the datasets. If you will append datasets, you want the variable names to be the same, but if you will merge datasets, you want the variable names to be different.

We need to make the variable names the same between the two datasets before appending them. We first rename the variables for moms1.dta and then save it as moms1temp.dta.

```
. use moms1
. rename mage age
. rename mrace race
. rename mhs hs
. save moms1temp
file moms1temp.dta saved
```

We then do the same kind of renaming for dads1.dta and save it as dads1temp.dta.

```
. use dads1
. rename dage age
. rename drace race
. rename dhs hs
. save dads1temp
file dads1temp.dta saved
```

Because moms1temp.dta shares the same variable names with dads1temp.dta, we can successfully append these datasets.

```
. clear
. append using moms1temp dads1temp
. list
```

	famid	age	race	hs
1.	1	33	2	1
2.	2	28	1	1
3.	3	24	2	1
4.	4	21	1	0
5.	1	21	1	0
6.	2	25	1	1
7.	3	31	2	1
8.	4	25	2	1

Conflicting variable labels

Consider `momslab.dta` and `dadslab.dta`. These datasets are described below.

```
. use momslab

. describe

Contains data from momslab.dta
  obs:             4
 vars:             4                          27 Dec 2009 21:47
 size:            80 (99.9% of memory free)
-------------------------------------------------------------------------------
              storage  display     value
variable name   type   format      label      variable label
-------------------------------------------------------------------------------
famid           float  %5.0g                  Family ID
age             float  %5.0g                  Mom's Age
race            float  %9.0g       eth        Mom's Ethnicity
hs              float  %15.0g      grad       Is Mom a HS Graduate?
-------------------------------------------------------------------------------
Sorted by:  famid

. use dadslab

. describe

Contains data from dadslab.dta
  obs:             4
 vars:             4                          27 Dec 2009 21:47
 size:            80 (99.9% of memory free)
-------------------------------------------------------------------------------
              storage  display     value
variable name   type   format      label      variable label
-------------------------------------------------------------------------------
famid           float  %5.0g                  Family ID
age             float  %5.0g                  Dad's Age
race            float  %9.0g       eth        Dad's Ethnicity
hs              float  %15.0g      hsgrad     Is Dad a HS Graduate?
-------------------------------------------------------------------------------
Sorted by:  famid
```

Note the variable labels used in each of these files. The variable labels in `momslab.dta` specifically identify the variables as belonging to the mom, and likewise, the labels in `dadslab.dta` describe the variables as belonging to the dad. These labels seem perfect. Let's see what happens when we **append** these two files.

```
. clear

. append using momslab dadslab
(label eth already defined)
```

(*Continued on next page*)

Now let's describe the combined file.

```
. describe
Contains data
  obs:             8
 vars:             4
 size:           160 (99.9% of memory free)
-------------------------------------------------------------------------------
              storage  display     value
variable name   type   format      label      variable label
-------------------------------------------------------------------------------
famid           float  %5.0g                  Family ID
age             float  %5.0g                  Mom's Age
race            float  %9.0g       eth        Mom's Ethnicity
hs              float  %15.0g      grad       Is Mom a HS Graduate?
-------------------------------------------------------------------------------
Sorted by:
     Note:  dataset has changed since last saved
```

The variable labels are based on the labels specified in `momslab.dta`. (The labels from `momslab.dta` were used because that file was specified earlier on the `append` command.) The labels that made so much sense when labeling the moms no longer make as much sense when applied to the combined file.

The solution is either to select more neutral labels in the original datasets or to use the `variable label` command to change the labels after appending the datasets.

Conflicting value labels

Let's again use `momslab.dta` and `dadslab.dta` to illustrate conflicts that can arise with value labels. Looking at the `describe` command for these datasets (on page 181), we can see that the variable `race` is labeled using a value label named `eth` in both datasets. In `momslab.dta`, the variable `hs` is labeled with a label named `grad`, while the same variable in `dadslab.dta` is labeled with a label named `hsgrad`.

Let's list the observations from each of these datasets.

```
. use momslab
. list

     +------------------------------------------------+
     | famid   age         race                   hs |
     |------------------------------------------------|
  1. |     1    33    Mom Black          Mom HS Grad |
  2. |     2    28    Mom White          Mom HS Grad |
  3. |     3    24    Mom Black          Mom HS Grad |
  4. |     4    21    Mom White      Mom Not HS Grad |
     +------------------------------------------------+
```

```
. use dadslab
. list
```

	famid	age	race	hs
1.	1	21	Dad White	Dad Not HS Grad
2.	2	25	Dad White	Dad HS Grad
3.	3	31	Dad Black	Dad HS Grad
4.	4	25	Dad Black	Dad HS Grad

Note how the labeled values for **race** and **hs** are different in the two datasets. Let's see what happens when we append these two files together.

```
. clear
. append using momslab dadslab, generate(datasrc)
(label eth already defined)
. list, sepby(datasrc)
```

	datasrc	famid	age	race	hs
1.	1	1	33	Mom Black	Mom HS Grad
2.	1	2	28	Mom White	Mom HS Grad
3.	1	3	24	Mom Black	Mom HS Grad
4.	1	4	21	Mom White	Mom Not HS Grad
5.	2	1	21	Mom White	Mom Not HS Grad
6.	2	2	25	Mom White	Mom HS Grad
7.	2	3	31	Mom Black	Mom HS Grad
8.	2	4	25	Mom Black	Mom HS Grad

Looking at the listing of **race** and **hs**, we can see that these variables are labeled using the value labels from **momslab.dta**.[1] This also applied to the definition of the value label named **eth**: the definition from **momslab.dta** took precedence (this is what the **(label eth already defined)** message meant).

This would not be such a problem if the labels from **momslab.dta** were written in a general way that could apply to both moms and dads. But as written, the labels are misleading. They imply that all the observations come from a mom.

We can either go back and change the labels in **momslab.dta** before merging the datasets or simply change the labels afterward. It is probably just as easy to change the labels afterward. See section 4.4 for more on how to do this.

1. Conflicts among value labels are resolved by giving precedence to the dataset that is referenced first. The master dataset takes precedence over the using dataset. If there are multiple using datasets, the earlier using datasets take precedence over the later using datasets.

Inconsistent variable coding

Suppose that you append two datasets, each of which uses different coding for the same variable. This can be hard to detect because each dataset is internally consistent but the coding is not consistent between the datasets. Let's illustrate this by appending a variation of `moms.dta`, named `momshs.dta`, with the dads dataset named `dads.dta`.

First, let's look at `momshs.dta`.

```
. use momshs
. list

     +---------------------------+
     | famid   age   race   hs   |
     |---------------------------|
  1. |     3    24      2    2   |
  2. |     2    28      1    2   |
  3. |     4    21      1    1   |
  4. |     1    33      2    1   |
     +---------------------------+
```

And then we look at `dads.dta`.

```
. use dads
. list

     +---------------------------+
     | famid   age   race   hs   |
     |---------------------------|
  1. |     1    21      1    0   |
  2. |     4    25      2    1   |
  3. |     2    25      1    1   |
  4. |     3    31      2    1   |
     +---------------------------+
```

Note the difference in the coding of `hs` in these two datasets. In `momshs.dta`, `hs` is coded using a 1 = no and 2 = yes coding scheme, but `dads.dta` uses dummy coding, i.e., 0 = no and 1 = yes. Let's pretend we did not yet notice this problem and observe the consequences of appending these two files together, as shown below.

```
. use momshs
. append using dads
```

The `append` command was successful and did not produce any errors. We can list the observations from the combined file, and there are no obvious errors.

```
. list

     ┌───────────────────────────┐
     │ famid   age   race    hs  │
     ├───────────────────────────┤
  1. │     3    24      2     2  │
  2. │     2    28      1     2  │
  3. │     4    21      1     1  │
  4. │     1    33      2     1  │
  5. │     1    21      1     0  │
     ├───────────────────────────┤
  6. │     4    25      2     1  │
  7. │     2    25      1     1  │
  8. │     3    31      2     1  │
     └───────────────────────────┘
```

Let's look at a tabulation of the variable **hs**. This is a yes/no variable indicating whether the person graduated high school, so it should only have two levels. But as we see below, this variable has three levels. This is often the first clue when you have appended two datasets that use a different coding scheme for the same variable.

```
. tabulate hs

         HS │
  Graduate? │      Freq.     Percent        Cum.
────────────┼───────────────────────────────────
          0 │          1       12.50       12.50
          1 │          5       62.50       75.00
          2 │          2       25.00      100.00
────────────┼───────────────────────────────────
      Total │          8      100.00
```

The solution, of course, is to ensure that the **hs** variable uses the same coding before appending the two datasets. Below we repeat the appending process, but we first recode **hs** in **momshs.dta** to use dummy coding (thus making it commensurate with the coding of **hs** in **dads.dta**).

```
. use momshs
. recode hs (1=0) (2=1)
(hs: 4 changes made)
. append using dads
```

With **hs** coded the same way in both datasets, the **hs** variable now has two levels. We can see in the combined dataset that three parents did not graduate high school and five parents did graduate high school.

```
. tabulate hs

         HS │
  Graduate? │      Freq.     Percent        Cum.
────────────┼───────────────────────────────────
          0 │          3       37.50       37.50
          1 │          5       62.50      100.00
────────────┼───────────────────────────────────
      Total │          8      100.00
```

Mixing variable types across datasets

Let's see what happens when you append datasets in which the variables have different data types. As section A.5 describes, Stata variables fall into two general data types: string types and numeric types. Let's start by examining what happens if we try to append two datasets in which one of the variables is stored as a numeric type and the other is stored as a string type. In `moms.dta`, the variable `hs` is stored as a numeric (`float`) variable, but in `dadstr.dta`, `hs` is stored as a string (`str3`) variable.

Below we can see what happens when we try to append these two datasets.

```
. use moms
. append using dadstr
hs is str3 in using data
r(106);
```

As the error message indicates, the variable `hs` is stored as a `str3` (a string with length 3) in the using dataset. Stata cannot reconcile this with `hs` in `moms.dta` because here it is a numeric (`float`), so `merge` reports an error. We need to make `hs` either numeric in both datasets (see section 5.13) or a string in both datasets (see section 5.14). Let's convert `hs` to numeric in `dadstr.dta` and then append that with `moms.dta`, as shown below.

```
. use dadstr
. destring hs, replace
hs has all characters numeric; replaced as byte
. append using moms
hs was byte now float
```

As we can see below, the combined dataset reflects the values for `hs` from each dataset and is stored as a numeric (`float`) data type.

```
. list

     +-------------------------+
     | famid   age   race   hs |
     |-------------------------|
  1. |     1    21      1    0 |
  2. |     4    25      2    1 |
  3. |     2    25      1    1 |
  4. |     3    31      2    1 |
  5. |     3    24      2    1 |
     |-------------------------|
  6. |     2    28      1    1 |
  7. |     4    21      1    0 |
  8. |     1    33      2    1 |
     +-------------------------+

. describe hs

              storage  display     value
variable name   type   format      label      variable label
--------------------------------------------------------------------
hs              float  %10.0g                 HS Graduate?
```

This illustrates the most serious conflict among variable types that can arise, when one variable is numeric and one variable is a string. However, there are other kinds of variable type conflicts that can arise. As we will see below, Stata resolves these other kinds of conflicts without our intervention.

As illustrated in section A.5, Stata permits us to store string variables using a length of 1 (i.e., `str1`) up to a length of 244 (i.e., `str244`). So how does Stata handle conflicting lengths for a string variable? In `momstr.dta`, the `hs` variable is stored as a string with length 1 (`str1`), whereas in `dadstr.dta`, the `hs` variable is stored as a string with length 3 (`str3`), as shown below.

```
. use momstr
. describe hs
              storage  display     value
variable name   type   format      label      variable label
-------------------------------------------------------------------------------
hs              str1   %9s                    HS Graduate?
. use dadstr
. describe hs
              storage  display     value
variable name   type   format      label      variable label
-------------------------------------------------------------------------------
hs              str3   %9s                    HS Graduate?
```

When appending these two datasets, Stata tells us that it is changing `hs` to be of type `str3`. It does this so that `hs` will be wide enough to accommodate the data from both datasets. Stata did this automatically for us, without any extra effort on our part.

```
. use momstr
. append using dadstr
hs was str1 now str3
```

What about differences among numeric variables? As section A.5 describes, Stata has five different numeric data storage types: `byte`, `int`, `long`, `float`, and `double`. When there are conflicts among numeric data types, Stata will automatically choose an appropriate data type for us.

(Continued on next page)

We can illustrate this automatic selection using `moms.dta` and `dadsdbl.dta`, described below.

```
. use moms

. describe

Contains data from moms.dta
  obs:             4
 vars:             4                          22 Dec 2009 20:07
 size:            80 (99.9% of memory free)
-------------------------------------------------------------------------------
              storage  display     value
variable name   type   format      label      variable label
-------------------------------------------------------------------------------
famid           float  %5.0g                  Family ID
age             float  %5.0g                  Age
race            float  %5.0g                  Ethnicity
hs              float  %7.0g                  HS Graduate?
-------------------------------------------------------------------------------
Sorted by:

. use dadsdbl

. describe

Contains data from dadsdbl.dta
  obs:             4
 vars:             4                          22 Dec 2009 20:02
 size:            76 (99.9% of memory free)
-------------------------------------------------------------------------------
              storage  display     value
variable name   type   format      label      variable label
-------------------------------------------------------------------------------
famid           int    %5.0g                  Family ID
age             byte   %5.0g                  Age
race            double %5.0g                  Ethnicity
hs              long   %7.0g                  HS Graduate?
-------------------------------------------------------------------------------
Sorted by:
```

Note how all the variables in `moms.dta` are stored as `float`, while in `dadsdbl.dta`, the variables are stored using four different data types (`int`, `byte`, `double`, and `long`). Let's see what happens when we append these two datasets.

```
. use moms

. append using dadsdbl
race was float now double
hs was float now double
```

During the appending process, Stata looks at each variable and chooses an appropriate data type for each. If the data type was changed from that specified in the master dataset (`moms.dta`), Stata displays a message. In this case, Stata tells us that it changed `race` from `float` to `double` and that it also changed `hs` from `float` to `double`. The important point is that Stata resolves any such discrepancies among numeric data types for you, selecting an appropriate data type that will ensure that no data are lost.

This concludes this section, about problems that can arise when appending datasets. For more information about appending datasets, see `help append`.

6.4 Merging: One-to-one match-merging

A match-merge combines two datasets using one (or more) key variables to link observations between the two datasets. In a one-to-one match-merge, the key variable(s) uniquely identifies each observation in each dataset. Consider the **moms1.dta** and **dads1.dta** datasets, below. The key variable, **famid**, uniquely identifies each observation in each dataset and can be used to link the observations from **moms.dta** with the observations from **dads.dta**. Because these datasets are so small, you can see that each observation from **moms.dta** has a match in **dads.dta** based on **famid**.

```
. use moms1
. list

     +-----------------------------+
     | famid   mage   mrace   mhs  |
     |-----------------------------|
  1. |     1     33       2     1  |
  2. |     2     28       1     1  |
  3. |     3     24       2     1  |
  4. |     4     21       1     0  |
     +-----------------------------+

. use dads1
. list

     +-----------------------------+
     | famid   dage   drace   dhs  |
     |-----------------------------|
  1. |     1     21       1     0  |
  2. |     2     25       1     1  |
  3. |     3     31       2     1  |
  4. |     4     25       2     1  |
     +-----------------------------+
```

We perform a **1:1** merge between **moms1.dta** and **dads1.dta**, linking them based on **famid**.

```
. use moms1
. merge 1:1 famid using dads1
    Result                           # of obs.
    -----------------------------------------
    not matched                             0
    matched                                 4  (_merge==3)
    -----------------------------------------
```

The output from the **merge** command confirms our expectations that each observation from **moms.dta** has a matched observation in **dads.dta** (and vice versa). We can see this for ourselves by listing the merged dataset.

(*Continued on next page*)

```
. list
```

	famid	mage	mrace	mhs	dage	drace	dhs	_merge
1.	1	33	2	1	21	1	0	matched (3)
2.	2	28	1	1	25	1	1	matched (3)
3.	3	24	2	1	31	2	1	matched (3)
4.	4	21	1	0	25	2	1	matched (3)

The listing shows the `famid` variable followed by the variables from `moms.dta` and then the variables from `dads.dta`. The last variable, `_merge`, was created by the `merge` command to show the matching status for each observation. In this example, every observation shows `matched (3)`, indicating that a match was found between the master and using dataset for every observation.

Tip! Merging jargon

In this example, `moms1.dta` is the *master* dataset because it is the dataset in memory when the `merge` command is issued. `dads1.dta` is called the *using* dataset because it is specified after the `using` keyword. The variable `famid` is called the *key variable* because it holds the key to linking the master and using files.

Let's consider a second example that involves some observations that do not match. Let's merge and inspect the datasets `moms2.dta` and `dads2.dta`.

```
. use moms2
. list
```

	famid	mage	mrace	mhs	fr_moms2
1.	1	33	2	1	1
2.	3	24	2	1	1
3.	4	21	1	0	1
4.	5	39	2	0	1

```
. use dads2
. list
```

	famid	dage	drace	dhs	fr_dads2
1.	1	21	1	0	1
2.	2	25	1	1	1
3.	4	25	2	1	1

Note how `moms2.dta` has an observation for family 3 and an observation for family 5 with no corresponding observations in `dads2.dta`. Likewise, `dads2.dta` has an observation for family 2, but there is no corresponding observation in `moms2.dta`. These observations will not be matched. When we merge these files, Stata will tell us about these nonmatched observations and help us track them, as we can see below.

```
. use moms2
. merge 1:1 famid using dads2
    Result                           # of obs.
    -----------------------------------------
    not matched                             3
        from master                         2  (_merge==1)
        from using                          1  (_merge==2)

    matched                                 2  (_merge==3)
    -----------------------------------------
```

The merge command summarizes how the matching went. Two observations were matched and three observations were not matched. Among the nonmatched observations, two observations originated from the master (moms2.dta) dataset, and one nonmatched observation originated from the using (dads2.dta) dataset. Let's now list the resulting merged dataset. (I first sorted the dataset on famid to make the listing easier to follow.)

```
. sort famid
. list famid mage mrace dage drace _merge
```

	famid	mage	mrace	dage	drace	_merge
1.	1	33	2	21	1	matched (3)
2.	2	.	.	25	1	using only (2)
3.	3	24	2	.	.	master only (1)
4.	4	21	1	25	2	matched (3)
5.	5	39	2	.	.	master only (1)

Families 3 and 5 have data from moms2.dta (master) but not dads2.dta (using). The _merge variable confirms this by displaying master only (1). Family 2 has data from dads2.dta (using) but not moms2.dta (master). The _merge variable informs us of this by displaying using only (2) for this observation. Families 1 and 4 had matched observations between the master and using datasets, and this is also indicated in the _merge variable, which shows matched (3).

Let's look more closely at the _merge variable. This variable, which tells us about the matching status for each observation, might appear to be a string variable, but it is a numeric variable. We can see this using the codebook command.

(Continued on next page)

```
. codebook _merge

_merge                                                        (unlabeled)

                  type:  numeric (byte)
                 label:  _merge

                 range:  [1,3]                        units:  1
         unique values:  3                        missing .:  0/5

           tabulation:  Freq.   Numeric  Label
                            2         1  master only (1)
                            1         2  using only (2)
                            2         3  matched (3)
```

The value for the _merge variable is just the number 1, 2, or 3 with a value label providing a more descriptive label. If we want to list just the matched observations, we can specify `if _merge == 3` with the `list` command, as shown below.

```
. list famid mage mrace dage drace _merge if _merge == 3

     famid   mage   mrace   dage   drace        _merge

  1.     1     33       2     21       1   matched (3)
  4.     4     21       1     25       2   matched (3)
```

Or we could list the observations that only originated from the master dataset (`moms2.dta`) like this:

```
. list famid mage mrace dage drace _merge if _merge == 1

     famid   mage   mrace   dage   drace            _merge

  3.     3     24       2      .       .   master only (1)
  5.     5     39       2      .       .   master only (1)
```

We could keep just the matched observations by using the `keep` command, as shown below.[2]

```
. keep if _merge == 3
(3 observations deleted)
. list famid mage mrace dage drace _merge

     famid   mage   mrace   dage   drace        _merge

  1.     1     33       2     21       1   matched (3)
  2.     4     21       1     25       2   matched (3)
```

When merging `moms2.dta` and `dads2.dta`, we called this a one-to-one merge because we assumed that `moms2.dta` contained one observation per `famid` and, likewise, `dads2.dta` contained one observation per `famid`. Suppose that one of the datasets

2. This could also be done using the `keep()` option, as illustrated in section 6.8.

had more than one observation per famid. momsdup.dta is such a dataset. This value of famid is accidentally repeated for the last observation (it shows as 4 for the last observation but should be 5).

```
. use momsdup
. list
```

	famid	mage	mrace	mhs	fr_moms2
1.	1	33	2	1	1
2.	3	24	2	1	1
3.	4	21	1	0	1
4.	4	39	2	0	1

This mistake should have been caught as a part of checking for duplicates (as described in section 3.8) on the famid variable, but suppose that we did not notice this. Fortunately, Stata catches this when we perform a one-to-one merge between momsdup.dta and dads2.dta, as shown below.

```
. use momsdup
. merge 1:1 famid using dads2
variable famid does not uniquely identify observations in the master data
r(459);
```

The error message is alerting us that famid does not uniquely identify observations in the master dataset (momsdup.dta). For a one-to-one merge, Stata checks both the master and the using datasets to make sure that the key variable(s) uniquely identifies the observations in each dataset. If not, an error message like the one above is displayed.

So far, all the examples have used one key variable for linking the master and using datasets, but it is possible to have two or more key variables that are used to link the master and using datasets. For example, consider kids1.dta, below.

```
. use kids1
. sort famid kidid
. list
```

	famid	kidid	kage	kfem
1.	1	1	3	1
2.	2	1	8	0
3.	2	2	3	1
4.	3	1	4	1
5.	3	2	7	0
6.	4	1	1	0
7.	4	2	3	0
8.	4	3	7	0

It takes two variables to identify each kid: famid and kidid. Let's merge this dataset with another dataset named kidname.dta (shown below).

```
. use kidname
. sort famid kidid
. list

     +----------------------+
     | famid   kidid  kname |
     |----------------------|
  1. |     1       1    Sue |
  2. |     2       1    Vic |
  3. |     2       2    Flo |
  4. |     3       1    Ivy |
  5. |     3       2    Abe |
     |----------------------|
  6. |     4       1    Tom |
  7. |     4       2    Bob |
  8. |     4       3    Cam |
     +----------------------+
```

The kids in these two files can be uniquely identified and linked based on the combination of `famid` and `kidid`. We can use these two variables together as the key variables for merging these two files, as shown below.

```
. use kids1
. merge 1:1 famid kidid using kidname
    Result                           # of obs.
    -----------------------------------------
    not matched                             0
    matched                                 8  (_merge==3)
    -----------------------------------------
```

The output from the `merge` command shows that all the observations in the merged file were matched. Below we can see the merged dataset.

```
. list

     +-----------------------------------------------------------+
     | famid   kidid   kage   kfem   kname                _merge |
     |-----------------------------------------------------------|
  1. |     1       1      3      1     Sue   matched (3)         |
  2. |     2       1      8      0     Vic   matched (3)         |
  3. |     2       2      3      1     Flo   matched (3)         |
  4. |     3       1      4      1     Ivy   matched (3)         |
  5. |     3       2      7      0     Abe   matched (3)         |
     |-----------------------------------------------------------|
  6. |     4       1      1      0     Tom   matched (3)         |
  7. |     4       2      3      0     Bob   matched (3)         |
  8. |     4       3      7      0     Cam   matched (3)         |
     +-----------------------------------------------------------+
```

This concludes this section on one-to-one merging. This section did not address any of the problems that can arise in such merges. Section 6.9 discusses problems that can arise when merging datasets, how to discover them, and how to deal with them.

6.5 Merging: One-to-many match-merging

Section 6.4 showed a `1:1` merge that merged moms with dads. This was called a `1:1` merge because the key variable(s) uniquely identified each observation within each dataset. By contrast, when matching moms to kids, a mom could match with more than one kid (a one-to-many merge). The moms dataset is the `1` dataset and the kids dataset is the `m` dataset. Despite this difference, the process of performing a `1:m` merge is virtually identical to the process of performing a `1:1` merge. This is illustrated by merging `moms1.dta` with `kids1.dta`. These two datasets are shown below.

```
. use moms1

. list

     +-----------------------------+
     | famid   mage   mrace   mhs  |
     |-----------------------------|
  1. |     1     33       2     1  |
  2. |     2     28       1     1  |
  3. |     3     24       2     1  |
  4. |     4     21       1     0  |
     +-----------------------------+

. use kids1

. list

     +-------------------------------+
     | famid   kidid   kage   kfem   |
     |-------------------------------|
  1. |     3       1      4      1   |
  2. |     3       2      7      0   |
  3. |     2       1      8      0   |
  4. |     2       2      3      1   |
  5. |     4       1      1      0   |
     |-------------------------------|
  6. |     4       2      3      0   |
  7. |     4       3      7      0   |
  8. |     1       1      3      1   |
     +-------------------------------+
```

The variable `famid` links the moms with the kids. You can see that the mom in family 1 will match to one child, but the mom in family 4 will match to three children. You can also see that for every mom, there is at least one matched child, and every child has a matching mom. We merge these two datasets below.

```
. use moms1

. merge 1:m famid using kids1
    Result                           # of obs.
    -----------------------------------------
    not matched                             0
    matched                                 8  (_merge==3)
    -----------------------------------------
```

The report shows that all observations were matched.

We can see the resulting merged dataset below. The dataset is sorted on `famid` and `kidid` to make the listing easier to follow.

```
. sort famid kidid
. list, sepby(famid)
```

	famid	mage	mrace	mhs	kidid	kage	kfem	_merge
1.	1	33	2	1	1	3	1	matched (3)
2.	2	28	1	1	1	8	0	matched (3)
3.	2	28	1	1	2	3	1	matched (3)
4.	3	24	2	1	1	4	1	matched (3)
5.	3	24	2	1	2	7	0	matched (3)
6.	4	21	1	0	1	1	0	matched (3)
7.	4	21	1	0	2	3	0	matched (3)
8.	4	21	1	0	3	7	0	matched (3)

In the listing above, note how the information for the moms with multiple children is repeated. For example, in family 4, the mom had three matching children. Her information (such as `mage` and `mrace`) appears three times, corresponding to each of the matching children.

Let's briefly consider an example where the observations do not match perfectly, by matching `moms2.dta` with `kids2.dta` (shown below).

```
. use moms2
. list
```

	famid	mage	mrace	mhs	fr_moms2
1.	1	33	2	1	1
2.	3	24	2	1	1
3.	4	21	1	0	1
4.	5	39	2	0	1

```
. use kids2
. list, sepby(famid)
```

	famid	kidid	kage	kfem
1.	2	2	3	1
2.	2	1	8	0
3.	3	2	7	0
4.	3	1	4	1
5.	4	2	3	0
6.	4	3	7	0
7.	4	1	1	0

`moms2.dta` has observations for family 1 and family 5, but there are no corresponding observations in `kids2.dta`. `kids2.dta` has an observation for family 2, but there is no corresponding observation in `moms2.dta`.

Let's now merge these datasets together.

```
. use moms2
. merge 1:m famid using kids2
    Result                           # of obs.
    -----------------------------------------
    not matched                             4
        from master                         2  (_merge==1)
        from using                          2  (_merge==2)

    matched                                 5  (_merge==3)
    -----------------------------------------
```

The report shows us the matching results. Five observations were matched and four observations were not matched. Among the nonmatched observations, two were from the master (moms2.dta) only and two were from the using (kids2.dta) only. Below we see the listing of the merged dataset.

```
. sort famid kidid
. list famid mage fr_moms2 kidid kage _merge, sepby(famid)
```

	famid	mage	fr_moms2	kidid	kage	_merge
1.	1	33	1	.	.	master only (1)
2.	2	.	.	1	8	using only (2)
3.	2	.	.	2	3	using only (2)
4.	3	24	1	1	4	matched (3)
5.	3	24	1	2	7	matched (3)
6.	4	21	1	1	1	matched (3)
7.	4	21	1	2	3	matched (3)
8.	4	21	1	3	7	matched (3)
9.	5	39	1	.	.	master only (1)

The _merge variable is master only (1) for families 1 and 5 because there was an observation for the moms but not the kids dataset. The _merge variable is using only (2) for family 2 because there were two observations in kids2.dta but no corresponding observation in moms2.dta.

Rather than listing the entire dataset, we could just list the nonmatched observations that originated from the master dataset (moms2.dta).

(Continued on next page)

```
. list famid mage fr_moms2 kidid kage _merge if _merge==1

     +------------------------------------------------------------+
     | famid   mage   fr_moms2   kidid   kage              _merge |
     |------------------------------------------------------------|
  1. |     1     33          1       .      .     master only (1) |
  9. |     5     39          1       .      .     master only (1) |
     +------------------------------------------------------------+
```

Or we could list the nonmatched observations that originated from the using dataset (`kids2.dta`).

```
. list famid mage fr_moms2 kidid kage _merge if _merge==2

     +----------------------------------------------------------+
     | famid   mage   fr_moms2   kidid   kage            _merge |
     |----------------------------------------------------------|
  2. |     2      .          .       1      8    using only (2) |
  3. |     2      .          .       2      3    using only (2) |
     +----------------------------------------------------------+
```

We could also list all the nonmatched observations.

```
. list famid mage fr_moms2 kidid kage _merge if _merge==1 | _merge==2

     +------------------------------------------------------------+
     | famid   mage   fr_moms2   kidid   kage              _merge |
     |------------------------------------------------------------|
  1. |     1     33          1       .      .     master only (1) |
  2. |     2      .          .       1      8      using only (2) |
  3. |     2      .          .       2      3      using only (2) |
  9. |     5     39          1       .      .     master only (1) |
     +------------------------------------------------------------+
```

So far, these examples have illustrated `1:m` (one-to-many) merging, but Stata also supports `m:1` (many-to-one) merging, in which the master dataset can have multiple observations that match to a using dataset in which the key variable(s) uniquely identifies each observation. For example, rather than merging the moms with the kids (as illustrated previously), we can merge the kids with the moms, as shown below.

```
. use kids1

. merge m:1 famid using moms1
    Result                           # of obs.
    -----------------------------------------
    not matched                             0
    matched                                 8  (_merge==3)
    -----------------------------------------
```

The output of the `merge` command shows that the merged dataset has eight matching observations and no nonmatching observations. We can see the resulting merged dataset below.

```
. sort famid kidid
. list, sepby(famid)
```

	famid	kidid	kage	kfem	mage	mrace	mhs	_merge
1.	1	1	3	1	33	2	1	matched (3)
2.	2	1	8	0	28	1	1	matched (3)
3.	2	2	3	1	28	1	1	matched (3)
4.	3	1	4	1	24	2	1	matched (3)
5.	3	2	7	0	24	2	1	matched (3)
6.	4	1	1	0	21	1	0	matched (3)
7.	4	2	3	0	21	1	0	matched (3)
8.	4	3	7	0	21	1	0	matched (3)

Note that the variables from `kids1.dta` appear before the variables from `moms1.dta` because `kids1.dta` was the master dataset and `moms1.dta` was the using dataset. Otherwise, the results of performing a `1:m` or an `m:1` merge are the same, so the choice of which to perform is up to you.

This section covered the basics of performing `1:m` and `m:1` merges. For more information, see section 6.9, which covers some of the problems that can arise when merging datasets.

6.6 Merging: Merging multiple datasets

In Stata version 11, the `merge` command was augmented with several new features and safeguards. As we have seen in the prior sections, you can perform `1:1`, `1:m`, and `m:1` merges using this new syntax. All of these are pairwise merges. This section illustrates how you can merge multiple datasets with a series of pairwise merges.[3]

Let's consider an example where we want to merge four datasets. We have seen two of the datasets before in this chapter, `moms2.dta` describing moms and `dads2.dta` describing dads. We also have a dataset named `momsbest2.dta`, which describes the mom's best friend, and a dataset named `dadsbest2.dta`, which describes the dad's best friend.

Let's approach this by merging the moms dataset with the dataset containing the moms' best friends and saving the resulting dataset. Then let's merge the dads dataset with the dataset containing the dads' best friends and save that dataset. We can then merge the two combined datasets. Below we start the process by merging the moms with their best friends.

3. The Stata 10 `merge` syntax continues to support merging multiple datasets, but I have always found this to be a perilous strategy and have avoided it, favoring the use of multiple pairwise merges.

```
. use moms2
. merge 1:1 famid using momsbest2, nogenerate
    Result                           # of obs.
    -----------------------------------------
    not matched                             3
        from master                         2
        from using                          1

    matched                                 2
    -----------------------------------------
```

The table shows us that there were two matched observations and three nonmatched observations. Normally, we would inspect the `_merge` variable in the merged dataset to identify the unmatched observations, but I added the `nogenerate` option to suppress the creation of the `_merge` variable. I did this because when I merge multiple datasets, I prefer to track the origin of the data using variables that I create in each dataset. The `moms2` dataset has a variable named `fr_moms2` (which contains a 1 for all observations), and `momsbest2` has a variable named `fr_momsbest2` (which contains a 1 for all observations). Let's look at the listing of all the variables, below (after sorting on `famid` to make the listing easier to follow).

```
. sort famid
. list, abb(20)
```

	famid	mage	mrace	mhs	fr_moms2	mbage	fr_momsbest2
1.	1	33	2	1	1	.	.
2.	2	.	.	.	.	29	1
3.	3	24	2	1	1	23	1
4.	4	21	1	0	1	37	1
5.	5	39	2	0	1	.	.

Looking at these `fr_` variables, we can see that in families 1 and 5, there was an observation for the mom but not her best friend. And in family 2, there was an observation for the mom's best friend but not one for the mom herself. Let's save this file, naming it `momsandbest.dta`.

```
. save momsandbest
file momsandbest.dta saved
```

`momsandbest.dta` can be merged with the file that merges the dads with their best friends. In fact, let's do that merge right now.

```
. use dads2
. merge 1:1 famid using dadsbest2, nogenerate
    Result                           # of obs.
    -----------------------------------------
    not matched                             1
        from master                         0
        from using                          1

    matched                                 3
    -----------------------------------------
```

```
. sort famid
. list, abb(20)
```

	famid	dage	drace	dhs	fr_dads2	dbage	fr_dadsbest2
1.	1	21	1	0	1	19	1
2.	2	25	1	1	1	28	1
3.	3	.	.	.	.	32	1
4.	4	25	2	1	1	38	1

```
. save dadsandbest
file dadsandbest.dta saved
```

In this merge, we can see that three cases matched and there was one that did not match. The dad from family 3 did not have an observation, although we did have an observation for his best friend. We save the merged dataset as `dadsandbest.dta`, which can be merged with `momsandbest.dta`, as shown below.

```
. use momsandbest
. merge 1:1 famid using dadsandbest, nogenerate
    Result                           # of obs.
    -----------------------------------------
    not matched                             1
        from master                         1
        from using                          0

    matched                                 4
    -----------------------------------------
```

The report tells us that the merging of these two datasets resulted in four matched observations and one observation that was not matched. At this stage, I find it more useful to focus on the `fr_` variables showing the origin of the observations from each of the four source files. Let's create a listing of `famid` with all the `fr_` variables, showing the final matching results for all four datasets.

```
. list famid fr_*, abb(20)
```

	famid	fr_moms2	fr_momsbest2	fr_dads2	fr_dadsbest2
1.	1	1	.	1	1
2.	2	.	1	1	1
3.	3	1	1	.	1
4.	4	1	1	1	1
5.	5	1	.	.	.

As we can see, family 1 missed data from the mom's best friend, while family 2 missed data from the mom. Family 3 missed data from the dad, while family 4 had complete data across all four datasets. Family 5 only had data from the moms. Let's save this dataset, naming it `momsdadsbest.dta`.

```
. save momsdadsbest
file momsdadsbest.dta saved
```

Suppose that we wanted to take this one step further and merge this combined file with kidname.dta, which contains the names of the kids from each family. Let's do this by reading in kidname.dta and creating a variable named fr_kids to identify observations as originating from kidname.dta.

```
. use kidname
. generate fr_kids = 1
```

We can now merge this file with momsdadsbest.dta via an m:1 merge, as shown below.

```
. merge m:1 famid using momsdadsbest
    Result                           # of obs.
    -----------------------------------------
    not matched                             1
        from master                         0  (_merge==1)
        from using                          1  (_merge==2)

    matched                                 8  (_merge==3)
    -----------------------------------------
```

Let's list the fr_ variables, showing the matching for the first five observations.

```
. list famid fr_* in 1/5, abb(20) sepby(famid)
```

	famid	fr_kids	fr_moms2	fr_momsbest2	fr_dads2	fr_dadsbest2
1.	1	1	1	.	1	1
2.	2	1	.	1	1	1
3.	2	1	.	1	1	1
4.	3	1	1	1	.	1
5.	3	1	1	1	.	1

This listing is useful for seeing the matching status for each observation, but if you have more than a handful of observations, this kind of listing would not be very useful. Instead, you might want a tabulation of the number of observations that corresponds to each pattern of matching results. When I was at UCLA, I wrote a program for situations like this called tablist,[4] which creates output that is a like a combination of the tabulate and list commands. tablist is used below to create a tabulation of all the fr_ variables.

4. You can download this program by using the command findit tablist (see section 10.2 for more details on using findit).

```
. tablist fr_*, abb(20) sort(v)
```

fr_kids	fr_moms2	fr_momsbest2	fr_dads2	fr_dadsbest2	Freq
1	1	1	1	1	3
1	1	1	.	1	2
1	1	.	1	1	1
1	.	1	1	1	2
.	1	.	.	.	1

This output shows that there are three observations that have data originating from all the source datasets. Then there are two observations that have data from all sources except the dads file. There is one observation with complete data except for the moms best friend. There are two observations that have data from all sources but the moms file, and one observation has only data from the moms file. I find this output useful for tracking observations that do not match, especially when merging many different files.

This section has shown how you can merge multiple files with a series of pairwise merges. The `fr_` variables that were included in each dataset helped us track the origin of the observations from each of the original datasets and show tabulations of the merging results. Finally, although not shown here, if you do merge multiple files, I would encourage you to create a diagram that illustrates each merging step. This can help clarify and document the merging process.

6.7 Merging: Update merges

There is a special kind of merge called an `update`. When performing such a merge, the using dataset provides changes that should be applied to the master dataset. For example, suppose that we had a version of the moms dataset that has some missing values and some errors in it. In addition, we have an update dataset that contains values to be used in place of the values in the master dataset. These updated values can be nonmissing values to be used in place of the missing values in the master dataset or corrections that should overwrite nonmissing values in the master dataset. By performing an update merge, we can take the master dataset and apply changes from the using dataset to create a corrected (or updated) dataset.

Let's illustrate this with a variation of the moms dataset named `moms5.dta`, which contains missing values and errors.

```
. use moms5
. list
```

	famid	mage	mrace	mhsgrad
1.	1	.	2	1
2.	2	82	.	1
3.	3	24	2	.
4.	4	21	1	0

moms5fixes.dta contains updated data for moms5.dta. Note that the observations where famid is 1 or 3 only contain updates for values that are missing in moms5.dta. The observation where famid is 2 contains an update for a missing value and contains a correction to the nonmissing value of mage.

```
. use moms5fixes
. list

     +--------------------------------+
     | famid   mage   mrace   mhsgrad |
     |--------------------------------|
  1. |     1     33       .         . |
  2. |     2     28       1         . |
  3. |     3      .       .         1 |
     +--------------------------------+
```

Let's try performing a regular merge between these two datasets.

```
. use moms5
. merge 1:1 famid using moms5fixes
    Result                           # of obs.
    -----------------------------------------
    not matched                             1
        from master                         1  (_merge==1)
        from using                          0  (_merge==2)

    matched                                 3  (_merge==3)
    -----------------------------------------
```

If you compare the merged results below with the contents of moms5.dta above, you can see that the original contents of the master dataset were all retained. This is the default behavior for the merge command.

```
. list

     +---------------------------------------------------------+
     | famid   mage   mrace   mhsgrad                   _merge |
     |---------------------------------------------------------|
  1. |     1      .       2         1              matched (3) |
  2. |     2     82       .         1              matched (3) |
  3. |     3     24       2         .              matched (3) |
  4. |     4     21       1         0          master only (1) |
     +---------------------------------------------------------+
```

If we add the update option, then the data in the using dataset are used to update the master dataset where the data in the master dataset are missing.

```
. use moms5
. merge 1:1 famid using moms5fixes, update
    Result                           # of obs.
    -----------------------------------------
    not matched                             1
        from master                         1  (_merge==1)
        from using                          0  (_merge==2)

    matched                                 3
        not updated                         0  (_merge==3)
        missing updated                     2  (_merge==4)
        nonmissing conflict                 1  (_merge==5)
    -----------------------------------------
```

The results from the merge command show that three values were matched between the master and using files. Two of those observations were **missing updated** and one was a **nonmissing conflict**. Let's look at the listing of the updated dataset to understand this better.

```
. sort famid
. list

     +-------------------------------------------------------------+
     | famid   mage   mrace   mhsgrad                       _merge |
     |-------------------------------------------------------------|
  1. |     1     33       2         1          missing updated (4) |
  2. |     2     82       1         1      nonmissing conflict (5) |
  3. |     3     24       2         1          missing updated (4) |
  4. |     4     21       1         0              master only (1) |
     +-------------------------------------------------------------+
```

The observations for families 1 and 3 contain corrections to missing data and these missing values were updated. The matching status for these observations is **missing updated** because the using dataset was used to update missing values in the master dataset. For family 2, the using file contained a correction for a nonmissing value of **mage**. The matching status for this observation is **nonmissing conflict** because the master dataset contains a nonmissing value that conflicts with the using dataset. This nonmissing value remained unchanged. (For example, in family 2, the mom's age remains 82).

If we want the nonmissing values in the master dataset to be replaced with the nonmissing values from the update dataset, we need to specify the **replace** option. When we use this option below, the resulting dataset reflects all the fixes contained in **moms5fixes.dta**, including the value of 82 that was replaced with 28.

(Continued on next page)

```
. use moms5
. merge 1:1 famid using moms5fixes, update replace
    Result                           # of obs.
    -----------------------------------------
    not matched                             1
        from master                         1  (_merge==1)
        from using                          0  (_merge==2)

    matched                                 3
        not updated                         0  (_merge==3)
        missing updated                     2  (_merge==4)
        nonmissing conflict                 1  (_merge==5)
    -----------------------------------------

. sort famid
. list

     +--------------------------------------------------------------+
     | famid   mage   mrace   mhsgrad                        _merge |
     |--------------------------------------------------------------|
  1. |     1     33       2         1           missing updated (4) |
  2. |     2     28       1         1       nonmissing conflict (5) |
  3. |     3     24       2         1           missing updated (4) |
  4. |     4     21       1         0               master only (1) |
     +--------------------------------------------------------------+
```

The utility of the update merge strategy becomes apparent if we imagine that `moms5.dta` had thousands or even millions of observations with some missing or incorrect values. We could then create a dataset with just the corrections (e.g., `moms5fixes.dta`). Used this way, the update merge can provide an efficient means of applying corrections to datasets. For more information, see `help merge`, especially the portions related to the `update` option.

6.8 Merging: Additional options when merging datasets

This section explores options that can be used with the `merge` command. The options `update` and `replace` were previously covered in section 6.7, so they are not covered here. Consider the example below, where `moms1.dta` is merged with `dads1.dta`.

```
. use moms1
. merge 1:1 famid using dads1
    Result                           # of obs.
    -----------------------------------------
    not matched                             0
    matched                                 4  (_merge==3)
    -----------------------------------------
```

The merged dataset contains all the variables from `moms1.dta` and `dads1.dta`, as shown below.

```
. list
```

	famid	mage	mrace	mhs	dage	drace	dhs	_merge
1.	1	33	2	1	21	1	0	matched (3)
2.	2	28	1	1	25	1	1	matched (3)
3.	3	24	2	1	31	2	1	matched (3)
4.	4	21	1	0	25	2	1	matched (3)

But perhaps we are only interested in keeping some of the variables from `dads1.dta` (e.g., `dage`). We can add the `keepusing(dage)` option to the `merge` command (below), and only the `dage` variable is kept from the using dataset (`dads1.dta`).

```
. use moms1
. merge 1:1 famid using dads1, keepusing(dage)
    Result                           # of obs.
    -----------------------------------------
    not matched                             0
    matched                                 4  (_merge==3)
    -----------------------------------------
```

The listing of the merged dataset (below) shows that `dage` is the only variable kept from the using dataset (`dads1.dta`). The `keepusing()` option is especially convenient if the using dataset has dozens or hundreds of variables but you want to keep just a few of them for the purpose of the merge.

```
. list
```

	famid	mage	mrace	mhs	dage	_merge
1.	1	33	2	1	21	matched (3)
2.	2	28	1	1	25	matched (3)
3.	3	24	2	1	31	matched (3)
4.	4	21	1	0	25	matched (3)

Let's next consider the `generate()` option. By default, Stata creates a variable named `_merge` that contains the matching result for each observation. The `generate()` option allows you to specify the name for this variable, allowing you to choose a more meaningful name than `_merge`.[5] The `generate(md)` option is used below to name the variable with the match results `md` instead of `_merge`.

5. `generate()` is also useful if `_merge` already exists from a prior merge. If you attempt another merge without specifying the `generate()` option, the merge will fail because `_merge` already exists and Stata will not replace it.

```
. use moms2
. merge 1:1 famid using dads2, generate(md)
    Result                           # of obs.
    -----------------------------------------
    not matched                             3
        from master                         2  (md==1)
        from using                          1  (md==2)

    matched                                 2  (md==3)
    -----------------------------------------
```

The listing of the merged dataset is shown below. The `md` variable tells us if the observation contains data from both the master and the using datasets (in which case the value is `matched (3)`), or contains data only from the master dataset (in which case the value is `master only (1)`), or contains data only from the using dataset (in which case the value is `using only (2)`).

```
. list famid mage dage fr_* md

     +-----------------------------------------------------------------+
     | famid   mage   dage   fr_moms2   fr_dads2                    md |
     |-----------------------------------------------------------------|
  1. |     1     33     21          1          1          matched (3) |
  2. |     3     24      .          1          .      master only (1) |
  3. |     4     21     25          1          1          matched (3) |
  4. |     5     39      .          1          .      master only (1) |
  5. |     2      .     25          .          1       using only (2) |
     +-----------------------------------------------------------------+
```

`moms2.dta` and `dads2.dta` contain a variable that can be used to determine the source of the data in the merged dataset. `moms2.dta` has the variable `fr_moms2`, which contains all ones, and similarly, `dads2.dta` contains the variable `fr_dads2`, which also contains all ones. As you can see above, these two variables can be used in lieu of the `_merge` variable to describe the results of the matching. We can also use these variables to create the summary table of match results automatically created by the `merge` command.

Let's merge these two files again, but this time let's specify the `nogenerate` option to omit creation of the `_merge` variable. Let's also suppress the summary table of match results by adding the `noreport` option.

```
. use moms2
. merge 1:1 famid using dads2, nogenerate noreport
```

The listing below shows some of the variables from the merged dataset along with the variables `fr_moms2` and `fr_dads2`. The `fr_moms2` and `fr_dads2` variables indicate the merging status of each observation. The observations from families (`famid`) 1 and 4 originate from both `moms2.dta` and `dads2.dta`. The observations from families 3 and 5 originate from only `moms2.dta`, while the observation from family 2 originates only from `dads2.dta`.

```
. sort famid
. list famid mage dage fr_*

     +------------------------------------------+
     | famid   mage   dage   fr_moms2   fr_dads2 |
     |------------------------------------------|
  1. |     1     33     21          1          1 |
  2. |     2      .     25          .          1 |
  3. |     3     24      .          1          . |
  4. |     4     21     25          1          1 |
  5. |     5     39      .          1          . |
     +------------------------------------------+
```

We can create a tabulation of fr_moms2 by fr_dads2 that summarizes the match results for the entire dataset. This corresponds to the information normally provided automatically by the merge command. Below we see that two observations matched (i.e., were from both moms2.dta and dads2.dta). Two observations were from moms2.dta but not dads2.dta, and one observation was from dads2.dta but not moms2.dta.

```
. tabulate fr_moms2 fr_dads2, missing
From moms2 |   Data from dads2?
  dataset? |         1          . |     Total
-----------+----------------------+----------
         1 |         2          2 |         4
         . |         1          0 |         1
-----------+----------------------+----------
     Total |         3          2 |         5
```

Let's now explore two other options, the keep() and assert() options. The keep() option allows you to specify which observations are to be kept in the merged file depending on their matched status (e.g., only keep observations that are matched between the master and using datasets). The assert() option permits you to test assertions regarding the matching of the master and using datasets (e.g., test whether all observations are matched between the two datasets). These options can be used separately or together. The behavior of these options depends on whether the update option is included. For the following examples, these options are discussed assuming the update option is not used.

Let's illustrate how the keep() option could be used when merging moms2.dta and dads2.dta. First, let's merge these two files without specifying the keep() option.

```
. use moms2, clear
. merge 1:1 famid using dads2
    Result                           # of obs.
    -----------------------------------------
    not matched                             3
        from master                         2  (_merge==1)
        from using                          1  (_merge==2)

    matched                                 2  (_merge==3)
    -----------------------------------------
```

The merge results above show that two observations are matched and three are not matched. Among the nonmatched observations, two were from the master dataset (`moms2.dta`) and one was from the using dataset (`dads2.dta`).

Let's add the `keep(match)` option. As we see below, only the two matched observations are now kept, and the report is based only on the observations that were kept.

```
. use moms2, clear
. merge 1:1 famid using dads2, keep(match)
    Result                           # of obs.
    -----------------------------------------
    not matched                             0
    matched                                 2  (_merge==3)
    -----------------------------------------
```

If you specify the `keep(match master)` option (shown below), only the matched observations (which are in both the master and the using datasets) and the unmatched observations from the master dataset (`moms2.dta`) are kept. The report reflects only the observations that are kept.

```
. use moms2, clear
. merge 1:1 famid using dads2, keep(match master)
    Result                           # of obs.
    -----------------------------------------
    not matched                             2
        from master                         2  (_merge==1)
        from using                          0  (_merge==2)

    matched                                 2  (_merge==3)
    -----------------------------------------
```

You could instead specify the `keep(match using)` option, which would keep only the matched observations (which are in both datasets) and the unmatched observations from the using dataset (`dads2.dta`).

```
. use moms2, clear
. merge 1:1 famid using dads2, keep(match using)
  (output omitted)
```

The `assert()` option works similarly to the `keep()` option but instead tests whether all observations meet the specified matching criteria. The `merge` command returns an error message if all the observations do not meet the specified criteria. For example, suppose that we merge `moms1.dta` and `dads1.dta` as shown below, including the `assert(match)` option on the `merge` command.

```
. use moms1
. merge 1:1 famid using dads1, assert(match)
    Result                           # of obs.
    -----------------------------------------
    not matched                             0
    matched                                 4  (_merge==3)
    -----------------------------------------
```

The report shows us that all four observations in the merged dataset were matched. This is confirmed because the `merge` command completed without an error. Had there been any nonmatched observations, the `assert(match)` option would have caused the `merge` command to give an error message. The next example shows what such an error message would look like.

As we have seen earlier in this section, `moms2.dta` and `dads2.dta` have some nonmatching observations. If we included the `assert(match)` option when merging those files, we would receive the following error message:

```
. use moms2
. merge 1:1 famid using dads2, assert(match)
merge:  after merge, all observations not matched
        (merged result left in memory)
r(9);
```

This error message clearly gets our attention, notifying us that not all observations were matched (i.e., there was at least one nonmatched observation).

Instead of specifying `assert(match)`, you can specify `assert(match master)` to make sure that each observation in the merged dataset either is a match or originates from the master dataset. You can also specify `assert(match using)` to ensure that each observation in the merged dataset is a match or originates from the using dataset. If the condition you specify on the `assert()` option is not met, an error will be given.

Careful! assert() and keep() with update

The previous examples of the `assert()` and `keep()` options assumed that the `update` option was not specified. Including the `update` option changes the meaning of `assert()` and `keep()`. For more information on using the `assert()` and `keep()` options with the `update` option, see `help merge`.

This section covered commonly used options available with the `merge` command. See `help merge` for more details about all the options available when merging datasets in Stata. The next section discusses problems that can arise when merging datasets.

6.9 Merging: Problems merging datasets

This section illustrates some of the common problems that can arise when merging datasets. Unless otherwise specified, these problems apply to all the kinds of merging illustrated in the previous sections (i.e., one-to-one merging, one-to-many merging, and update merging).

Common variable names

When merging two datasets, all the variable names in the two datasets should be different, except for the key variable(s).[6] But sometimes, we may name variables in different datasets without considering the implications for what will happen when the datasets are merged. Consider moms3.dta and dads3.dta, shown below.

```
. use moms3
. list
```

	famid	age	race	hs
1.	1	33	2	1
2.	2	28	1	1
3.	3	24	2	1
4.	4	21	1	0

```
. use dads3
. list
```

	famid	age	eth	gradhs
1.	1	21	1	0
2.	2	25	1	1
3.	3	31	2	1
4.	4	25	2	1

Note how both datasets have the variable age in common, although in moms3.dta it refers to the mom's age and in dads3.dta it refers to the dad's age. By contrast, the variables about race/ethnicity and graduating high school have different names in the two datasets. Let's see what happens when we merge these datasets.

```
. use moms3
. merge 1:1 famid using dads3
    Result                           # of obs.
    -----------------------------------------
    not matched                             0
    matched                                 4  (_merge==3)
    -----------------------------------------
```

In the listing below, note how the age variable reflects the age from moms3.dta (master). When the master and using datasets share the same variable name, the values from the master dataset will be retained, while the values from the using dataset are discarded.

6. If the update option is used, then the master and using datasets can share the same variable names.

```
. list

     ┌───────────────────────────────────────────────────────┐
     │ famid   age   race   hs   eth   gradhs         _merge │
     ├───────────────────────────────────────────────────────┤
  1. │     1    33      2    1     1        0   matched (3) │
  2. │     2    28      1    1     1        1   matched (3) │
  3. │     3    24      2    1     2        1   matched (3) │
  4. │     4    21      1    0     2        1   matched (3) │
     └───────────────────────────────────────────────────────┘
```

When many variables are in each dataset, this can be a tricky problem to detect. We can use the `cf` (compare files) command to determine if there are variables (aside from the key variables) in common between the master and using datasets.

```
. use moms3

. cf _all using dads3, all
           famid:  match
             age:  4 mismatches
            race:  does not exist in using
              hs:  does not exist in using
r(9);
```

The output shows that these two datasets have two variables in common: `famid` and `age`. We can deduce this because the `cf` command shows results comparing these two variables between the datasets. The `famid` variable is the key variable for merging these files, so it is supposed to have the same name in the two datasets. But the `age` variable should be different in the two files.

Even setting aside the problem with `age`, these variable names are confusing because I cannot tell which variables describe the mom and which variables describe the dad.

Let's contrast this with the variable names in `moms1.dta` and `dads1.dta`. Let's first merge these datasets.

```
. use moms1

. merge 1:1 famid using dads1
    Result                           # of obs.
    -----------------------------------------
    not matched                             0
    matched                                 4  (_merge==3)
    -----------------------------------------
```

When these two datasets are merged, it is clear which variables describe the mom and which variables describe the dad, as shown in the listing of the merged dataset below. Note how the mom variables are prefixed with `m` (to indicate they are from `moms1.dta`) and the dad variables are prefixed with `d` (to indicate they are from `dads1.dta`).

```
. list

     +----------------------------------------------------------------------+
     | famid   mage   mrace   mhs   dage   drace   dhs               _merge |
     |----------------------------------------------------------------------|
  1. |     1     33       2     1     21       1     0          matched (3) |
  2. |     2     28       1     1     25       1     1          matched (3) |
  3. |     3     24       2     1     31       2     1          matched (3) |
  4. |     4     21       1     0     25       2     1          matched (3) |
     +----------------------------------------------------------------------+
```

In summary, when you plan to merge datasets, you should pick names that will still make sense after the datasets are merged.

Datasets share same names for value labels

Let's see what happens when you merge datasets that have identical value labels.

Below we can see that `moms4.dta` has the variable `mrace`, which is labeled with a value label named `race`. `dads4.dta` has the variable `drace`, which is also labeled using a value label named `race`. This will cause a conflict when these two datasets are merged. By comparison, the variable `mhs` in `moms4.dta` has a different value label than the corresponding variable `dhs` in `dads4.dta`. There should be no merge problem with respect to these variables.

```
. use moms4

. describe mrace mhs

              storage  display     value
variable name   type   format      label      variable label
------------------------------------------------------------------
mrace           float  %9.0g       race       Ethnicity
mhs             float  %15.0g      mhsgrad    HS Graduate?

. use dads4

. describe drace dhs

              storage  display     value
variable name   type   format      label      variable label
------------------------------------------------------------------
drace           float  %9.0g       race       Ethnicity
dhs             float  %15.0g      dhsgrad    HS Graduate?
```

Let's now merge these two datasets together.

```
. use moms4

. merge 1:1 famid using dads4
(label race already defined)
    Result                           # of obs.
    -----------------------------------------
    not matched                             0
    matched                                 4  (_merge==3)
    -----------------------------------------
```

Note how Stata gives the message `(label race already defined)`. This tiny little message is warning us that the value label `race` is used in both datasets.

Looking at the listing below, we can see the consequences of both datasets sharing the value label race. The mrace and drace variables are both labeled using the value labels originating from moms4.dta (master dataset). By contrast, the mhs and dhs variables are labeled appropriately because the value labels for these variables had a different name in the master and the using datasets.

```
. list famid mrace drace mhs dhs

     +-------------------------------------------------------------------------+
     | famid       mrace       drace                 mhs                 dhs   |
     |-------------------------------------------------------------------------|
  1. |     1   Mom Black   Mom White          Mom HS Grad    Dad Not HS Grad   |
  2. |     2   Mom White   Mom White          Mom HS Grad        Dad HS Grad   |
  3. |     3   Mom Black   Mom Black          Mom HS Grad        Dad HS Grad   |
  4. |     4   Mom White   Mom Black      Mom Not HS Grad        Dad HS Grad   |
     +-------------------------------------------------------------------------+
```

When you specify the names for value labels, you want to choose names that will be unique when datasets are merged. If the names for the value labels are the same in the master and the using datasets, the value label from the master dataset will take precedence. The solution to this problem in our example is to give the value labels unique names in moms4.dta and dads4.dta before merging the two datasets.

Conflicts in the key variables

The key variables should have the same variable name in the master and the using datasets and should be of the same general type (i.e., either numeric in both datasets or string in both datasets).

Sometimes the key variables are named differently in the two datasets. For example, in one dataset the variable might be named famid, and in the other dataset the variable might be named id. The solution in such a case is simple: rename one of the variables so that they both have the same name.

Suppose that the key variable was stored as a numeric variable in one dataset and as a string variable in the other dataset. Before you can merge the datasets together, the key variables need to be either both numeric or both string. Section 5.13 illustrates how to convert string variables to numeric, and section 5.14 illustrates how to convert numeric variables to string.

Summary

There are several problems that can arise when merging datasets. Some problems produce no error messages and some produce innocent-looking messages. Only upon deeper inspection are these underlying problems revealed. Knowing about these problems can help you anticipate and avoid them.

6.10 Joining datasets

The previous sections have illustrated how to perform a `1:1` merge, a `1:m` merge, and an `m:1` merge. In rare cases, you want to perform an `m:m` merge (also known as a many-to-many merge). For example, we have a dataset called **parname.dta**, which has eight observations on parents from four families. In family 1, Sam is married to Lil; in family 2, Nik is married to Ula; and so forth.

```
. use parname
. sort famid mom
. list, sepby(famid)
```

	famid	mom	age	race	pname
1.	1	0	21	1	Sam
2.	1	1	33	2	Lil
3.	2	0	25	1	Nik
4.	2	1	28	1	Ula
5.	3	0	31	2	Al
6.	3	1	24	2	Ann
7.	4	0	25	2	Ted
8.	4	1	21	1	Bev

We also have a dataset called **kidname.dta** with the names of the kids in these families. Family 1 has one kid named Sue, family 2 has two kids named Vic and Flo, and so forth.

```
. use kidname
. sort famid kidid
. list, sepby(famid)
```

	famid	kidid	kname
1.	1	1	Sue
2.	2	1	Vic
3.	2	2	Flo
4.	3	1	Ivy
5.	3	2	Abe
6.	4	1	Tom
7.	4	2	Bob
8.	4	3	Cam

Suppose that we want to merge the parents with each of their kids. So for family 2, the parent Nik would match with his two kids, Vic and Flo. Also for family 2, the parent Ula would match with her two kids, Vic and Flo. Thus for family 2, there would be four resulting observations (two parents times two children). As you can see, this is a many-to-many merge, where many parents from each family (e.g., Nik and Ula) can be merged with many children from each family (e.g., Vic and Flo).

To prepare to join these two datasets, we first need to sort them on family ID, as shown below.

```
. use parname
. sort famid
. save parnamesort
file parnamesort.dta saved
. use kidname
. sort famid
. save kidnamesort
file kidnamesort.dta saved
```

Now we can join the datasets by using the `joinby` command.

```
. use parnamesort
. joinby famid using kidnamesort
```

The resulting dataset is listed below, after first sorting on `famid kidid pname`. This listing shows that each parent was matched with his or her child. For example, Nik and Ula from family 2 were matched with their first child, Vic, and Nik and Ula were also matched with their second child, Flo.

```
. sort famid kidid pname
. list famid kidid pname age kname, sepby(famid kidid)
```

	famid	kidid	pname	age	kname
1.	1	1	Lil	33	Sue
2.	1	1	Sam	21	Sue
3.	2	1	Nik	25	Vic
4.	2	1	Ula	28	Vic
5.	2	2	Nik	25	Flo
6.	2	2	Ula	28	Flo
7.	3	1	Al	31	Ivy
8.	3	1	Ann	24	Ivy
9.	3	2	Al	31	Abe
10.	3	2	Ann	24	Abe
11.	4	1	Bev	21	Tom
12.	4	1	Ted	25	Tom
13.	4	2	Bev	21	Bob
14.	4	2	Ted	25	Bob
15.	4	3	Bev	21	Cam
16.	4	3	Ted	25	Cam

Unlike the merge command, the joinby command defaults to keeping just the matched observations. But you can control this with the unmatched() option. Specifying unmatched(both) retains observations from both the master and the using datasets. Specifying unmatched(master) retains observations from the master dataset only or unmatched(using) to retain observations from the using dataset only.

Like the merge command, joinby permits the update and replace options, allowing joinby to update the master dataset from data contained in the using dataset. When used with joinby, these options behave in the same way as when used with merge; see section 6.7 for examples of their usage with merge.

The _merge() option permits you to specify the name of the variable used to indicate whether the observation came from the master, the using, or both datasets. This is similar to the generate() option illustrated in section 6.8 using the merge command.

For more information about performing these kinds of m:m merges, see help joinby.

6.11 Crossing datasets

I have seldom seen the cross command used, but it is handy when you need it. The cross command is kind of like a merge, except that every observation from the master dataset is merged with every observation from the using dataset. For example, consider moms1.dta and dads1.dta, below.

```
. use moms1
. list
```

	famid	mage	mrace	mhs
1.	1	33	2	1
2.	2	28	1	1
3.	3	24	2	1
4.	4	21	1	0

```
. use dads1
. list
```

	famid	dage	drace	dhs
1.	1	21	1	0
2.	2	25	1	1
3.	3	31	2	1
4.	4	25	2	1

If we wanted to form a dataset that contained every possible pairing between every mom and every dad, we could cross the datasets, as shown below.

```
. use moms1
. cross using dads1
```

Crossing these two datasets creates a dataset where every mom is matched with every dad, yielding 16 observations (4×4). If you needed a dataset that showed every possible marriage between every mom and every dad, you can see that the `cross` command did this easily.

```
. list
```

	famid	mage	mrace	mhs	dage	drace	dhs
1.	1	33	2	1	21	1	0
2.	2	28	1	1	21	1	0
3.	3	24	2	1	21	1	0
4.	4	21	1	0	21	1	0
5.	1	33	2	1	25	1	1
6.	1	33	2	1	31	2	1
7.	1	33	2	1	25	2	1
8.	2	28	1	1	25	1	1
9.	2	28	1	1	31	2	1
10.	2	28	1	1	25	2	1
11.	3	24	2	1	25	1	1
12.	3	24	2	1	31	2	1
13.	3	24	2	1	25	2	1
14.	4	21	1	0	25	1	1
15.	4	21	1	0	31	2	1
16.	4	21	1	0	25	2	1

For more information about crossing datasets, see `help cross`.

7 Processing observations across subgroups

The same set of statistics can produce opposite conclusions at different levels of aggregation.

—Thomas Sowell

7.1 Introduction

Our datasets are often not completely flat but instead have some kind of grouping structure. Perhaps the groupings reflect subgroups (e.g., race or gender) or nested structure (e.g., kids within a family or multiple observations on one person). Stata refers to these groupings as by-groups. This chapter discusses the tools that Stata offers for specialized processing of by-groups.

Section 7.2 introduces the `by` prefix command, which allows you to repeat a Stata command for each level of a by-group, such as performing correlations among variables for each level of race. Then section 7.3 shows how you can combine by-group processing with the `egen` command to easily create variables that contain summary statistics for each by-group. The heart of this chapter spans sections 7.4–7.7, which describe how to combine the `by` prefix with the `generate` command to perform complex and useful computations within by-groups. For example, you can compute differences between adjacent observations within a by-group, fill in missing values using the last valid observation within a group, identify singleton groups, or identify changes across time within a group. This chapter concludes with section 7.8, which compares `by` with `tsset`.

7.2 Obtaining separate results for subgroups

This section illustrates the use of the `by` prefix for repeating a command on groups of observations. For example, using `wws2.dta`, say that we want to summarize the wages separately for women who are married and unmarried. The `tabulate` command with the `summarize()` option can do this for us.

```
. use wws2
(Working Women Survey w/fixes)
. tabulate married, summarize(wage)

            |     Summary of hourly wage
    married |        Mean   Std. Dev.       Freq.
------------+------------------------------------
          0 |   8.0920006    6.354849         804
          1 |   7.6319496   5.5017864        1440
------------+------------------------------------
      Total |   7.7967807   5.8245895        2244
```

If we wish, we could instead use the `by` prefix before the `summarize` command. Note that the data are first sorted on `married`, and then the `summarize` command is prefaced with `by married:` to indicate that Stata is to run the `summarize wage` command separately for every level of `married` (i.e., for those who are unmarried and for those who are married).

```
. sort married
. by married: summarize wage

-------------------------------------------------------------------------------
-> married = 0
    Variable |       Obs        Mean    Std. Dev.       Min        Max
-------------+--------------------------------------------------------
        wage |       804    8.092001    6.354849          0   40.19808

-------------------------------------------------------------------------------
-> married = 1
    Variable |       Obs        Mean    Std. Dev.       Min        Max
-------------+--------------------------------------------------------
        wage |      1440     7.63195    5.501786   1.004952   40.74659
```

Each level of married (unmarried and married) is referred to as a by-group, and when a command is prefaced with `by`, that command is performed for each level of the by-group. If we were confident that the dataset was already sorted by `married`, then we could have omitted the `sort married` command. However, if we were wrong and the data were not sorted, we would get the following error:

```
. by married: summarize wage
not sorted
r(5);
```

The `bysort` command can be used to combine the `sort married` command and the `by married:` prefix in one step.

```
. bysort married: summarize wage
  (output omitted)
```

Or we can shorten `bysort` to just `bys`.

```
. bys married: summarize wage
  (output omitted)
```

The `by` prefix becomes more important when used with commands that have no other means for obtaining results separately for each by-group. For example, to run correlations separately for those who are married and unmarried, we would need to run two separate `correlate` commands.

```
. correlate wage age if married==0
  (output omitted)
. correlate wage age if married==1
  (output omitted)
```

Instead, we can simply use `by`.

```
. by married: correlate wage age
  (output omitted)
```

Let's consider another example, obtaining separate correlations for each level of `race`.

```
. bysort race: correlate wage age
------------------------------------------------------------------------------

-> race = 1
(obs=1637)

             |     wage      age
-------------+------------------
        wage |   1.0000
         age |   0.0017   1.0000

------------------------------------------------------------------------------

-> race = 2
(obs=581)

             |     wage      age
-------------+------------------
        wage |   1.0000
         age |  -0.0331   1.0000

------------------------------------------------------------------------------

-> race = 3
(obs=26)

             |     wage      age
-------------+------------------
        wage |   1.0000
         age |  -0.2194   1.0000
```

Note how the sample size is getting a little bit small for the third group. Although an analysis may be completely satisfactory with the overall sample, problems can possibly arise for a subsample implied by one or more of the by-groups.

For more information about the `by` prefix, see `help by`.

7.3 Computing values separately by subgroups

The `by` prefix can be combined with the `egen` command to combine the power of by-group processing with the `generate` command to perform computations across by-groups.[1] Let's illustrate this using `tv1.dta`, which has information on the weight, TV-watching time, and vacation status of four kids over one or more days per kid.

1. For more about `egen`, see sections 5.10, 5.11, and 5.12.

```
. use tv1
. list, sepby(kidid)
```

	kidid	dt	female	wt	tv	vac
1.	1	07jan2002	1	53	1	1
2.	1	08jan2002	1	55	3	1
3.	2	16jan2002	1	58	8	1
4.	3	18jan2002	0	60	2	0
5.	3	19jan2002	0	63	5	1
6.	3	21jan2002	0	66	1	1
7.	3	22jan2002	0	64	6	0
8.	4	10jan2002	1	62	7	0
9.	4	11jan2002	1	58	1	0
10.	4	13jan2002	1	55	4	0

The `egen` command makes it easy to create a variable that contains the average TV-watching time for each kid. The `mean()` function is used to get the mean, and when prefixed with `bysort kidid`, the mean is computed separately for each kid, as shown below.

```
. bysort kidid: egen avgtv = mean(tv)
. list kidid tv avgtv, sepby(kidid)
```

	kidid	tv	avgtv
1.	1	1	2
2.	1	3	2
3.	2	8	8
4.	3	2	3.5
5.	3	5	3.5
6.	3	1	3.5
7.	3	6	3.5
8.	4	7	4
9.	4	1	4
10.	4	4	4

In the same way, the `sd()` function can be used to get the standard deviation of TV-watching time within each kid. Kid number 2 has a missing value for the standard deviation because she had only one value. We then compute a z-score, standardized according to each kid's mean and standard deviation, as shown below.

```
. bysort kidid: egen sdtv = sd(tv)
(1 missing value generated)
. generate ztv = (tv - avgtv)/sdtv
(1 missing value generated)
```

```
. list kidid tv avgtv sdtv ztv, sepby(kidid)
```

	kidid	tv	avgtv	sdtv	ztv
1.	1	1	2	1.414214	-.7071068
2.	1	3	2	1.414214	.7071068
3.	2	8	8	.	.
4.	3	2	3.5	2.380476	-.630126
5.	3	5	3.5	2.380476	.630126
6.	3	1	3.5	2.380476	-1.05021
7.	3	6	3.5	2.380476	1.05021
8.	4	7	4	3	1
9.	4	1	4	3	-1
10.	4	4	4	3	0

Consider the variable `vac`, which is a dummy variable that is 0 if the kid was not on vacation and 1 if the kid was on vacation. Let's use the following functions with this variable and see what happens: `total()`, `sd()`, `min()`, and `max()`.

```
. bysort kidid: egen vac_total = total(vac)
. bysort kidid: egen vac_sd = sd(vac)
(1 missing value generated)
. bysort kidid: egen vac_min = min(vac)
. bysort kidid: egen vac_max = max(vac)
. list kidid vac*, sepby(kidid) abb(10)
```

	kidid	vac	vac_total	vac_sd	vac_min	vac_max
1.	1	1	2	0	1	1
2.	1	1	2	0	1	1
3.	2	1	1	.	1	1
4.	3	0	2	.5773503	0	1
5.	3	1	2	.5773503	0	1
6.	3	1	2	.5773503	0	1
7.	3	0	2	.5773503	0	1
8.	4	0	0	0	0	0
9.	4	0	0	0	0	0
10.	4	0	0	0	0	0

The variable `vac_total` represents the number of days the kid was on vacation. The variable `vac_sd` gives the standard deviation, which can be used to determine if the kid's vacation status changed (for kid 3, the standard deviation is nonzero, so the vacation status changed). The variable `vac_min` is 1 if the kid was always on vacation, and the variable `vac_max` is 1 if the kid was ever on vacation.

Let's apply these same tricks to the time spent watching TV. Suppose that we are interested in trying to get kids to watch less than four hours of TV per day, and falling below that threshold is our research interest. We can compute the dummy variable `tvlo` to be 1 if TV-watching time is less than four hours per day and 0 if it is more (see section 5.7 for more on how this dummy variable was created).

```
. generate tvlo = (tv < 4) if ! missing(tv)
```

We can use the same tricks we used for `vac` on `tvlo`, yielding some useful results, as shown below.

```
. egen tvlocnt = count(tvlo), by(kidid)
. egen tvlototal = total(tvlo), by(kidid)
. egen tvlosame = sd(tvlo), by(kidid)
(1 missing value generated)
. egen tvloall = min(tvlo), by(kidid)
. egen tvloever = max(tvlo), by(kidid)
. list kidid tv tvlo*, sepby(kidid) abb(20)
```

	kidid	tv	tvlo	tvlocnt	tvlototal	tvlosame	tvloall	tvloever
1.	1	1	1	2	2	0	1	1
2.	1	3	1	2	2	0	1	1
3.	2	8	0	1	0	.	0	0
4.	3	2	1	4	2	.5773503	0	1
5.	3	5	0	4	2	.5773503	0	1
6.	3	1	1	4	2	.5773503	0	1
7.	3	6	0	4	2	.5773503	0	1
8.	4	7	0	3	1	.5773503	0	1
9.	4	1	1	3	1	.5773503	0	1
10.	4	4	0	3	1	.5773503	0	1

The variable `tvlocnt` is the number of valid observations each kid had for `tvlo`. The variable `tvlototal` is the number of days the kid watched less than four hours of TV. The variable `tvlosame` is 0 if the kid had multiple observations always with the same value of `tvlo` and is greater than 0 if the kid had differing values on `tvlo` (and is missing if the kid had only one observation). A kid who always watched less than four hours of TV has a 1 for `tvloall`, and a kid who ever watched less than four hours of TV has a 1 for `tvloever`.

There are many more `egen` functions beyond the ones illustrated here. There are functions for measures of central tendency, including the `mean()`, `median()`, and `mode()` functions, as well as functions for measures of variability, such as the `iqr()`, `mad()`, `mdev()`, `kurt()`, and `skew()` functions. Rather than try to list all these functions, let's look at how to read the help file for `egen`. I typed `help egen` and below show the help for the `mean()`, `skew()`, and `rowmean()` functions.

```
mean(exp)                                        (allows by varlist:)
       creates a constant (within varlist) containing the mean of exp.

skew(varname)                                    (allows by varlist:)
     returns the skewness (within varlist) of varname.

rowmean(varlist)
     may not be combined with by.  It creates the (row) means of the
     variables in varlist, ignoring missing values;
     (rest of help omitted)
```

For the `mean()` function, the help tells us that it allows `by` *varlist*:, which means that we can use the `by` prefix with this function. The `mean()` function also takes an expression (*exp*), which means that you can insert either one variable name or a logical/mathematical expression (see section 5.3 for more on mathematical expressions and section A.6 for more on logical expressions).

The `skew()` function also allows the `by` prefix, but it only accepts a *varname*. So for the `skew()` function, you can only supply the name of one variable.

Unlike the two previous examples, the `rowmean()` function does not allow the `by` prefix. This function is for performing computations across variables, not across observations (see section 5.10 for more information about performing computations across variables). That is why it accepts a *varlist*, meaning that you can supply one or more variables to the `rowmean()` function, and it will compute the mean across the row of variables.

> **Warning! varlist versus expressions with egen**
>
> Note how the `mean()` function takes an expression. This means that you can type something like `var1-var5`, and it will interpret this expression as `var1` minus `var5`. You might think you are supplying a variable list, like `var1 var2 var3 var4 var5`, but `egen` will only interpret `var1-var5` as a variable list when the `egen` function accepts a *varlist* (not an *exp*).

The `egen` command is both powerful and simple to use. For more information about it, see `help egen`.

7.4 Computing values within subgroups: Subscripting observations

This section illustrates subscripting observations. I think this is a concept that is easier to show than to explain, so let's dive right into an example using `tv1.dta`, shown below.

```
. use tv1
. list, sepby(kidid)
```

	kidid	dt	female	wt	tv	vac
1.	1	07jan2002	1	53	1	1
2.	1	08jan2002	1	55	3	1
3.	2	16jan2002	1	58	8	1
4.	3	18jan2002	0	60	2	0
5.	3	19jan2002	0	63	5	1
6.	3	21jan2002	0	66	1	1
7.	3	22jan2002	0	64	6	0
8.	4	10jan2002	1	62	7	0
9.	4	11jan2002	1	58	1	0
10.	4	13jan2002	1	55	4	0

Although it is not apparent, each variable is a vector, and you can access any particular observation from that vector. Below we use the **display** command to show the TV-watching time for the first observation by specifying `tv[1]`. Likewise, we display the TV-watching time for the second observation by specifying `tv[2]`. Stata calls this subscripting.

```
. display tv[1]
1
. display tv[2]
3
```

We can display the difference in TV-watching times between observation 2 and observation 1, as shown below.

```
. display tv[2] - tv[1]
2
```

Below we display the TV-watching time for the last observation by specifying `tv[_N]`. `tv[_N]` represents the last observation because the variable `_N` represents the number of observations in the dataset (in this case, 10). We also display the second to last observation in the dataset by displaying `tv[_N-1]`; Stata permits us to supply an expression within the brackets (e.g., `_N-1`).

```
. display tv[_N]
4
. display tv[_N-1]
1
```

Let's issue the following series of **generate** commands to see how they behave:

```
. generate n = _n
. generate N = _N
. generate tvn = tv[_n]
```

```
. generate tvp = tv[_n-1]
(1 missing value generated)
. generate tvs = tv[_n+1]
(1 missing value generated)
. generate tv1 = tv[1]
. generate tvN = tv[_N]
```

The variables created by these generate commands are shown below.

```
. list kidid tv n N tvn tvp tvs tv1 tvN
```

	kidid	tv	n	N	tvn	tvp	tvs	tv1	tvN
1.	1	1	1	10	1	.	3	1	4
2.	1	3	2	10	3	1	8	1	4
3.	2	8	3	10	8	3	2	1	4
4.	3	2	4	10	2	8	5	1	4
5.	3	5	5	10	5	2	1	1	4
6.	3	1	6	10	1	5	6	1	4
7.	3	6	7	10	6	1	7	1	4
8.	4	7	8	10	7	6	1	1	4
9.	4	1	9	10	1	7	4	1	4
10.	4	4	10	10	4	1	.	1	4

The meanings of the newly created variables are described in the following bullet points:

- The variable n contains the observation number (_n represents the observation number).
- The variable N contains the number of observations.
- The variable tvn is the same as tv; the addition of the [_n] is superfluous.
- The variable tvp was assigned tv[_n-1], the value of tv for the previous observation (and missing for the first observation because there is no previous observation).
- The variable tvs was assigned tv[_n+1], the value of tv for the subsequent observation (and missing for the last observation because there is no subsequent observation).
- The variable tv1 was assigned the value of tv for the first observation.
- The variable tvN was assigned the value of tv for the last observation.

Consider the following commands:

```
. use tv1
. generate newvar = tv[X]
```

Table 7.1 below shows the different meanings that newvar would have depending on the value we insert for *X*.

Table 7.1. Meanings of **newvar** depending on the value inserted for X

Value of X	Meaning of newvar
_n	The value of tv for the *current* observation (i.e., a copy of tv).
_n-1	The value of tv for the *previous* observation. This would be missing for the first observation (because there is no previous observation).
_n+1	The value of tv for the *subsequent* observation. This would be missing for the last observation (because there is no subsequent observation).
1	The value of tv for the *first* observation.
_N	The value of tv for the *last* observation.

Let's issue basically the same set of commands but this time prefixed with `by kidid:` to see how the results are affected by the presence of the `by` prefix.

```
. sort kidid dt
. by kidid: generate byn = _n
. by kidid: generate byN = _N
. by kidid: generate bytv = tv
. by kidid: generate bytvn = tv[_n]
. by kidid: generate bytvp = tv[_n-1]
(4 missing values generated)
. by kidid: generate bytvs = tv[_n+1]
(4 missing values generated)
. by kidid: generate bytv1 = tv[1]
. by kidid: generate bytvN = tv[_N]
. list kidid tv byn byN bytv bytvn bytvp bytvs bytv1 bytvN, sepby(kidid)
```

	kidid	tv	byn	byN	bytv	bytvn	bytvp	bytvs	bytv1	bytvN
1.	1	1	1	2	1	1	.	3	1	3
2.	1	3	2	2	3	3	1	.	1	3
3.	2	8	1	1	8	8	.	.	8	8
4.	3	2	1	4	2	2	.	5	2	6
5.	3	5	2	4	5	5	2	1	2	6
6.	3	1	3	4	1	1	5	6	2	6
7.	3	6	4	4	6	6	1	.	2	6
8.	4	7	1	3	7	7	.	1	7	4
9.	4	1	2	3	1	1	7	4	7	4
10.	4	4	3	3	4	4	1	.	7	4

In the presence of the `by` prefix, the subscripts change meaning to reflect the observation number in relation to each by-group (in this case, relative to each kid). So `tv[1]` means the TV-watching time for the first observation (within the kid), `tv[_N]` means the TV-watching time for the last observation (within the kid), and `tv[_n-1]` means the TV-watching time for the previous observation (within the kid); if it is the first observation within the kid, then it is undefined (missing). This is described in more detail below for each variable:

- The variable `byn` contains the observation number *within each kid* (each `kidid`). `byn` restarts the count of observations for each new `kidid`.
- The variable `byN` contains the number of observations within each kid.
- The variable `bytv` is simply a copy of the variable `tv`, so the presence of the `by` prefix had no impact. Likewise for the variable `bytvn`.
- The variable `bytvp` contains the TV-watching time for the previous observation within each kid (and is a missing value for the kid's first observation because there is no previous observation for that kid).
- The variable `bytvs` is the TV-watching time for the subsequent observation within each kid (and is missing on the last observation for each kid because there is no subsequent observation for that kid).
- The variable `bytv1` has the TV-watching time for the first observation within each kid.
- The variable `bytvN` has the TV-watching time for the last observation within each kid.

Consider the following commands:

```
. use tv1
. sort kidid dt
. by kidid: generate newvar = tv[X]
```

Table 7.2 below shows the values that would be assigned to `newvar` based on the value we insert for *X*.

Table 7.2. Values assigned to `newvar` based on the value inserted for X

Value of X	Meaning of `newvar`
_n	The value of `tv` for the *current* observation for each kid (i.e., a copy of `tv`).
_n-1	The value of `tv` for the *previous* observation for each kid. `newvar` would be missing for the kid's first observation because there is no previous observation.
_n+1	The value of `tv` for the *subsequent* observation for each kid. `newvar` would be missing for the kid's last observation because there is no subsequent observation.
1	The value of `tv` for the *first* observation for each kid.
_N	The value of `tv` for the *last* observation for each kid.

In the previous examples, we correctly used the `sort` command to sort the data on `kidid` and `dt`. Suppose that we had instead just sorted the data on `kidid` only, as shown in the two commands below.

```
. sort kidid
. by kidid: generate bytv1 = tv[1]
```

Or equivalently, you might combine these commands using `bysort`, as shown below (see section A.3 for more on `bysort`).

```
. bysort kidid: generate bytv1 = tv[1]
```

Either way, you might be tempted to assume that `tv[1]` refers to the TV-watching time for the first date that the kid was observed. The `tv[1]` variable refers to the first observation for each kid within the dataset, which is not necessarily the same as the first date that the kid was observed. Unless we sort the data on both `kidid` and `dt`, we do not know how the observations are ordered within each kid. Because it is important in these examples to know how the observations are sorted, it is important to first use the `sort` command to sort the data on both `kidid` and `dt`. Or if you want to use the `bysort` command, you would use the following syntax:

```
. bysort kidid (dt): generate bytv1 = tv[1]
```

This syntax specifies that the data should first be sorted on `kidid` and `dt`. Then `kidid` is used in combination with the `by` prefix when the `generate` command is issued. Expressed this way, `tv[1]` represents the TV-watching time for the first date that the kid was observed.

This section began to explore what can be accomplished when subscripting observations in Stata, especially when combined with the `by` prefix. The following section builds upon this, showing how you can perform computations across observations when combining the `by` prefix with the `generate` command.

7.5 Computing values within subgroups: Computations across observations

This section builds upon the previous section, illustrating how you can combine the `by` prefix with the `generate` command to perform computations across observations within by-groups. Consider the following commands, noting that the dataset is sorted on `kidid` and `dt`.

```
. use tv1
. sort kidid dt
. by kidid: generate newvar = X
```

Table 7.3 below shows some expressions that could go in place of X and the meaning that `newvar` would have.

Table 7.3. Expressions to replace X and the meaning that `newvar` would have

Value of X	Meaning of `newvar`
`tv - tv[_n-1]`	The difference in TV-watching times between the *current* and *previous* observations, within each kid. This is missing for the kid's first observation.
`tv - tv[_n+1]`	The difference in TV-watching times between the *current* and *subsequent* observations, within each kid. This is missing for the kid's last observation.
`tv - tv[1]`	The difference in TV-watching times between the *current* and *first* observations, within each kid.
`tv - tv[_N]`	The difference in TV-watching times between the *current* and *last* observations, within each kid.
`(tv[_n-1] + tv + tv[_n+1])/3`	The average TV-watching time for the *previous*, *current*, and *subsequent* observations. This is missing for the kid's first and last observations.

Each of these computations is illustrated below.

```
. use tv1
. sort kidid dt
. by kidid: generate tvdfp = tv - tv[_n-1]
(4 missing values generated)
```

```
. by kidid: generate tvdfs = tv - tv[_n+1]
(4 missing values generated)
. by kidid: generate tvdff = tv - tv[1]
. by kidid: generate tvdfl = tv - tv[_N]
. by kidid: generate tvavg = (tv[_n-1] + tv + tv[_n+1])/3
(7 missing values generated)
```

The results of these computations are shown below.

```
. list kidid dt tv tvdfp tvdfs tvdff tvdfl tvavg, sepby(kidid)
```

	kidid	dt	tv	tvdfp	tvdfs	tvdff	tvdfl	tvavg
1.	1	07jan2002	1	.	-2	0	-2	.
2.	1	08jan2002	3	2	.	2	0	.
3.	2	16jan2002	8	.	.	0	0	.
4.	3	18jan2002	2	.	-3	0	-4	.
5.	3	19jan2002	5	3	4	3	-1	2.666667
6.	3	21jan2002	1	-4	-5	-1	-5	4
7.	3	22jan2002	6	5	.	4	0	.
8.	4	10jan2002	7	.	6	0	3	.
9.	4	11jan2002	1	-6	-3	-6	-3	4
10.	4	13jan2002	4	3	.	-3	0	.

- The variable `tvdfp` contains the kid's current TV-watching time minus the kid's TV-watching time for the previous observation. In other words, this is a change score compared with the previous observation for the kid. This is missing for the kid's first observation because there is no previous observation for the kid.
- The variable `tvdfs` contains the kid's current TV-watching time minus the kid's TV-watching time for the subsequent observation. This is missing on the last observation for the kid because there is no subsequent observation for the kid.
- The variable `tvdff` contains the kid's current TV-watching time minus the TV-watching time for the kid's first observation. This can be thought of as a change score compared with the first (baseline) observation.
- The variable `tvdfl` contains the kid's current TV-watching time minus the TV-watching time for the kid's last observation.
- Finally, the variable `tvavg` contains the average of the kid's previous, current, and subsequent observations on TV-watching time.

This section and the previous section illustrated how the `by` prefix can be combined with `generate` and the subscripting of observations to perform some sophisticated computations among observations within by-groups. For more information about subscripting, see `help subscripting`. The next section illustrates what can be performed using running sums when computed alone and when combined with by-groups.

7.6 Computing values within subgroups: Running sums

The `sum()` function allows you to create running sums across observations. Consider this simple example using `tv1.dta`.

```
. use tv1
. generate tvsum = sum(tv)
. list kidid tv tvsum

     +---------------------+
     | kidid   tv   tvsum  |
     |---------------------|
  1. |     1    1       1  |
  2. |     1    3       4  |
  3. |     2    8      12  |
  4. |     3    2      14  |
  5. |     3    5      19  |
     |---------------------|
  6. |     3    1      20  |
  7. |     3    6      26  |
  8. |     4    7      33  |
  9. |     4    1      34  |
 10. |     4    4      38  |
     +---------------------+
```

As you can see, the variable `tvsum` contains the running sum, across observations, for the variable `tv`. Let's apply this to the dummy variable `vac`, which is 1 if the kid is on vacation and 0 if the kid is not on vacation.

```
. generate vacsum = sum(vac)
. list kidid vac vacsum

     +------------------------+
     | kidid   vac   vacsum   |
     |------------------------|
  1. |     1     1        1   |
  2. |     1     1        2   |
  3. |     2     1        3   |
  4. |     3     0        3   |
  5. |     3     1        4   |
     |------------------------|
  6. |     3     1        5   |
  7. |     3     0        5   |
  8. |     4     0        5   |
  9. |     4     0        5   |
 10. |     4     0        5   |
     +------------------------+
```

The `sum()` function works the same way for this dummy variable. The variable `vacsum` is basically a running count of the number of vacation days.

Let's see what happens if we combine these commands with the `by kidid:` prefix. Based on the previous sections, we would expect that the running sums would be performed separately for each kid.

```
. use tv1
. sort kidid dt
. by kidid: generate bytvsum = sum(tv)
. by kidid: generate byvacsum = sum(vac)
```

Indeed, as we see below, the variable `bytvsum` is the running sum of `tv` performed separately for each kid. Likewise, `byvacsum` is the running sum (count) of the number of vacation days for each kid.

```
. list kidid tv vac bytvsum byvacsum, sepby(kidid)

     +-------------------------------------------+
     | kidid   tv   vac   bytvsum   byvacsum |
     |-------------------------------------------|
  1. |     1    1     1         1          1 |
  2. |     1    3     1         4          2 |
     |-------------------------------------------|
  3. |     2    8     1         8          1 |
     |-------------------------------------------|
  4. |     3    2     0         2          0 |
  5. |     3    5     1         7          1 |
  6. |     3    1     1         8          2 |
  7. |     3    6     0        14          2 |
     |-------------------------------------------|
  8. |     4    7     0         7          0 |
  9. |     4    1     0         8          0 |
 10. |     4    4     0        12          0 |
     +-------------------------------------------+
```

Consider the following incomplete command, noting that the dataset is sorted on `kidid` and `dt`.

```
. use tv1
. sort kidid dt
. by kidid: generate newvar = X
```

We could replace X with a variety of expressions that involve the `sum()` function. Table 7.4 below shows some expressions that could go in place of X and the meaning that `newvar` would have.

Table 7.4. Expressions to replace X and the meaning that `newvar` would have

Value of X	Meaning of `newvar`
`sum(tv)`	The running sum of TV-watching time within each kid.
`sum(tv)/_n`	The running average of TV-watching time within each kid.
`sum(vac)`	The running sum of `vac`. Because `vac` is a 0/1 variable indicating if the kid is on vacation, the running sum of `vac` is basically a running count of the number of days the kid was on vacation.
`sum(vac)/_n`	The running proportion of the number of days the kid was on vacation.

The variables described above are computed using the generate commands below.

```
. use tv1
. sort kidid dt
. by kidid: generate tvrunsum = sum(tv)
. by kidid: generate tvrunavg = sum(tv)/_n
. by kidid: generate vacruncnt = sum(vac)
. by kidid: generate vacrunprop = sum(vac)/_n
```

The variable **tvrunsum** is the running sum of TV-watching time for each kid, and **tvrunavg** is the running average of TV-watching time for each kid. Likewise, **vacruncnt** is the running count of vacation days for each kid, and **vacrunprop** is the running proportion of vacation days for each kid.

```
. list kidid tv* vac*, sepby(kidid) abb(20)
```

	kidid	tv	tvrunsum	tvrunavg	vac	vacruncnt	vacrunprop
1.	1	1	1	1	1	1	1
2.	1	3	4	2	1	2	1
3.	2	8	8	8	1	1	1
4.	3	2	2	2	0	0	0
5.	3	5	7	3.5	1	1	.5
6.	3	1	8	2.666667	1	2	.6666667
7.	3	6	14	3.5	0	2	.5
8.	4	7	7	7	0	0	0
9.	4	1	8	4	0	0	0
10.	4	4	12	4	0	0	0

This and the previous sections illustrated many of the tools that Stata provides for performing computations across observations. The following section combines these elements to show more examples of how you can use these tools together to perform useful tasks.

7.7 Computing values within subgroups: More examples

In the previous sections, we have seen that when we combine by with the generate command, we can obtain many things, including the following:

- The observation number within the group, _n.
- The number of observations in the group, _N.
- The value of any particular observation within the group via subscripting (e.g., x[1], x[_n-1], x[_n+1], or x[_N]).
- Computations between and among observations within the group (e.g., x[_n] - x[_n-1]).
- The ability to create running sums within groups with the sum() function.

These fundamental tools can be combined in many unique ways. This section illustrates some examples of combining these tools together using tv1.dta as an example.

Number per group, singleton groups, first in group, last in group

When using the by prefix, we have seen that _n represents the observation number within the by-group and _N represents the number of observations within the by-group. A variable named singleton is created below that is 1 if the value of _N is 1 and is 0 otherwise. If _N is 1, then there is only one observation in the group. A dummy variable called first is created that is 1 if it is the first observation for the kid. We know an observation is the first for a kid because the value of _n is equal to 1. Finally, a dummy variable called last is created that represents the last observation for a kid. We can identify the last observation because the value of the current observation for the kid (_n) is equal to the total number of observations for the kid (_N). These computations are illustrated below.

```
. use tv1
. sort kidid dt
. by kidid: generate n = _n
. by kidid: generate N = _N
. by kidid: generate singleton = (_N==1)
. by kidid: generate first = (_n==1)
. by kidid: generate last = (_n==_N)
```

The variables created by these generate commands are listed below.

```
. list kidid n N singleton first last, sepby(kidid) abb(20)
```

	kidid	n	N	singleton	first	last
1.	1	1	2	0	1	0
2.	1	2	2	0	0	1
3.	2	1	1	1	1	1
4.	3	1	4	0	1	0
5.	3	2	4	0	0	0
6.	3	3	4	0	0	0
7.	3	4	4	0	0	1
8.	4	1	3	0	1	0
9.	4	2	3	0	0	0
10.	4	3	3	0	0	1

Changing states: starting and ending vacation

The variable vac contains 0 if the kid is not on vacation and 1 if the kid is on vacation. You might be interested in seeing how starting vacation and ending vacation impacts TV-watching time, so you might want to have a variable that indicates the day the kid started vacation and the last day of vacation. Starting vacation means that the current value of vacation is 1, while the previous value of vacation is 0. Likewise, the last day of vacation means that the current value of vacation is 1, while the next value is 0. This is illustrated below.

```
. use tv1
. sort kidid dt
. by kidid: generate vacstart = (vac==1) & (vac[_n-1]==0)
. by kidid: generate vacend = (vac==1) & (vac[_n+1]==0)
. list kidid vac*, sepby(kidid) abb(20)
```

	kidid	vac	vacstart	vacend
1.	1	1	0	0
2.	1	1	0	0
3.	2	1	0	0
4.	3	0	0	0
5.	3	1	1	0
6.	3	1	0	1
7.	3	0	0	0
8.	4	0	0	0
9.	4	0	0	0
10.	4	0	0	0

Fill in missing values

tv2.dta is like tv1.dta except that it has a couple of missing values on tv and wt, as shown below.

```
. use tv2
. sort kidid dt
. list, sepby(kidid)
```

	kidid	dt	female	wt	tv	vac
1.	1	07jan2002	1	53	1	1
2.	1	08jan2002	1	55	3	1
3.	2	16jan2002	1	58	8	1
4.	3	18jan2002	0	60	2	0
5.	3	19jan2002	0	.	.	.
6.	3	21jan2002	0	66	.	1
7.	3	22jan2002	0	64	6	0
8.	4	10jan2002	1	62	7	0
9.	4	11jan2002	1	58	.	.
10.	4	13jan2002	1	.	4	0

Suppose that we wanted to fill in a missing value of `tv` with the last nonmissing value. We first make a copy of `tv`, calling it `tvimp1`. Then we replace it with the previous value of `tv` if `tv` is missing. This intentionally only carries a valid value forward for the first missing value (e.g., for kid number 3, the second missing `tv` value remains missing).

```
. generate tvimp1 = tv
(3 missing values generated)
. by kidid: replace tvimp1 = tv[_n-1] if missing(tv)
(2 real changes made)
. list kidid tv tvimp1, sepby(kidid)

     +----------------------+
     | kidid   tv   tvimp1  |
     |----------------------|
  1. |     1    1        1  |
  2. |     1    3        3  |
     |----------------------|
  3. |     2    8        8  |
     |----------------------|
  4. |     3    2        2  |
  5. |     3    .        2  |
  6. |     3    .        .  |
  7. |     3    6        6  |
     |----------------------|
  8. |     4    7        7  |
  9. |     4    .        7  |
 10. |     4    4        4  |
     +----------------------+
```

Instead, you might want to continue to carry forward the last known valid value for all consecutive missing values. This strategy starts the same way as the one just mentioned in that we create a new variable (in this case, `tvimp2`) that is a copy of `tv`. But in the `replace` command, the `tvimp2` variable appears both on the left side and on the right side of the equal sign. The value of `tvimp2` is replaced with the prior value when a missing value is present. For the case where there are two or more consecutive missing values, the first missing value is replaced with the valid value of `tvimp2` from the prior observation, which then becomes the source for replacing the next missing value.

(*Continued on next page*)

```
. generate tvimp2 = tv
(3 missing values generated)

. by kidid: replace tvimp2 = tvimp2[_n-1] if missing(tvimp2)
(3 real changes made)

. list kidid tv tvimp2, sepby(kidid)

     +----------------------+
     | kidid   tv   tvimp2  |
     |----------------------|
  1. |     1    1        1  |
  2. |     1    3        3  |
     |----------------------|
  3. |     2    8        8  |
     |----------------------|
  4. |     3    2        2  |
  5. |     3    .        2  |
  6. |     3    .        2  |
  7. |     3    6        6  |
     |----------------------|
  8. |     4    7        7  |
  9. |     4    .        7  |
 10. |     4    4        4  |
     +----------------------+
```

Instead, you might prefer to interpolate the missing value, replacing it with the average of the previous and next values. This is illustrated below.

```
. generate tvimp3 = tv
(3 missing values generated)

. by kidid: replace tvimp3 = (tv[_n-1]+tv[_n+1])/2 if missing(tv)
(1 real change made)

. list kidid tv tvimp3, sepby(kidid)

     +----------------------+
     | kidid   tv   tvimp3  |
     |----------------------|
  1. |     1    1        1  |
  2. |     1    3        3  |
     |----------------------|
  3. |     2    8        8  |
     |----------------------|
  4. |     3    2        2  |
  5. |     3    .        .  |
  6. |     3    .        .  |
  7. |     3    6        6  |
     |----------------------|
  8. |     4    7        7  |
  9. |     4    .      5.5  |
 10. |     4    4        4  |
     +----------------------+
```

The missing `tv` value for the fourth kid is replaced with 5.5 (i.e., $(7+4)/2$), but the missing `tv` values for the third kid are not replaced because there were two consecutive missing values (thus the interpolation yielded a missing value).

Changes in TV-watching time

You might be interested in focusing on the changes in TV-watching time from one observation to the next. We can compute a variable that represents the change in TV-watching time by taking the current TV-watching time (`tv[_n]`) minus the previous TV-watching time (`tv[_n-1]`), as shown below.

```
. use tv1
. sort kidid dt
. by kidid: generate tvchange = tv[_n] - tv[_n-1]
(4 missing values generated)
. list kidid tv*, sepby(kidid)
```

	kidid	tv	tvchange
1.	1	1	.
2.	1	3	2
3.	2	8	.
4.	3	2	.
5.	3	5	3
6.	3	1	-4
7.	3	6	5
8.	4	7	.
9.	4	1	-6
10.	4	4	3

Of course, the first value is always missing because for the first observation, the value of `tv[_n-1]` resolves to `tv[0]`, which is a missing value. But all the change values are exactly as we would expect.

Perhaps you might want to create an indicator variable that notes when the change scores are less than −2 (meaning that TV-watching time went down by two or more hours). You can do that like this:

(*Continued on next page*)

```
. generate tvch2 = (tvchange <= -2) if ! missing(tvchange)
(4 missing values generated)

. list kidid tv*, sepby(kidid)
```

	kidid	tv	tvchange	tvch2
1.	1	1	.	.
2.	1	3	2	0
3.	2	8	.	.
4.	3	2	.	.
5.	3	5	3	0
6.	3	1	-4	1
7.	3	6	5	0
8.	4	7	.	.
9.	4	1	-6	1
10.	4	4	3	0

As you can see, Stata offers a powerful suite of tools for combining the elements of `by` and `generate` to perform a wide array of computations across observations and within by-groups. For more information about using `by` with `generate`, see `help by`.

7.8 Comparing the by and tsset commands

Section 7.5 illustrated how `by` could be used for performing computations across observations within groups. For example, using `tv1.dta`, we saw how we could obtain the TV-watching time for the prior observation like this:

```
. use tv1

. sort kidid dt

. by kidid: generate ltv = tv[_n - 1]
(4 missing values generated)

. list, sepby(kidid)
```

	kidid	dt	female	wt	tv	vac	ltv
1.	1	07jan2002	1	53	1	1	.
2.	1	08jan2002	1	55	3	1	1
3.	2	16jan2002	1	58	8	1	.
4.	3	18jan2002	0	60	2	0	.
5.	3	19jan2002	0	63	5	1	2
6.	3	21jan2002	0	66	1	1	5
7.	3	22jan2002	0	64	6	0	1
8.	4	10jan2002	1	62	7	0	.
9.	4	11jan2002	1	58	1	0	7
10.	4	13jan2002	1	55	4	0	1

The value of ltv is the time spent watching TV from the prior observation for the given kid. Note that it is not necessarily the time spent watching TV on the prior day. For example, for observation 6 on Jan 21, the value of ltv is 5, which represents the value from Jan 19. Consider this alternative method of performing this computation:

```
. tsset kidid dt, daily
       panel variable:  kidid (unbalanced)
        time variable:  dt, 07jan2002 to 22jan2002, but with gaps
                delta:  1 day

. generate ltv2 = L.tv
(6 missing values generated)
```

The tsset command is used to tell Stata that the data are grouped by kidid and that dt determines the date of observation. The daily option indicates that dt represents days (as opposed to weeks, months, or years). Having issued this command, Stata understands that prefacing a variable with L. (that is an L as in lag) means that you want the value of tv from the prior period (the prior day).[2] We can see the results below.

```
. list kidid dt wt tv ltv ltv2, sepby(kidid)
```

	kidid	dt	wt	tv	ltv	ltv2
1.	1	07jan2002	53	1	.	.
2.	1	08jan2002	55	3	1	1
3.	2	16jan2002	58	8	.	.
4.	3	18jan2002	60	2	.	.
5.	3	19jan2002	63	5	2	2
6.	3	21jan2002	66	1	5	.
7.	3	22jan2002	64	6	1	1
8.	4	10jan2002	62	7	.	.
9.	4	11jan2002	58	1	7	7
10.	4	13jan2002	55	4	1	.

Note the cases where ltv (TV-watching time for the previous observation) differs from ltv2 (TV-watching time for the previous day). For example, on Jan 21, the prior day is Jan 20, and there is no value of tv for that day; this is why the value of ltv2 is missing on that day. Consider these additional examples:

2. Contrast this with tv[_n-1], which represents the prior observation regardless of how long ago that prior observation occurred.

```
. generate ftv = F.tv
(6 missing values generated)

. generate dtv = D.tv
(6 missing values generated)

. list kidid dt wt tv ftv dtv, sepby(kidid)
```

	kidid	dt	wt	tv	ftv	dtv
1.	1	07jan2002	53	1	3	.
2.	1	08jan2002	55	3	.	2
3.	2	16jan2002	58	8	.	.
4.	3	18jan2002	60	2	5	.
5.	3	19jan2002	63	5	.	3
6.	3	21jan2002	66	1	6	.
7.	3	22jan2002	64	6	.	5
8.	4	10jan2002	62	7	1	.
9.	4	11jan2002	58	1	.	-6
10.	4	13jan2002	55	4	.	.

Note how `F.tv` (F as in forward) represents the value of `tv` in the next period, and `D.tv` (D as in difference) represents the current value of `tv` minus the previous value of `tv`. When we specify `F.tv`, it is as though we are specifying `F1.tv`, explicitly indicating that we want to move in advance one period. You can specify whatever period you wish; for example, `F2.tv` would refer to the value of `tv` two periods in the future. This can be equally applied with `L.tv` and `D.tv`.

For more information about using and performing computations involving time-series data, see `help tsset`.

8 Changing the shape of your data

To call in the statistician after the experiment is done may be no more than asking him to perform a post-mortem examination: he may be able to say what the experiment died of.

—Ronald Fisher

8.1 Introduction

For most datasets, there is no question about the shape of the data. Every observation is represented by a different row, and every variable is represented by a different column. However, some datasets have a nested structure that can be represented in more than one way. For example, when there are multiple measurements per person, those multiple measurements could be represented as additional columns (in a format that is called a wide dataset) or as additional rows (in a format that is called a long dataset).

This chapter considers these kinds of datasets that could be stored using more than one shape. Section 8.2 describes wide and long datasets, and illustrates the situations in which you would want to use each. Then sections 8.3 and 8.4 illustrate how you can convert long datasets to wide datasets, and sections 8.5 and 8.6 illustrate how you can convert wide datasets to long datasets. Section 8.7 describes multilevel datasets, which combine information measured at more than one level (like those used in multilevel modeling). Finally, section 8.8 discusses collapsed datasets and how you can create them.

8.2 Wide and long datasets

This section focuses on two shapes you can use for structuring your data: wide and long. This section illustrates these two shapes and the advantages of each.

Let's start our examination of long and wide datasets by looking at an example of a wide dataset named `cardio_wide.dta`.

```
. use cardio_wide

. describe

Contains data from cardio_wide.dta
  obs:             6
 vars:            12                          22 Dec 2009 20:43
 size:           144 (99.9% of memory free)
-------------------------------------------------------------------------------
              storage  display     value
variable name   type   format      label      variable label
-------------------------------------------------------------------------------
id              byte   %3.0f                  ID of person
age             byte   %3.0f                  Age of person
bp1             int    %3.0f                  Blood pressure systolic Trial 1
bp2             int    %3.0f                  Blood pressure systolic Trial 2
bp3             int    %3.0f                  Blood pressure systolic Trial 3
bp4             int    %3.0f                  Blood pressure systolic Trial 4
bp5             int    %3.0f                  Blood pressure systolic Trial 5
pl1             int    %3.0f                  Pulse: Trial 1
pl2             byte   %3.0f                  Pulse: Trial 2
pl3             int    %3.0f                  Pulse: Trial 3
pl4             int    %3.0f                  Pulse: Trial 4
pl5             byte   %3.0f                  Pulse: Trial 5
-------------------------------------------------------------------------------
Sorted by:
```

```
. list
```

	id	age	bp1	bp2	bp3	bp4	bp5	pl1	pl2	pl3	pl4	pl5
1.	1	40	115	86	129	105	127	54	87	93	81	92
2.	2	30	123	136	107	111	120	92	88	125	87	58
3.	3	16	124	122	101	109	112	105	97	128	57	68
4.	4	23	105	115	121	129	137	52	79	71	106	39
5.	5	18	116	128	112	125	111	70	64	52	68	59
6.	6	27	108	126	124	131	107	74	78	92	99	80

This dataset contains information about six people who took part in a study regarding exertion and its effects on systolic blood pressure and pulse rate. The blood pressure and pulse rate were measured over five trials: the first time at rest, the next two times while working out, and the final two times when recovering from the workout. The time interval between trials was 10 minutes for all people.

This dataset contains one observation per person, and the blood pressure measurements for the five trials are named **bp1**–**bp5**. Likewise, the five pulse measures are named **pl1**–**pl5**. The age of the participant was stored in the variable **age**. This is called a wide dataset because as more trials are added, the dataset gets wider.

The long counterpart of this dataset is `cardio_long.dta`, shown below.

```
. use cardio_long
. describe
Contains data from cardio_long.dta
  obs:            30
 vars:             5                          21 Dec 2009 22:08
 size:           330 (99.9% of memory free)
-------------------------------------------------------------------------------
              storage  display     value
variable name   type   format      label      variable label
-------------------------------------------------------------------------------
id              byte   %3.0f                  ID of person
trial           byte   %9.0g                  Trial number
age             byte   %3.0f                  Age of person
bp              int    %3.0f                  Blood pressure (systolic)
pl              int    %3.0f                  Pulse
-------------------------------------------------------------------------------
Sorted by:  id  trial
```

(*Continued on next page*)

```
. list
```

	id	trial	age	bp	pl
1.	1	1	40	115	54
2.	1	2	40	86	87
3.	1	3	40	129	93
4.	1	4	40	105	81
5.	1	5	40	127	92
6.	2	1	30	123	92
7.	2	2	30	136	88
8.	2	3	30	107	125
9.	2	4	30	111	87
10.	2	5	30	120	58
11.	3	1	16	124	105
12.	3	2	16	122	97
13.	3	3	16	101	128
14.	3	4	16	109	57
15.	3	5	16	112	68
16.	4	1	23	105	52
17.	4	2	23	115	79
18.	4	3	23	121	71
19.	4	4	23	129	106
20.	4	5	23	137	39
21.	5	1	18	116	70
22.	5	2	18	128	64
23.	5	3	18	112	52
24.	5	4	18	125	68
25.	5	5	18	111	59
26.	6	1	27	108	74
27.	6	2	27	126	78
28.	6	3	27	124	92
29.	6	4	27	131	99
30.	6	5	27	107	80

As you can see, this long dataset contains the same information as its wide counterpart, but instead of storing the information for each additional trial in a new variable, each additional trial is stored in a new observation (row). This is called a long dataset because for every additional trial, the dataset gets longer.

The wide dataset has one observation per person, and the long dataset has five observations per person (one for each trial). The blood pressure measurements are stored in the variable `bp`, and the pulse measures are stored in `pl`. The variable `trial` identifies the trial number for the blood pressure and pulse measurements. Like the wide dataset, the long dataset contains the age of the person, but this information is repeated for each trial.

Note how everyone has the same number of trials (five). In such cases, the choice of whether to use a wide dataset or a long dataset depends on what you want to do with the dataset. Let's consider the different kinds of analyses we can perform on the wide and long versions of these datasets. `cardio_wide.dta` makes it easy to obtain the correlations among the blood pressure measurements across time points.

```
. use cardio_wide

. correlate bp1 bp2 bp3 bp4 bp5
(obs=6)

             |      bp1      bp2      bp3      bp4      bp5
-------------+---------------------------------------------
         bp1 |   1.0000
         bp2 |   0.2427   1.0000
         bp3 |  -0.7662  -0.6657   1.0000
         bp4 |  -0.7644   0.3980   0.2644   1.0000
         bp5 |  -0.3643  -0.4984   0.3694  -0.0966   1.0000
```

By contrast, `cardio_long.dta` makes it a bit simpler to assess the correlations between pulse and blood pressure at each of the time points. This extends to any analysis that you might want to perform separately at each time point, not just correlations.

```
. use cardio_long

. bysort trial: correlate bp pl

-------------------------------------------------------------------------------
-> trial = 1
(obs=6)

             |       bp       pl
-------------+------------------
          bp |   1.0000
          pl |   0.7958   1.0000

-------------------------------------------------------------------------------
-> trial = 2
(obs=6)

             |       bp       pl
-------------+------------------
          bp |   1.0000
          pl |  -0.1985   1.0000

-------------------------------------------------------------------------------
-> trial = 3
(obs=6)

             |       bp       pl
-------------+------------------
          bp |   1.0000
          pl |  -0.4790   1.0000

-------------------------------------------------------------------------------
-> trial = 4
(obs=6)

             |       bp       pl
-------------+------------------
          bp |   1.0000
          pl |   0.5639   1.0000
```

```
-> trial = 5
(obs=6)

             |       bp       pl
-------------+------------------
          bp |   1.0000
          pl |  -0.3911   1.0000
```

The wide format is amenable to multivariate analysis of the multiple observations across trials with commands such as mvreg and factor. For example, if we had more observations, we could use the mvreg command on this wide dataset to examine the impact of age on the five blood pressure measurements.

```
. use cardio_wide

. mvreg bp1 bp2 bp3 bp4 bp5 = age
  (output omitted)
```

The long format is required for using the xt suite of commands. For example, with cardio_long.dta, we could perform a random-intercepts regression predicting blood pressure from age.

```
. use cardio_long

. xtset id trial
       panel variable:  id (strongly balanced)
        time variable:  trial, 1 to 5
                delta:  1 unit

. xtreg bp age
  (output omitted)
```

The long format is also more amenable to graphing the data across time with the xtline command.

```
. xtline bp, overlay
  (output omitted)
```

Let's compare these two formats from a data-management perspective. Suppose that we are entering data, and there are many person-level variables (not only age but also the person's gender, height, weight, education, race, and so forth). With a wide dataset, such information needs to be entered only once. But in a long dataset, such information needs to be entered five times, corresponding to the five trials. In such a case, the wide format would be more advantageous (also see section 8.7, which illustrates another means of handling such cases by using a multilevel structure).

Likewise, suppose that we are merging cardio_wide.dta with other datasets that are stored at the person level. When merging such datasets, it would be easier to assess the adequacy of the matching in the wide dataset, where there is one observation per person, compared with the long dataset, where there are multiple observations per person.

Now let's consider the differences between the wide and long forms of the data when performing computations involving the data across trials. Suppose that we want to recode the pulse variable to be 0 if the pulse was below 90 and 1 if the pulse was 90 or above. With the long dataset, we can issue one `recode` command, as shown below.

```
. use cardio_long
. recode pl (min/89=0) (90/max=1), generate(plhi)
(30 differences between pl and plhi)
. list id trial pl* in 1/10
```

	id	trial	pl	plhi
1.	1	1	54	0
2.	1	2	87	0
3.	1	3	93	1
4.	1	4	81	0
5.	1	5	92	1
6.	2	1	92	1
7.	2	2	88	0
8.	2	3	125	1
9.	2	4	87	0
10.	2	5	58	0

By contrast, using the wide dataset, this recoding would need to be done repeatedly for each trial. The long version makes such recoding much easier because one `recode` command does the recoding across all trials; in the wide version, you need to issue one `recode` command for every trial. (Section 9.8 illustrates how to reduce such repetition by using `foreach` loops.)

```
. use cardio_wide
. recode pl1 (min/99=0) (100/max=1), generate(plhi1)
(6 differences between pl1 and plhi1)
. recode pl2 (min/99=0) (100/max=1), generate(plhi2)
(6 differences between pl2 and plhi2)
. recode pl3 (min/99=0) (100/max=1), generate(plhi3)
(6 differences between pl3 and plhi3)
. recode pl4 (min/99=0) (100/max=1), generate(plhi4)
(6 differences between pl4 and plhi4)
. recode pl5 (min/99=0) (100/max=1), generate(plhi5)
(6 differences between pl5 and plhi5)
. list id pl*, noobs
```

id	pl1	pl2	pl3	pl4	pl5	plhi1	plhi2	plhi3	plhi4	plhi5
1	54	87	93	81	92	0	0	0	0	0
2	92	88	125	87	58	0	0	1	0	0
3	105	97	128	57	68	1	0	1	0	0
4	52	79	71	106	39	0	0	0	1	0
5	70	64	52	68	59	0	0	0	0	0
6	74	78	92	99	80	0	0	0	0	0

The wide version of the data is used below to compute the overall average pulse rate. The `egen` command is used with the `rowmean()` function to compute the mean across the five measures of pulse.

```
. use cardio_wide
. egen pl_avg = rowmean(pl*)
. list id pl*
```

	id	pl1	pl2	pl3	pl4	pl5	pl_avg
1.	1	54	87	93	81	92	81.4
2.	2	92	88	125	87	58	90
3.	3	105	97	128	57	68	91
4.	4	52	79	71	106	39	69.4
5.	5	70	64	52	68	59	62.6
6.	6	74	78	92	99	80	84.6

With the long dataset, this computation is equally easy. Here we use the `egen` command with the `mean()` function combined with the `bysort id:` prefix to compute the means across the levels of `id`. (See section 7.3 for more on combining the `by` prefix with the `egen` command.)

```
. use cardio_long
. bysort id: egen pl_avg = mean(pl)
. list id trial pl* in 1/10
```

	id	trial	pl	pl_avg
1.	1	1	54	81.4
2.	1	2	87	81.4
3.	1	3	93	81.4
4.	1	4	81	81.4
5.	1	5	92	81.4
6.	2	1	92	90
7.	2	2	88	90
8.	2	3	125	90
9.	2	4	87	90
10.	2	5	58	90

Finally, let's compare the wide and long dataset formats when performing computations among time points. The wide version of the data below is used to compute changes in pulse rates across adjacent trials. The computations themselves are simple and easy to understand, but they need to be repeated multiple times across the trials (section 9.9 shows how a `foreach` loop could be used for this kind of problem).

```
. use cardio_wide
. generate pldiff2 = pl2 - pl1
. generate pldiff3 = pl3 - pl2
```

```
. generate pldiff4 = pl4 - pl3
. generate pldiff5 = pl5 - pl4
```

Below we can see the original variables as well as the differences between the pulse rates in the adjacent trials.

```
. list pl*
```

	pl1	pl2	pl3	pl4	pl5	pldiff2	pldiff3	pldiff4	pldiff5
1.	54	87	93	81	92	33	6	-12	11
2.	92	88	125	87	58	-4	37	-38	-29
3.	105	97	128	57	68	-8	31	-71	11
4.	52	79	71	106	39	27	-8	35	-67
5.	70	64	52	68	59	-6	-12	16	-9
6.	74	78	92	99	80	4	14	7	-19

In the long format, we need to draw upon the logic of by-groups (as illustrated in section 7.5). The `generate` command is prefaced with `by id:` and computes the pulse rate for the current trial (`pl`) minus the pulse rate of the previous trial (`pl[_n-1]`) within each person. This requires fewer commands but is conceptually a bit trickier.

```
. use cardio_long
. sort id trial
. by id: generate pldiff = pl - pl[_n-1]
(6 missing values generated)
. list id trial pl* in 1/10
```

	id	trial	pl	pldiff
1.	1	1	54	.
2.	1	2	87	33
3.	1	3	93	6
4.	1	4	81	-12
5.	1	5	92	11
6.	2	1	92	.
7.	2	2	88	-4
8.	2	3	125	37
9.	2	4	87	-38
10.	2	5	58	-29

As noted before, this example dataset has an equal number of time points (trials) per person and an equal spacing between the time points (10 minutes). In such a case, the choice between a wide or a long format can depend on things like the kind of analysis task that you want to perform or the way you would like to approach certain data-management tasks.

But imagine we have a dataset with hospital admissions, where we measure blood pressure for each person when he or she is admitted to the hospital. A person could have one or two admissions, or perhaps she might have dozens of admissions, and perhaps a couple of people have two hundred or more admissions. Storing this as a wide dataset could grow unwieldy because the width of the dataset would be determined by the maximum number of admissions. For example, if one person has 210 admissions, then we would have to have the variables bp1–bp210 for all people in the dataset, even though most of these values would be missing. Further, if a new person who has even more admissions was added to the dataset, the dataset would need to be further widened to accommodate this new person.

Compare this scenario with storing the dataset in a long format, in which there would be one observation for each admission. A person with 1 admission would have 1 observation (recording their blood pressure), and a person with 210 admissions would have 210 observations (each with their blood pressure upon admission). If a person who has more admissions is added, the structure of the dataset remains the same—that person simply contributes more observations to the dataset. In general, when the number of observations per person varies considerably or is unpredictable, a long format can be much more advantageous than a wide format.

Similarly, if the multiple observations are unevenly spaced across time (and you care about the amount of time between observations), a long dataset can be much better than a wide dataset. For example, imagine that you are collecting data studying weight loss over time. You plan to measure a person's weight every seven days, but in practice, the time between weight measurements is erratic. For such a dataset, I highly recommend using the long format for storing the data, as illustrated below in weights_long.dta.

```
. use weights_long
. list, sepby(id)

     +----------------+
     | id   days   wt |
     |----------------|
  1. |  1      7  166 |
  2. |  1     14  163 |
  3. |  1     21  164 |
  4. |  1     28  162 |
     |----------------|
  5. |  2      9  188 |
  6. |  2     13  184 |
  7. |  2     22  185 |
  8. |  2     27  182 |
     |----------------|
  9. |  3      6  158 |
 10. |  3     12  155 |
 11. |  3     31  157 |
     |----------------|
 12. |  4      8  192 |
 13. |  4     17  190 |
 14. |  4     22  191 |
 15. |  4     30  193 |
     |----------------|
 16. |  5      5  145 |
 17. |  5     11  142 |
 18. |  5     20  140 |
 19. |  5     26  137 |
     +----------------+
```

For each weight measurement, we enter an observation with the `id` variable for the person, when the observation was recorded (`days`), and the person's weight (`wt`). Such long datasets can be analyzed using commands like `xtset` followed by `xtreg` or `xtline`, allowing us to examine weight loss as a function of time, as shown below.

```
. xtset id days
       panel variable:  id (unbalanced)
        time variable:  days, 5 to 31, but with gaps
                delta:  1 unit
. xtreg wt days
  (output omitted)
. xtline wt, overlay
  (output omitted)
```

I have often seen people get into trouble by trying to enter such data as a wide dataset, as illustrated in `weights_wide.dta`.

```
. use weights_wide
. list
```

	id	days1	wt1	days2	wt2	days3	wt3	days4	wt4
1.	1	7	166	14	163	21	164	28	162
2.	2	9	188	13	184	22	185	27	182
3.	3	6	158	12	155	31	157	.	.
4.	4	8	192	17	190	22	191	30	193
5.	5	5	145	11	142	20	140	26	137

The variables `days1`–`days4` reflect the number of days (since the start of the study) when the weight was measured for each, and `wt1`–`wt4` represent the weight on that day. Some people (like person 1) came in exactly on schedule, but others (like person 4) deviated considerably from coming every 7 days, and person 3 altogether missed the weigh-in that was supposed to occur at 21 days, thus this person's third weight measurement aligns more (in time) with the weight measurements of the others' fourth measurements. A long data structure avoids these kinds of problems and yields a dataset that is ready for analysis with the `xt` suite of commands.

This section illustrated the long and wide formats for storing data. In some cases, where you have equally spaced time points and an equal (or roughly equal) number of observations per person, you can choose between the wide and long data formats depending on your analysis and data-management preferences. In such cases, the `reshape` command allows you to easily switch from long to wide (see sections 8.3 and 8.4) and from wide to long (see sections 8.5 and 8.6). But when the number of observations per person can vary substantially (as described in the hospital admissions example) or when the time between observations can vary across persons (as described in the weight loss example), I highly recommend choosing and sticking with the long format for structuring your data.

8.3 Introduction to reshaping long to wide

Section 8.2 described wide and long datasets and illustrated situations where a wide format might be preferable to a long format. This section illustrates how you can use the `reshape` command to convert your data from a long format to a wide format. Let's briefly look at the first 10 observations from `cardio_long.dta` before showing how to reshape it into a wide format.

```
. use cardio_long

. describe

Contains data from cardio_long.dta
  obs:            30
 vars:             5                          21 Dec 2009 22:08
 size:           330 (99.9% of memory free)
-------------------------------------------------------------------------------
              storage  display     value
variable name   type   format      label      variable label
-------------------------------------------------------------------------------
id              byte   %3.0f                  ID of person
trial           byte   %9.0g                  Trial number
age             byte   %3.0f                  Age of person
bp              int    %3.0f                  Blood pressure (systolic)
pl              int    %3.0f                  Pulse
-------------------------------------------------------------------------------
Sorted by:  id  trial

. list in 1/10

     +-------------------------------+
     | id   trial   age    bp    pl  |
     |-------------------------------|
  1. |  1       1    40   115    54  |
  2. |  1       2    40    86    87  |
  3. |  1       3    40   129    93  |
  4. |  1       4    40   105    81  |
  5. |  1       5    40   127    92  |
     |-------------------------------|
  6. |  2       1    30   123    92  |
  7. |  2       2    30   136    88  |
  8. |  2       3    30   107   125  |
  9. |  2       4    30   111    87  |
 10. |  2       5    30   120    58  |
     +-------------------------------+
```

This dataset contains observations of blood pressure (`bp`) and pulse (`pl`) across five trials for each person. We can reshape this dataset into a wide format as shown below.

```
. reshape wide bp pl, i(id) j(trial)
(note: j = 1 2 3 4 5)
Data                               long   ->   wide
-----------------------------------------------------------------------------
Number of obs.                       30   ->      6
Number of variables                   5   ->     12
j variable (5 values)             trial   ->   (dropped)
xij variables:
                                     bp   ->   bp1 bp2 ... bp5
                                     pl   ->   pl1 pl2 ... pl5
-----------------------------------------------------------------------------
```

After the `reshape wide` command, we specified three pieces of information. First, we specified the names of the variables that had multiple measurements per observation (i.e., `bp pl`). Note that `age` was excluded from this list because age is constant within a person. Second, we supplied the `i(id)` option to specify the variable(s) that defines an observation in the wide version of the dataset; in the wide dataset, each observation is defined by the variable `id`. Finally, we specified the `j(trial)` option to specify the variable that identifies the repeated observations for each person; in the wide version of the file, the values of `trial` will be used as the numeric suffix for the multiple measurements of `bp` and `pl`. The reshaped wide version of this dataset is shown below.

```
. describe
Contains data
  obs:             6
 vars:            12
 size:           156 (99.9% of memory free)
--------------------------------------------------------------------------------
              storage  display     value
variable name   type   format      label      variable label
--------------------------------------------------------------------------------
id              byte   %3.0f                  ID of person
bp1             int    %3.0f                  1 bp
pl1             int    %3.0f                  1 pl
bp2             int    %3.0f                  2 bp
pl2             int    %3.0f                  2 pl
bp3             int    %3.0f                  3 bp
pl3             int    %3.0f                  3 pl
bp4             int    %3.0f                  4 bp
pl4             int    %3.0f                  4 pl
bp5             int    %3.0f                  5 bp
pl5             int    %3.0f                  5 pl
age             byte   %3.0f                  Age of person
--------------------------------------------------------------------------------
Sorted by:  id
     Note:  dataset has changed since last saved
. list
```

	id	bp1	pl1	bp2	pl2	bp3	pl3	bp4	pl4	bp5	pl5	age
1.	1	115	54	86	87	129	93	105	81	127	92	40
2.	2	123	92	136	88	107	125	111	87	120	58	30
3.	3	124	105	122	97	101	128	109	57	112	68	16
4.	4	105	52	115	79	121	71	129	106	137	39	23
5.	5	116	70	128	64	112	52	125	68	111	59	18
6.	6	108	74	126	78	124	92	131	99	107	80	27

We can reshape the data back into long format simply by issuing the `reshape long` command.

(*Continued on next page*)

```
. reshape long
(note: j = 1 2 3 4 5)
Data                                 wide   ->   long
-----------------------------------------------------------------------------
Number of obs.                          6   ->      30
Number of variables                    12   ->       5
j variable (5 values)                       ->   trial
xij variables:
                          bp1 bp2 ... bp5   ->   bp
                          pl1 pl2 ... pl5   ->   pl
-----------------------------------------------------------------------------
```

As you can see below, after reshaping this dataset back into its long form, it is nearly identical to the original long dataset. The variable names, labels, and values are all the same. The only difference is in the order of the variables: age now appears at the end of the dataset.

```
. describe
Contains data
  obs:            30
 vars:             5
 size:           330 (99.9% of memory free)
-------------------------------------------------------------------------------
              storage  display     value
variable name   type   format      label      variable label
-------------------------------------------------------------------------------
id              byte   %3.0f                  ID of person
trial           byte   %9.0g                  Trial number
bp              int    %3.0f                  Blood pressure (systolic)
pl              int    %3.0f                  Pulse
age             byte   %3.0f                  Age of person
-------------------------------------------------------------------------------
Sorted by:  id  trial
     Note:  dataset has changed since last saved
. list in 1/10

     +---------------------------+
     | id   trial    bp    pl   age |
     |---------------------------|
  1. |  1       1   115    54    40 |
  2. |  1       2    86    87    40 |
  3. |  1       3   129    93    40 |
  4. |  1       4   105    81    40 |
  5. |  1       5   127    92    40 |
     |---------------------------|
  6. |  2       1   123    92    30 |
  7. |  2       2   136    88    30 |
  8. |  2       3   107   125    30 |
  9. |  2       4   111    87    30 |
 10. |  2       5   120    58    30 |
     +---------------------------+
```

For many, or perhaps most, cases, this is all that you need to know about reshaping data from a long format to a wide format. However, there are some complications that can arise. Those are discussed in the next section.

8.4 Reshaping long to wide: Problems

This section considers some complications that can arise when reshaping data from a long format to a wide format. Consider `cardio_long2.dta`. Note how the value of `age` is wrong for trial 3: it is 44 instead of 40.

```
. use cardio_long2
. list in 1/10

     +-----------------------------+
     | id   trial   age    bp    pl |
     |-----------------------------|
  1. |  1       1    40   115    54 |
  2. |  1       2    40    86    87 |
  3. |  1       3    44   129    93 |
  4. |  1       4    40   105    81 |
  5. |  1       5    40   127    92 |
     |-----------------------------|
  6. |  2       1    30   123    92 |
  7. |  2       2    30   136    88 |
  8. |  2       3    30   107   125 |
  9. |  2       4    30   111    87 |
 10. |  2       5    30   120    58 |
     +-----------------------------+
```

Let's see what happens when we try to reshape these data to a wide format.

```
. reshape wide bp pl, i(id) j(trial)
(note: j = 1 2 3 4 5)
age not constant within id
Type "reshape error" for a listing of the problem observations.
r(9);
```

The error message is telling us that the `age` variable is supposed to be the same for each person (`id`), but the values of `age` are not constant (the same) within `id`. Although we already know where the problem lies, we can use the `reshape error` command to display the problem.

```
. reshape error
(note: j = 1 2 3 4 5)
i (id) indicates the top-level grouping such as subject id.
j (trial) indicates the subgrouping such as time.
xij variables are bp pl.
Thus the following variable(s) should be constant within i:
      age
age not constant within i (id) for 1 value of i:

     +------------------+
     | id   trial   age |
     |------------------|
  1. |  1       1    40 |
  2. |  1       2    40 |
  3. |  1       3    44 |
  4. |  1       4    40 |
  5. |  1       5    40 |
     +------------------+

(data now sorted by id trial)
```

The `reshape error` command not only tells us that the `age` variable is changing over time but also shows us the offending observations from the long dataset to help us diagnose and fix the problem. The solution is ensure that the values for `age` are all consistent for a given person. Here we need to make sure that `age` is 40 for all observations associated with an `id` value of 1.

This same error can arise if we forget to mention one of the variables that changes over time in the `reshape wide` command. Using the original `cardio_long.dta`, we try to reshape the data into wide format but mention only the `bp` variable (forgetting the `pl` variable). The `reshape wide` command assumes that `pl` is constant within `id` (which it is not) and gives us an error, as shown below.

```
. use cardio_long

. reshape wide bp, i(id) j(trial)
(note: j = 1 2 3 4 5)
pl not constant within id
Type "reshape error" for a listing of the problem observations.
r(9);
```

In this case, the remedy is simply to repeat the command, remembering to include `pl`.

```
. use cardio_long

. reshape wide bp pl, i(id) j(trial)
(note: j = 1 2 3 4 5)
Data                               long   ->   wide
-----------------------------------------------------------------------------
Number of obs.                       30   ->      6
Number of variables                   5   ->     12
j variable (5 values)             trial   ->   (dropped)
xij variables:
                                     bp   ->   bp1 bp2 ... bp5
                                     pl   ->   pl1 pl2 ... pl5
-----------------------------------------------------------------------------
```

For more information about reshaping datasets from long to wide, see `help reshape`.

8.5 Introduction to reshaping wide to long

This section introduces how to use the `reshape` command to reshape a wide dataset into a long dataset. Let's illustrate this using `cardio_wide.dta` (shown below).

```
. use cardio_wide

. describe

Contains data from cardio_wide.dta
  obs:             6
 vars:            12                          22 Dec 2009 20:43
 size:           144 (99.9% of memory free)
-------------------------------------------------------------------------------
              storage  display     value
variable name   type   format      label      variable label
-------------------------------------------------------------------------------
id              byte   %3.0f                  ID of person
age             byte   %3.0f                  Age of person
bp1             int    %3.0f                  Blood pressure systolic Trial 1
bp2             int    %3.0f                  Blood pressure systolic Trial 2
bp3             int    %3.0f                  Blood pressure systolic Trial 3
bp4             int    %3.0f                  Blood pressure systolic Trial 4
bp5             int    %3.0f                  Blood pressure systolic Trial 5
pl1             int    %3.0f                  Pulse: Trial 1
pl2             byte   %3.0f                  Pulse: Trial 2
pl3             int    %3.0f                  Pulse: Trial 3
pl4             int    %3.0f                  Pulse: Trial 4
pl5             byte   %3.0f                  Pulse: Trial 5
-------------------------------------------------------------------------------
Sorted by:

. list
```

	id	age	bp1	bp2	bp3	bp4	bp5	pl1	pl2	pl3	pl4	pl5
1.	1	40	115	86	129	105	127	54	87	93	81	92
2.	2	30	123	136	107	111	120	92	88	125	87	58
3.	3	16	124	122	101	109	112	105	97	128	57	68
4.	4	23	105	115	121	129	137	52	79	71	106	39
5.	5	18	116	128	112	125	111	70	64	52	68	59
6.	6	27	108	126	124	131	107	74	78	92	99	80

This dataset contains six observations with an ID variable (id), the person's age (age), five measurements of blood pressure (bp1–bp5), and five measurements of pulse (pl1–pl5). We see how to reshape this dataset into a long format below.

```
. reshape long bp pl, i(id) j(trialnum)
(note: j = 1 2 3 4 5)
Data                               wide   ->   long
-----------------------------------------------------------------------------
Number of obs.                        6   ->      30
Number of variables                  12   ->       5
j variable (5 values)                     ->   trialnum
xij variables:
                        bp1 bp2 ... bp5   ->   bp
                        pl1 pl2 ... pl5   ->   pl
-----------------------------------------------------------------------------
```

We specified three chunks of information with the `reshape long` command. We first specified the prefix of the variables that we wanted to be reshaped long (i.e., `bp pl`). Next we specified the `i(id)` option, indicating the variable that defines the observations in the wide dataset. Finally, we specified the `j(trialnum)` option to provide the name that we wanted for the variable that identifies the multiple measurements per person in the long dataset. In this case, we called it **trialnum**, but you can call this variable anything you like. Below we see what the long version of this dataset looks like, showing the first 10 observations.

```
. describe
Contains data
  obs:            30
 vars:             5
 size:           330 (99.9% of memory free)
-------------------------------------------------------------------------------
              storage  display     value
variable name   type   format      label      variable label
-------------------------------------------------------------------------------
id              byte   %3.0f                  ID of person
trialnum        byte   %9.0g
age             byte   %3.0f                  Age of person
bp              int    %3.0f
pl              int    %3.0f
-------------------------------------------------------------------------------
Sorted by:  id  trialnum
     Note:  dataset has changed since last saved
. list in 1/10
```

	id	trialnum	age	bp	pl
1.	1	1	40	115	54
2.	1	2	40	86	87
3.	1	3	40	129	93
4.	1	4	40	105	81
5.	1	5	40	127	92
6.	2	1	30	123	92
7.	2	2	30	136	88
8.	2	3	30	107	125
9.	2	4	30	111	87
10.	2	5	30	120	58

In the long version, there is one observation per trial for each person. Each observation contains a variable with the trial number (**trialnum**), the person's age (**age**), the blood pressure measurement for that trial (**bp**), and the pulse measurement for that trial (**pl**).

If we want to reshape this back into a wide format, we can do this using the **reshape wide** command, as shown below.

```
. reshape wide
(note: j = 1 2 3 4 5)
Data                                long   ->   wide
-----------------------------------------------------------------------------
Number of obs.                        30   ->      6
Number of variables                    5   ->     12
j variable (5 values)           trialnum   ->   (dropped)
xij variables:
                                      bp   ->   bp1 bp2 ... bp5
                                      pl   ->   pl1 pl2 ... pl5
-----------------------------------------------------------------------------
```

This dataset is the same as the original wide dataset except that the variable labels have been replaced with more generic labels.

```
. describe
Contains data
  obs:             6
 vars:            12
 size:           156 (99.9% of memory free)
-------------------------------------------------------------------------------
              storage  display     value
variable name   type   format      label      variable label
-------------------------------------------------------------------------------
id              byte   %3.0f                  ID of person
bp1             int    %3.0f                  1 bp
pl1             int    %3.0f                  1 pl
bp2             int    %3.0f                  2 bp
pl2             int    %3.0f                  2 pl
bp3             int    %3.0f                  3 bp
pl3             int    %3.0f                  3 pl
bp4             int    %3.0f                  4 bp
pl4             int    %3.0f                  4 pl
bp5             int    %3.0f                  5 bp
pl5             int    %3.0f                  5 pl
age             byte   %3.0f                  Age of person
-------------------------------------------------------------------------------
Sorted by:  id
     Note:  dataset has changed since last saved
. list
```

	id	bp1	pl1	bp2	pl2	bp3	pl3	bp4	pl4	bp5	pl5	age
1.	1	115	54	86	87	129	93	105	81	127	92	40
2.	2	123	92	136	88	107	125	111	87	120	58	30
3.	3	124	105	122	97	101	128	109	57	112	68	16
4.	4	105	52	115	79	121	71	129	106	137	39	23
5.	5	116	70	128	64	112	52	125	68	111	59	18
6.	6	108	74	126	78	124	92	131	99	107	80	27

This section covered the basics of reshaping data from wide to long. For details on complications that can arise in such reshaping, see section 8.6.

8.6 Reshaping wide to long: Problems

This section illustrates some of the problems that can arise when reshaping data from wide to long. Let's start by using `cardio_wide.dta`, which was also used in the previous section.

```
. use cardio_wide
```

Let's reshape this dataset into long format but purposefully forget to specify the prefix for the pulse measurements (i.e., `pl`). As you can see, the `reshape long` command treats the variables `pl1`–`pl5` as variables that do not change over time. Just like the variable `age`, all five pulse values are repeated for each observation within a person.

```
. reshape long bp, i(id) j(trialnum)
(note: j = 1 2 3 4 5)
Data                               wide   ->   long
-----------------------------------------------------------------------
Number of obs.                        6   ->     30
Number of variables                  12   ->      9
j variable (5 values)                     ->   trialnum
xij variables:
                        bp1 bp2 ... bp5   ->   bp
-----------------------------------------------------------------------
. list in 1/10
```

	id	trialnum	age	bp	pl1	pl2	pl3	pl4	pl5
1.	1	1	40	115	54	87	93	81	92
2.	1	2	40	86	54	87	93	81	92
3.	1	3	40	129	54	87	93	81	92
4.	1	4	40	105	54	87	93	81	92
5.	1	5	40	127	54	87	93	81	92
6.	2	1	30	123	92	88	125	87	58
7.	2	2	30	136	92	88	125	87	58
8.	2	3	30	107	92	88	125	87	58
9.	2	4	30	111	92	88	125	87	58
10.	2	5	30	120	92	88	125	87	58

The remedy for this, of course, is to include the omitted variable, `pl`.

```
. use cardio_wide
. reshape long bp pl, i(id) j(trialnum)
  (output omitted)
```

Consider a variation of `cardio_wide.dta` named `cardio_wide2.dta`.

```
. use cardio_wide2
. describe
Contains data from cardio_wide2.dta
  obs:             6
 vars:            12                          31 Dec 2009 15:46
 size:           144 (99.9% of memory free)
-------------------------------------------------------------------------------
              storage  display     value
variable name   type   format      label      variable label
-------------------------------------------------------------------------------
id              byte   %3.0f                  ID of person
age             byte   %3.0f                  Age of person
t1bp            int    %3.0f                  Blood pressure systolic Trial 1
t2bp            int    %3.0f                  Blood pressure systolic Trial 2
t3bp            int    %3.0f                  Blood pressure systolic Trial 3
t4bp            int    %3.0f                  Blood pressure systolic Trial 4
t5bp            int    %3.0f                  Blood pressure systolic Trial 5
t1pl            int    %3.0f                  Pulse: Trial 1
t2pl            byte   %3.0f                  Pulse: Trial 2
t3pl            int    %3.0f                  Pulse: Trial 3
t4pl            int    %3.0f                  Pulse: Trial 4
t5pl            byte   %3.0f                  Pulse: Trial 5
-------------------------------------------------------------------------------
Sorted by:
. list *bp *pl
```

	t1bp	t2bp	t3bp	t4bp	t5bp	t1pl	t2pl	t3pl	t4pl	t5pl
1.	115	86	129	105	127	54	87	93	81	92
2.	123	136	107	111	120	92	88	125	87	58
3.	124	122	101	109	112	105	97	128	57	68
4.	105	115	121	129	137	52	79	71	106	39
5.	116	128	112	125	111	70	64	52	68	59
6.	108	126	124	131	107	74	78	92	99	80

As you can see, rather than having blood pressure measurements named `bp1`–`bp5`, they are named `t1bp` (time 1 blood pressure) to `t5bp`. The pulse variables are named using the same structure, `t1pl`–`t5pl`.

The `reshape long` command can handle variable names like this, but you must use the `@` symbol to indicate where the numbers 1–5 can be found in the variables. As you can see in the example below, specifying `t@bp` indicates that the number associated with `trialnum` is in between `t` and `bp`. The pulse measurements are specified using the same strategy, specifying `t@pl`.

(Continued on next page)

```
. reshape long t@bp t@pl, i(id) j(trialnum)
(note: j = 1 2 3 4 5)

Data                               wide   ->   long
-----------------------------------------------------------------
Number of obs.                        6   ->     30
Number of variables                  12   ->      5
j variable (5 values)                     ->   trialnum
xij variables:
                     t1bp t2bp ... t5bp   ->   tbp
                     t1pl t2pl ... t5pl   ->   tpl
-----------------------------------------------------------------
```

The resulting long variables are named `tbp` and `tpl`.

```
. describe

Contains data
  obs:            30
 vars:             5
 size:           330 (99.9% of memory free)
--------------------------------------------------------------------
              storage  display     value
variable name   type   format      label      variable label
--------------------------------------------------------------------
id              byte   %3.0f                  ID of person
trialnum        byte   %9.0g
age             byte   %3.0f                  Age of person
tbp             int    %3.0f
tpl             int    %3.0f
--------------------------------------------------------------------
Sorted by:  id  trialnum
     Note:  dataset has changed since last saved

. list in 1/10

     +-------------------------------+
     | id   trialnum   age   tbp   tpl |
     |-------------------------------|
  1. |  1          1    40   115    54 |
  2. |  1          2    40    86    87 |
  3. |  1          3    40   129    93 |
  4. |  1          4    40   105    81 |
  5. |  1          5    40   127    92 |
     |-------------------------------|
  6. |  2          1    30   123    92 |
  7. |  2          2    30   136    88 |
  8. |  2          3    30   107   125 |
  9. |  2          4    30   111    87 |
 10. |  2          5    30   120    58 |
     +-------------------------------+
```

Let's have a look at a variation on `cardio_wide.dta` named `cardio_wide3.dta`. In this dataset, a variable named `bp2005` reflects the person's blood pressure as measured by their doctor in the year 2005, and `pl2005` is their pulse rate as measured by their doctor in 2005.

```
. use cardio_wide3

. clist bp* pl*, noobs

bp1  bp2  bp3  bp4  bp5      bp2005  pl1  pl2  pl3  pl4  pl5      pl2005
115   86  129  105  127         112   54   87   93   81   92          81
123  136  107  111  120         119   92   88  125   87   58          90
124  122  101  109  112         113  105   97  128   57   68          91
105  115  121  129  137         121   52   79   71  106   39          69
116  128  112  125  111         118   70   64   52   68   59          62
108  126  124  131  107         119   74   78   92   99   80          84
```

Although the **reshape long** command is smart, it only does exactly what you say (which might not be what you mean). In this case, when you try to reshape this dataset into long format, it will treat the blood pressure and pulse readings from 2005 as though they were just another trial from this experiment (like they were the 2,005th measurement).

```
. reshape long bp pl, i(id) j(trialnum)
(note: j = 1 2 3 4 5 2005)

Data                               wide   ->   long
-----------------------------------------------------------------------------
Number of obs.                        6   ->     36
Number of variables                  14   ->      5
j variable (6 values)                     ->   trialnum
xij variables:
                      bp1 bp2 ... bp2005  ->   bp
                      pl1 pl2 ... pl2005  ->   pl
-----------------------------------------------------------------------------

. list in 1/12, sepby(id)

     +----------------------------------+
     | id   trialnum   age    bp    pl  |
     |----------------------------------|
  1. |  1          1    40   115    54  |
  2. |  1          2    40    86    87  |
  3. |  1          3    40   129    93  |
  4. |  1          4    40   105    81  |
  5. |  1          5    40   127    92  |
  6. |  1       2005    40   112    81  |
     |----------------------------------|
  7. |  2          1    30   123    92  |
  8. |  2          2    30   136    88  |
  9. |  2          3    30   107   125  |
 10. |  2          4    30   111    87  |
 11. |  2          5    30   120    58  |
 12. |  2       2005    30   119    90  |
     +----------------------------------+
```

What we intend is for `bp2005` and `pl2005` to be treated like `age`, as constant variables that do not change over the trials within a person. We could rename these variables using the **rename** command, or we can specify `j(trialnum 1 2 3 4 5)`[1] to indicate that the values for **trialnum** range from 1 to 5. The **reshape long** command then changes the shape of the variables named `bp1`–`bp5` and `pl1`–`pl5`, and treats any other variables as time constant.

1. This can also be abbreviated using the dash, `j(trialnum 1-5)`.

```
. use cardio_wide3
. reshape long bp pl, i(id) j(trialnum 1-5)
Data                               wide   ->   long
-----------------------------------------------------------------------------
Number of obs.                        6   ->      30
Number of variables                  14   ->       7
j variable (5 values)                     ->   trialnum
xij variables:
                        bp1 bp2 ... bp5   ->   bp
                        pl1 pl2 ... pl5   ->   pl
-----------------------------------------------------------------------------
```

Now the resulting long dataset is what we intended:

```
. list in 1/10, sepby(id)
```

	id	trialnum	age	bp	bp2005	pl	pl2005
1.	1	1	40	115	112	54	81
2.	1	2	40	86	112	87	81
3.	1	3	40	129	112	93	81
4.	1	4	40	105	112	81	81
5.	1	5	40	127	112	92	81
6.	2	1	30	123	119	92	90
7.	2	2	30	136	119	88	90
8.	2	3	30	107	119	125	90
9.	2	4	30	111	119	87	90
10.	2	5	30	120	119	58	90

This section covered most of the sticky situations that can arise when reshaping your data from wide to long. For more information, see `help reshape`.

Tip! No ID variable

Suppose that your wide dataset is lacking an ID variable, a variable that identifies each observation in the dataset. When you reshape a dataset from wide to long, you need to specify a variable that identifies each wide observation. If you do not have such a variable, you can easily create one that contains the observation number by typing

```
generate id = _n
```

You can then use the `i(id)` option with the `reshape long` command.

8.7 Multilevel datasets

Toward the end of section 8.2, a weight loss example was used that contained measurements of weight over time. The time between the measurements was planned to be 7 days but in reality was more erratic. Because of the unequal spacing of time between weight measurements, I recommended a long format for storing the data. However, consider this variation on `weights_long.dta`, named `weights_long2.dta`. This dataset contains additional information about each person, namely, his or her gender, age, race, and education.

```
. list, sepby(id)

     +-------------------------------------------+
     | id   female   age   race   ed   days   wt |
     |-------------------------------------------|
  1. |  1        1    22      1    9      7  166 |
  2. |  1        1    22      1    9     14  163 |
  3. |  1        1    22      1    9     21  164 |
  4. |  1        1    22      1    9     28  162 |
     |-------------------------------------------|
  5. |  2        0    43      2   13      9  188 |
  6. |  2        0    43      2   13     13  184 |
  7. |  2        0    43      2   13     22  185 |
  8. |  2        0    43      2   13     27  182 |
     |-------------------------------------------|
  9. |  3        0    63      3   11      6  158 |
 10. |  3        0    63      3   11     12  155 |
 11. |  3        0    63      3   11     31  157 |
     |-------------------------------------------|
 12. |  4        1    26      2   15      8  192 |
 13. |  4        1    26      2   15     17  190 |
 14. |  4        1    26      2   15     22  191 |
 15. |  4        1    26      2   15     30  193 |
     |-------------------------------------------|
 16. |  5        1    29      1   12      5  145 |
 17. |  5        1    29      1   12     11  142 |
 18. |  5        1    29      1   12     20  140 |
 19. |  5        1    29      1   12     26  137 |
     +-------------------------------------------+
```

If you were entering the data for this type of dataset, you would notice that the information about the person needs to be entered multiple times. This not only adds more work but also increases the chances for data-entry mistakes. For example, the same person may be entered as a female on one observation (day) and as a male on another observation.

Using the nomenclature from multilevel modeling, this dataset contains information at two different levels, a person level (level 2) and times within the person (level 1). The variables `female`, `age`, `race`, and `ed` are all person-level (level 2) variables, while the variables `days` and `wt` are time-level (level 1) variables.

In such a case, the data entry and data verification can be much simpler by creating two datasets, a level-2 (person) dataset and a level-1 (time) dataset. These two datasets can then be merged together with a `1:m` merge to create the combined multilevel dataset. Let's see how this works.

First, we can see the person-level (level-2) data stored in `weights_level2.dta`.

```
. use weights_level2
. list

     +--------------------------------+
     | id   female   age   race   ed |
     |--------------------------------|
  1. |  1        1    22      1    9 |
  2. |  2        0    43      2   13 |
  3. |  3        0    63      3   11 |
  4. |  4        1    26      2   15 |
  5. |  5        1    29      1   12 |
     +--------------------------------+
```

The level-1 data (namely, `days` and `wt`) are stored in `weights_level1.dta`.

```
. use weights_level1
. list, sepby(id)

     +------------------+
     | id   days    wt |
     |------------------|
  1. |  1      7   166 |
  2. |  1     14   163 |
  3. |  1     21   164 |
  4. |  1     28   162 |
     |------------------|
  5. |  2      9   188 |
  6. |  2     13   184 |
  7. |  2     22   185 |
  8. |  2     27   182 |
     |------------------|
  9. |  3      6   158 |
 10. |  3     12   155 |
 11. |  3     31   157 |
     |------------------|
 12. |  4      8   192 |
 13. |  4     17   190 |
 14. |  4     22   191 |
 15. |  4     30   193 |
     |------------------|
 16. |  5      5   145 |
 17. |  5     11   142 |
 18. |  5     20   140 |
 19. |  5     26   137 |
     +------------------+
```

We can now perform a `1:m` merge to merge these two datasets (see section 6.5 for more information about `1:m` merging).

```
. use weights_level2
. merge 1:m id using weights_level1
    Result                           # of obs.
    -----------------------------------------
    not matched                             0
    matched                                19  (_merge==3)
    -----------------------------------------
```

Now we have one multilevel dataset that combines the person-level and time-level information together, as shown below.

```
. sort id days
. list, sepby(id)
```

	id	female	age	race	ed	days	wt	_merge
1.	1	1	22	1	9	7	166	matched (3)
2.	1	1	22	1	9	14	163	matched (3)
3.	1	1	22	1	9	21	164	matched (3)
4.	1	1	22	1	9	28	162	matched (3)
5.	2	0	43	2	13	9	188	matched (3)
6.	2	0	43	2	13	13	184	matched (3)
7.	2	0	43	2	13	22	185	matched (3)
8.	2	0	43	2	13	27	182	matched (3)
9.	3	0	63	3	11	6	158	matched (3)
10.	3	0	63	3	11	12	155	matched (3)
11.	3	0	63	3	11	31	157	matched (3)
12.	4	1	26	2	15	8	192	matched (3)
13.	4	1	26	2	15	17	190	matched (3)
14.	4	1	26	2	15	22	191	matched (3)
15.	4	1	26	2	15	30	193	matched (3)
16.	5	1	29	1	12	5	145	matched (3)
17.	5	1	29	1	12	11	142	matched (3)
18.	5	1	29	1	12	20	140	matched (3)
19.	5	1	29	1	12	26	137	matched (3)

This dataset can now be used with commands like `xtmixed`, allowing us to study the impact of level-1 variables (such as `days`) and level-2 variables (such as `age`).

```
. xtmixed wt days age || id:
  (output omitted)
```

Although the example illustrated here used time as a level-1 variable and person as a level-2 variable, the same principles would apply to any two-level structure, such as students (level 1) nested within schools (level 2).

What if you have three levels of data, say, students (level 1) nested within schools (level 2) nested within school districts (level 3)? In such a case, I would recommend creating three datasets: `districts.dta` (level 3), `schools.dta` (level 2), and `students.dta` (level 1). The districts and schools would be linked by the district ID, and the schools and students would be linked by the school ID. These hypothetical datasets could then be merged as shown below.

```
. use districts
. merge 1:m districtid using schools, generate(merge1)
. merge 1:m schoolid using students, generate(merge2)
```

The multilevel data structure relies heavily on the use of `1:m` dataset merges. You can see section 6.5 for more information about `1:m` merges.

8.8 Collapsing datasets

This final section of this chapter discusses how you can collapse datasets. For example, consider `cardio_long.dta`, which we have seen before in this chapter. This dataset contains multiple observations per person with the measurements of their blood pressure and pulse across several trials.

```
. use cardio_long

. list, sepby(id)

     +-------------------------------+
     | id   trial   age    bp     pl |
     |-------------------------------|
  1. |  1       1    40   115     54 |
  2. |  1       2    40    86     87 |
  3. |  1       3    40   129     93 |
  4. |  1       4    40   105     81 |
  5. |  1       5    40   127     92 |
     |-------------------------------|
  6. |  2       1    30   123     92 |
  7. |  2       2    30   136     88 |
  8. |  2       3    30   107    125 |
  9. |  2       4    30   111     87 |
 10. |  2       5    30   120     58 |
     |-------------------------------|
 11. |  3       1    16   124    105 |
 12. |  3       2    16   122     97 |
 13. |  3       3    16   101    128 |
 14. |  3       4    16   109     57 |
 15. |  3       5    16   112     68 |
     |-------------------------------|
 16. |  4       1    23   105     52 |
 17. |  4       2    23   115     79 |
 18. |  4       3    23   121     71 |
 19. |  4       4    23   129    106 |
 20. |  4       5    23   137     39 |
     |-------------------------------|
 21. |  5       1    18   116     70 |
 22. |  5       2    18   128     64 |
 23. |  5       3    18   112     52 |
 24. |  5       4    18   125     68 |
 25. |  5       5    18   111     59 |
     |-------------------------------|
 26. |  6       1    27   108     74 |
 27. |  6       2    27   126     78 |
 28. |  6       3    27   124     92 |
 29. |  6       4    27   131     99 |
 30. |  6       5    27   107     80 |
     +-------------------------------+
```

Suppose that we wanted to collapse this dataset across observations within a person, creating a mean blood pressure score and mean pulse score across the observations for each person. We could do that with the `collapse` command, as shown below.

```
. collapse bp pl, by(id)
. list
```

	id	bp	pl
1.	1	112.40	81.40
2.	2	119.40	90.00
3.	3	113.60	91.00
4.	4	121.40	69.40
5.	5	118.40	62.60
6.	6	119.20	84.60

The dataset in memory has been replaced with this new dataset, where `bp` now contains the mean blood pressure across the observations within the person and `pl` has the mean pulse averaged across the observations within the person.

Unless we specify otherwise, the `collapse` command computes the mean of the specified variables. But the `collapse` command is not limited to just computing means. We can manually specify other statistics that we want to compute. Below we indicate that we want to compute the minimum and maximum values of `bp` and `pl`, specified by the keywords `min` and `max` surrounded by parentheses. Note how on the right side of the equal sign is the original variable name and on the left side of the equal sign is the new variable name.

```
. use cardio_long
. collapse (min) bpmin=bp plmin=pl (max) bpmax=bp plmax=pl, by(id)
. list
```

	id	bpmin	plmin	bpmax	plmax
1.	1	86	54	129	93
2.	2	107	58	136	125
3.	3	101	57	124	128
4.	4	105	39	137	106
5.	5	111	52	128	70
6.	6	107	74	131	99

Here is one more example, showing the computation of the mean and standard deviation of the blood pressure and pulse scores.

(*Continued on next page*)

```
. use cardio_long
. collapse (mean) bpmean=bp plmean=pl (sd) bpsd=bp plsd=pl, by(id)
. list
```

	id	bpmean	plmean	bpsd	plsd
1.	1	112	81	18	16
2.	2	119	90	11	24
3.	3	114	91	10	29
4.	4	121	69	12	26
5.	5	118	63	8	7
6.	6	119	85	11	10

The `collapse` command supports many other statistics beyond `mean`, `sd`, `min`, and `max`. It also can compute the median (`median`), sums (`sum`), number of nonmissing observations (`count`), and many others. See `help collapse` for a comprehensive list and for more details.

9 Programming for data management

For a long time it puzzled me how something so expensive, so leading edge, could be so useless. And then it occurred to me that a computer is a stupid machine with the ability to do incredibly smart things, while computer programmers are smart people with the ability to do incredibly stupid things. They are, in short, a perfect match.

—Bill Bryson

9.1 Introduction

The word programming can be a loaded word. I use it here to describe the creation of a series of commands that can be easily repeated to perform a given task. As such, this chapter is about how to create a series of Stata commands that be easily repeated to perform data-management and data analysis tasks. But you might say that you already know how to use Stata for your data management and data analysis. Why spend time learning about programming? My colleague at UCLA, Phil Ender, had a wise saying that I loved: "There is the short road that is long and the long road that is short." Investing time in learning and applying programming strategies may seem like it will cost you extra time, but at the end of your research project, you will find that it is part of the "long road that is short".

Sections 9.2–9.5 focus on long-term strategies to save you time and effort in your research project. Section 9.2 describes long-term goals for data management and the benefits of investing a little bit of time in long-term planning even in the midst of short-term deadlines. Then section 9.3 discusses how you can combine Stata commands together into do-files that allow you to easily reproduce data-management and data analysis tasks. This section also illustrates how to save the contents of the Results window into a log file, saving a record of your Stata commands and output. Section 9.4 illustrates how you can automate the process of data checking by using the `assert` command. Such automation can reduce the need to scrutinize tables for out-of-range or impossible data values. The concepts from the previous two sections are then brought together into section 9.5, which shows how you can combine do-files into one master do-file that automates your data management, data checking, and data analysis.

Section 9.6 illustrates one of the most fundamental concepts in Stata programming, the concept of a Stata macro, which is expanded upon in section 9.7. The real power of Stata macros comes out when they are combined with other commands as, for example, in section 9.8, which shows how you can perform repetitive tasks over a series of variables by combining the `foreach` command with a Stata macro to loop over variables. This discussion continues in section 9.9, which illustrates how you can use a Stata macro to loop over numbers, allowing you to perform repetitive tasks where only the numeric part of a variable name changes. Then section 9.10 shows how the `foreach` command can loop over any arbitrary list, extending the range of tasks that you can automate with looping.

Section 9.11 illustrates how many Stata commands create "saved results", which are variables that reflect values displayed in output of the command. Section 9.12 illustrates how these saved results can be captured and stored as a Stata dataset.

Many of the elements of this chapter are brought together in section 9.13, which introduces how to write Stata programs. Such programs combine the concept of do-files with Stata macros to allow you to create and execute a series of commands that behave differently based on the information (variables) that you pass into the program.

9.2 Tips on long-term goals in data management

There was an episode of Seinfeld where Jerry stayed up late partying and, when asked about the consequences for himself the next day, he said, "Let tomorrow guy deal with it." The joke, of course, is that Jerry is showing denial of the obvious fact that he will become "tomorrow guy". This same logic often applies in research projects. Such projects contain a mixture of short-term goals and long-term goals that are at odds with each other. The short-term goal of finishing an analysis quickly can be at odds with the long-term goal of having well-documented and reproducible results six months in the future when it will be time to respond to reviewer comments. Dealing with the reviewer comments is the problem of "tomorrow guy" or "tomorrow gal". This section contains tips to help you focus on some of the long-term goals of data management. The nuts and bolts of implementing many of these tips are covered in the following sections of this chapter.

Reviewer comments

Say that you are analyzing data for a research publication. Although the short-term goal is to produce results that go into a manuscript for publication, the real goal is get that manuscript published. That means responding to comments from reviewers, which might necessitate further data management and data analysis. By using do-files and log files (as described in section 9.3), you will have a record of all the steps you performed for managing and analyzing your data in do-files and a record of the results in log files. Further, by combining all your individual do-files into one master do-file (as described in section 9.5), your entire data-management and data analysis process is completely clear and completely automated. That one master do-file permits you to easily reproduce your original results and make changes in response to reviewer comments.

Check your do-files

Speaking of manuscripts, think of all the time and effort we spend to review and check them for accuracy and to eliminate errors. We reread them, we spell-check them, and we ask friends to review them. Just as we engage in various ways to check and double-check our manuscripts, I recommend doing the same for your do-files. Even after you have read over and have double-checked your own do-file, I would recommend showing your do-files to a friend or colleague and asking him or her to look them over just as you might ask for a friendly review of a manuscript. Your friend could point out potential mistakes and review your do-file for clarity. If a stranger can understand your do-file, then you can feel more confident that you will understand it when you revisit it in the future.

Documentation

It is not sufficient to write accurate do-files if they are not understandable and clear to you in the future. To help with this, I recommend creating documentation for your project. As we work on a project, I think we all have the feeling that we will remember what the files are named, why we chose to create a variable one way instead of another, and so forth. A good research project will be returned to in the future, to respond to reviewer comments or for creating additional future publications. The usefulness of your data in the future may hinge on whether you documented key information that you may forget over time. Certainly, labeling your data (as described in chapter 4) is a start, but labeling can be supplemented with comments in your do-files, a journal, flow diagrams, and so on. The type of system that you choose for documenting your do-files and your project is much less important than picking something that works for you and sticking with it.

Writing do-files that work in future versions

Stata is a dynamic program, and each release not only adds substantial new features but also often includes some changes to the way some commands work. To ensure that your do-files work the same, regardless of how Stata changes in the future, you should include the `version` command at the top of each do-file (as illustrated in section 9.3). For example, including `version 11.0` at the top of your do-file means that all the commands contained in that do-file will be executed according to the syntax and rules that were in force in Stata 11.0, even if you are using a future version of Stata.

Separating intermediate work from final work

As you work on a project, you probably save alternate versions of your do-files and datasets. Maybe you decide to make substantial changes to a do-file and save it as a new file in case the substantial changes are undesirable. Likewise, you might do the same thing with respect to datasets, saving a new version after a substantial change. In the short term, this strategy is beneficial by allowing you to revert back to a previous version of a file in case your work makes things worse instead of better. At some point, you outgrow these intermediate versions and their presence becomes more of a nuisance than a help, especially if at a later time you become confused about which file represents the most up-to-date version. Rather than deleting these intermediate files, you can simply move them into a separate folder that you create and designate for such files.[1] Whatever system that you choose, periodically take steps to prune dead versions of files so that in the future, you do not accidentally confuse an old version of a file with the most up-to-date version.

1. Or you can use *version control* software that tracks different versions of your files.

Develop routines

When I was a teenager, I frequently lost my keys at home. When I would get home, I would lay them down in the first open space that I found. When it was time to leave, I had to think of where I had put down my keys. I finally learned to place my keys in the same place every time I got home. Just like I finally developed a routine for my keys, so too should we all develop routines for our data management. Although each project is different, projects often contain many of the same features. When you find yourself doing the same task as you have done in prior projects, you might see if you want to adopt the same routine as you did before. Developing such routines means that you not only know how to approach a problem in the future but also can predict how you approached it from past projects.

Back up your data

Okay, everyone tells you this: back up your data! Without backups, your do-files and data can be completely lost because of hard drive failure, theft, fire, earthquake, or gremlins. There are many ways you can back up your data: via external hard drives, CD/DVDs, Internet storage, and even automated online backup services. When you think of the hundreds or thousands of hours that can be lost, taking a little bit of time to devise and adhere to a back-up strategy only makes good sense.

Tip! Remote storage of backups

Suppose that you routinely back up your files, but those backups are stored near your computer (as a CD, DVD, or external hard drive). Your original data, along with the backups, could be lost to theft, fire, or earthquake. Storing backups online or taking backup copies on CD/DVD to a different location helps protect you against such threats.

Protecting your participants

So far, all the considerations discussed focus on preserving and protecting your data, but we have not addressed an even more important concern, protecting the participants in your study. Ask yourself if your data contain confidential information. If so, you should investigate the obligations that you have regarding securing your data. This is especially true if your data contain information like social security numbers, phone numbers, or any other information that easily links the information directly to the person. Consider whether you need to take steps to de-identify your datasets (by removing variables that could possibly identify participants) or encrypt your datasets so that only those who have the password to the data can decrypt them. Having a data breach not only would harm your participants but also could have serious consequences for the future progress of your study.

Summary

When working on a research project, it is so easy to get caught up in short-term goals that you can easily forget the dangers of ignoring the ultimate long-term goals. Focusing too much on the quick attainment of such short-term goals could lead to analyses that cannot be reproduced (because of the lack of documented do-files), incorrect results that need to be retracted (because a do-file was not checked and had errors in it), complete loss of data because of a hard-disk crash (because a backup was not performed), and so forth. The recommendations presented in this section are like insurance for your research project, trying to protect your research project against these various dangers. Like any insurance policy, there is a small cost involved, but consider that cost against the hundreds of hours you invest in your project. Sometimes the effort invested in such insurance seems wasted until we need to break out that insurance policy, and then all that effort is well rewarded.

9.3 Executing do-files and making log files

Throughout this book, commands have been discussed as though you were typing them at the Stata Command window, issuing them one at a time. However, there are problems with using this method as the only means for your data management and data analysis. If you issue all the commands in the Command window, there is no record of what you did. You will not remember how you recoded a particular variable or how you merged certain datasets together. Or imagine that you found out that the original raw dataset was corrupt. You would need to manually repeat all the commands to read in the data, label the data, create the variables, and so forth. Please trust me when I tell you that this sort of thing happens all the time.

Instead, you can save your commands that read the data, label the data, create variables, and such, into one file called a do-file. When you execute the do-file, all the commands within it are quickly and easily executed. This section shows how you can create do-files and how you can create log files, which save the results of your do-files, providing you with a saved record of your data management and data analysis. Below is an example of a small do-file named `example1.do`.

```
. type example1.do
use wws2, clear
summarize age wage hours
tabulate married
```

We can execute the commands contained in this do-file by typing `do example1`. By doing this, `wws2.dta` is used and the `summarize` and `tabulate` commands are issued, and then the program is complete. By using do-files, you can easily run the same sequence of commands by just typing one command, in this case `do example1`.

```
. do example1

. use wws2, clear
(Working Women Survey w/fixes)
```

```
. summarize age wage hours

    Variable |       Obs        Mean    Std. Dev.       Min        Max
-------------+--------------------------------------------------------
         age |      2246    36.22707    5.337859         21         48
        wage |      2244    7.796781     5.82459          0   40.74659
       hours |      2242    37.21811    10.50914          1         80

. tabulate married

    married |      Freq.     Percent        Cum.
------------+-----------------------------------
          0 |        804       35.80       35.80
          1 |      1,442       64.20      100.00
------------+-----------------------------------
      Total |      2,246      100.00

.
end of do-file
```

You can create your own do-files by using the `doedit` command, which opens the Stata Do-file Editor. It works just like any text editor that you may have used (e.g., Notepad), but the Do-file Editor includes syntax highlighting, bookmarking, and the ability to run Stata commands from within the Do-file Editor. You can try it for yourself by typing `doedit` and entering the commands as shown in `example1.do`. You can then save the file as, say, `myexample.do`. And then you could type `do myexample` and Stata would execute it. Alternatively, you could click on one of the **Execute** icons to execute the contents of the commands in the Do-file Editor. For more information about using the Do-file Editor, see `help doedit` and be sure to see the link to the tutorial for the Do-file Editor in the *Getting Started with Stata* manual.

Returning to `example1.do`, there is one thing that bothers me about this do-file: it does not save the results in any permanent way. Once we close Stata, the contents of the Results window vanish. Consider an improved version, `example2.do`, that saves the results in a Stata log file.

```
. type example2.do
log using example2
use wws2, clear
summarize age wage hours
tabulate married
log close
```

When `example2.do` is executed, a log file is opened and the results are saved in `example2.smcl`.[2] After the main commands are executed, the `log close` command closes the log file, and we can then view the contents of the log. Let's execute `example2.do` below.

```
. do example2
  (output omitted)
```

2. `smcl` stands for Stata Markup and Control Language. It is a markup language, like HTML, that can control the display of the output, e.g., making the output bold, italic, underlined. SMCL also allows links.

Let's use the `type` command to view the results stored in `example2.smcl`.[3] This shows us the results formatted using SMCL codes to make the output display nicely.

```
. type example2.smcl

      name:  <unnamed>
       log:  C:\data\example2.smcl
  log type:  smcl
 opened on:  16 Nov 2009, 15:00:42
. use wws2, clear
(Working Women Survey w/fixes)
. summarize age wage hours
    Variable |       Obs        Mean    Std. Dev.       Min        Max
-------------+--------------------------------------------------------
         age |      2246    36.22707    5.337859         21         48
        wage |      2244    7.796781     5.82459          0   40.74659
       hours |      2242    37.21811    10.50914          1         80
. tabulate married
   Is woman |
  currently |
   married? |      Freq.     Percent        Cum.
------------+-----------------------------------
          0 |        804       35.80       35.80
          1 |      1,442       64.20      100.00
------------+-----------------------------------
      Total |      2,246      100.00
. log close
      name:  <unnamed>
       log:  C:\data\example2.smcl
  log type:  smcl
 closed on:  16 Nov 2009, 15:00:42
```

However, if you view this file using another editor program or even using the Stata Do-file Editor, you will see all the SMCL code in the file. Let's look at this file with the SMCL codes by using the `type` command with the `asis` option.

3. You can also use the Stata Viewer window to view such files by typing, for example, `view example2.smcl`.

```
. type example2.smcl, asis
{smcl}
{com}{sf}{ul off}{txt}{.-}
      name:  {res}<unnamed>
       {txt}log:  {res}C:\data\example2.smcl
  {txt}log type:  {res}smcl
 {txt}opened on:  {res}16 Nov 2009, 15:00:42
{txt}
{com}. use wws2, clear
(Working Women Survey w/fixes)
{txt}
{com}. summarize age wage hours
{txt}    Variable {c |}       Obs        Mean    Std. Dev.       Min        Max
{hline 13}{c +}{hline 56}
{space 9}age {c |}{res}      2246    36.25111    5.437983         21         83
{txt}{space 8}wage {c |}{res}      2244    7.796781     5.82459          0
> 40.74659
{txt}{space 7}hours {c |}{res}      2242    37.21811    10.50914          1
>      80
{txt}
{com}. tabulate married
   {txt}Is woman {c |}
  currently {c |}
   married? {c |}      Freq.     Percent        Cum.
{hline 12}{c +}{hline 35}
          0 {c |}{res}        804       35.80       35.80
{txt}          1 {c |}{res}      1,442       64.20      100.00
{txt}{hline 12}{c +}{hline 35}
      Total {c |}{res}      2,246      100.00
{txt}
{com}. log close
      {txt}name:  {res}<unnamed>
       {txt}log:  {res}C:\data\example2.smcl
  {txt}log type:  {res}smcl
 {txt}closed on:  {res}16 Nov 2009, 15:00:42
{txt}{.-}
{smcl}
{txt}{sf}{ul off}
```

You might want to create a log file that excludes the SMCL language that contains just plain text. As shown in **example3.do**, below, the **translate** command is used to convert **example3.smcl** into **example3.log**. Specifying the **.log** extension tells the **translate** command that we want **example3.log** to be a plain-text file.

```
. type example3.do
log using example3
use wws2, clear
summarize age wage hours
tabulate married
log close
translate example3.smcl example3.log
```

(*Continued on next page*)

Let's now execute this do-file and inspect `example3.log`.

```
. do example3
  (output omitted)
. type example3.log
--------------------------------------------------------------------------------
      name:  <unnamed>
       log:  C:\data\example3.log
  log type:  text
 opened on:  16 Nov 2009, 15:00:42
. use wws2, clear
(Working Women Survey w/fixes)
. summarize age wage hours
    Variable |       Obs        Mean    Std. Dev.       Min        Max
-------------+--------------------------------------------------------
         age |      2246    36.22707    5.337859         21         48
        wage |      2244    7.796781     5.82459          0   40.74659
       hours |      2242    37.21811    10.50914          1         80
. tabulate married
   Is woman |
  currently |
   married? |      Freq.     Percent        Cum.
------------+-----------------------------------
          0 |        804       35.80       35.80
          1 |      1,442       64.20      100.00
------------+-----------------------------------
      Total |      2,246      100.00
. log close
      name:  <unnamed>
       log:  C:\data\example3.log
  log type:  text
 closed on:  16 Nov 2009, 15:00:42
--------------------------------------------------------------------------------
```

Although this output is not as aesthetically pleasing as `example3.smcl`, it will be much easier to paste such results into a word processing program. However, the results will look better if you use a fixed-width font (like Courier). The columns will be out of alignment if you use a proportional font, like Arial or Times Roman.

Note! Directly making plain-text logs with log using

By default, the `log using` command creates a `.smcl`-style log file. As `example3.do` showed, the `translate` command can be used to convert a `.smcl` file to a plain-text log file. We can, instead, directly create such a plain-text log file by typing `log using example3.log`. The `.log` extension automatically results in a plain-text log. An alternative to specifying the `.log` extension would be to use the `text` option of the `log` command, as in `log using example3, text`. Because both of these methods automatically create a plain-text log file, you no longer need the `translate` command.

Say that we decide to execute the file `example3.do` again. I ran this again on my computer and received this error:

```
. do example3

. log using example3
file C:\DATA\example3.smcl already exists
r(602);
```

This error is similar to the kind of error you would get if you tried to use the `save` command to overwrite an existing Stata `.dta` file. Then you need to add the `replace` option to let Stata know it is okay to overwrite the existing file. Likewise, we need to add the `replace` option to the `log using` and `translate` commands if we want to run these commands and overwrite the existing log files. In `example4.do` (below), the `replace` option is added to the `log using` command so that Stata knows it is okay to overwrite `example4.smcl`, and the `replace` option is added to the `translate` command to indicate that it is okay to overwrite `example4.log`.

```
. type example4.do
log using example4, replace
use wws2, clear
summarize age wage hours
tabulate married
log close
translate example4.smcl example4.log, replace
```

Now we can execute `example4.do`; each time it runs, it will overwrite `example4.smcl` and `example4.log` (if they exist).

```
. do example4
  (output omitted)
```

This would seem to be the perfect prototype for a do-file, but there is one problem. Suppose that there is an error inside the do-file. Stata will quit processing the do-file and the log file will remain open (because the program never reached the `log close` command to close the log file). If we try to run the do-file again (without closing the log file) the `log using` command will give an error like this:

```
. do example4.do

. log using example4
log file already open
r(604);
```

To address this, we add the command `capture log close` at the top of `example5.do`. This closes the log, whether it was open or not.

```
. type example5.do
capture log close
log using example5, replace
use wws2, clear
summarize age wage hours
tabulate married
log close
translate example5.smcl example5.log, replace
```

We can execute `example5.do` over and over, and it will properly close the log file if needed and overwrite any existing log files.

Below we see an example skeleton do-file that builds upon `example5.do`. Comments are included to describe the commands, but you do not need to include them. (You can see section A.4 for more about the different ways to add comments to Stata do-files.) Following the do-file is an explanation of the newly added commands.

```
capture log close          // Close the log in case it was open
log using myfile, replace // Open the log (change myfile to the name for your log)
version 11.0               // Set the version of Stata you are using
set more off               // Run program without showing -more-
clear all                  // Remove any prior data from memory
set memory 200m            // Set size of memory as needed
* your commands here
log close                  // close the log when you are done
                           // Optional: convert myfile.smcl to myfile.log
translate myfile.smcl myfile.log, replace
```

As we have seen before, the do-file begins with commands to close the log and establish a log file.

The `version` command tells Stata to run the program based on the rules that applied for the version of Stata that you specify. Even years in the future, when you run this program using something like Stata version 14, including the `version 11.0` command would request that the do-file be executed using the syntax that applied back in Stata version 11.0.

The `set more off` command suppresses the —`more`— message that prompts you to press a key after a screen full of output. By inserting this command, the program will run to the end without presenting you with the —`more`— prompt.

The `clear all` command clears data (and many other things) to give you a fresh, clean slate for executing your do-file.

The `set memory` command is used to allocate enough memory for you to be able to read your dataset. If your dataset is larger than the amount of memory allocated, then Stata will give you an `r(901)` error saying that there is no more room to add more observations. By issuing the `set memory` command inside of your do-file, you can be sure that enough memory is allocated for your dataset.

After that, you can insert your commands and then conclude the do-file with the `log close` command that closes the log. If you wish, you could add the `translate` command to convert `myfile.smcl` to `myfile.log`.

As you develop your own style of writing do-files, you can create your own skeleton do-file that includes the commands you want to use at the start of a do-file.

For more information about do-files, see `help do`, and for more information about the Do-file Editor, see `help doedit`. The next section will build upon this section by illustrating how you can automate the process of data checking using do-files.

9.4 Automating data checking

An important part of data management is data checking. Throughout the book, I have tried to emphasize not only how to perform a task but also how to visually check that it was performed correctly. This section extends such checking one step further by showing how you can automate the process of data checking within your do-files. Let's illustrate this process by creating a small do-file that reads the raw dataset `wws.csv` and checks some of the variables in that dataset.

```
. insheet using wws.csv
(30 vars, 2246 obs)
```

Consider the variable `race`. This variable should have values of 1, 2, or 3.

```
. tabulate race

       race |      Freq.     Percent        Cum.
------------+-----------------------------------
          1 |      1,636       72.84       72.84
          2 |        583       25.96       98.80
          3 |         26        1.16       99.96
          4 |          1        0.04      100.00
------------+-----------------------------------
      Total |      2,246      100.00
```

The `tabulate` command shows us that one woman has a value of 4, which is not a valid value for `race`. If this command were embedded inside of a do-file, it is possible that we might overlook this error. Checking data in this way means that we need to carefully read the results in the log file to detect problems.

Stata has another way of checking for such problems: using the `assert` command. After the `assert` command, you provide a logical expression that should always be true of your data. For example, the `inlist(race,1,2,3)` function will be true if the variable `race` is 1, 2, or 3 and will be false otherwise. If the expression is ever false, the `assert` command tells us the number of observations for which the assertion was false and the command returns an error. Let's use the `assert` command to determine if `race` is coded correctly (i.e., that `race` is 1, 2, or 3). (See section A.6 for more on logical expressions in Stata.)

```
. assert inlist(race,1,2,3)
1 contradiction in 2246 observations
assertion is false
r(9);
```

Now consider how this would work in the context of a do-file. Say that we had a do-file called `wwscheck1.do` with the following contents:[4]

```
. type wwscheck1.do
* first attempt at wwscheck1.do
insheet using wws.csv, clear
assert inlist(race,1,2,3)
```

4. For simplicity, the example do-files in this chapter omit commands that create a log, set memory, and such. In practice, you would want to include these omitted commands as part of your do-files.

When we type do wwscheck1, we get the following results:

```
. do wwscheck1
. insheet using wws.csv, clear
(30 vars, 2246 obs)
. assert inlist(race,1,2,3)
1 contradiction in 2246 observations
assertion is false
r(9);
```

The assert command stops the do-file and clobbers us over the head telling us that race takes on values aside from 1, 2, or 3. Although we might have overlooked the invalid value of race from the tabulate command, it is difficult to ignore errors identified by the assert command. The assert command tells us that there is one observation that failed the assertion, but we do not know which particular observation failed the assertion. The list command below shows us that observation by listing the observations where race is not 1, 2, or 3, i.e., observations where ! inlist(race,1,2,3).

```
. list idcode race age married if ! inlist(race,1,2,3)
```

	idcode	race	age	married
2013.	543	4	39	0

This woman has an idcode of 543 and happens to be 39 years old and not married. Pretend that I went back to the survey she filled out and saw that the value for race was 1, so a data-entry error was made in wws.csv. You might think that we should fix this by changing the raw data, but then that would conceal the correction we have made. I suggest making the correction right inside of our do-file as shown below.

```
. type wwscheck1.do
* second attempt at wwscheck1.do
insheet using wws.csv, clear
replace race = 1 if idcode == 543
assert inlist(race,1,2,3)
```

When we execute this do-file (see below), it completes without error. Because the file completed, we can assume that our assertions were met and that the data have passed all the checks we created for them.

```
. do wwscheck1
. insheet using wws.csv, clear
(30 vars, 2246 obs)
. replace race = 1 if idcode == 543
(1 real change made)
. assert inlist(race,1,2,3)
end of do-file
```

Let's extend this do-file to check the variable age as well. Based on knowledge of this sample, the ages should range from 21 to no more than 50 years. Let's add this assertion to the wwscheck1.do file, placing it after the race assertion.

```
. type wwscheck1.do
* third attempt at wwscheck1.do
insheet using wws.csv, clear

* correct race and then check race
replace race = 1 if idcode == 543
assert inlist(race,1,2,3)

* check age
assert (age >= 21) & (age <= 50)
```

When we run this do-file, the assertion with respect to `race` passes, but the assertion with respect to `age` fails.

```
. do wwscheck1.do

. * third attempt at wwscheck1.do
. insheet using wws.csv, clear
(30 vars, 2246 obs)

. * correct race and then check race
. replace race = 1 if idcode == 543
(1 real change made)

. assert inlist(race,1,2,3)

. * check age
. assert (age >= 21) & (age <= 50)
2 contradictions in 2246 observations
assertion is false
r(9);

end of do-file
r(9);
```

Below we list the observations where the assertion for `age` was not true.

```
. list idcode age if ! ((age >= 21) & (age <= 50))
```

	idcode	age
2205.	80	54
2219.	51	83

I looked back at the survey, and it turns out that the woman with the `idcode` of 51 should have been 38 and the woman with the `idcode` of 80 should have been 45. We can make these corrections and repeat the process of checking `race` and `age`.

```
. type wwscheck1.do

* fourth attempt at wwscheck1.do
insheet using wws.csv, clear

* correct race and then check race
replace race = 1 if idcode == 543
assert inlist(race,1,2,3)
```

```
* correct age and then check age
replace age = 38 if idcode == 51
replace age = 45 if idcode == 80
assert (age >= 21) & (age <= 50)
```

When we run this updated do-file, it runs until completion (as shown below), indicating that both `race` and `age` meet the assertions we specified.

```
. do wwscheck1

. * fourth attempt at wwscheck1.do
. insheet using wws.csv, clear
(30 vars, 2246 obs)
.
. * correct race and then check race
. replace race = 1 if idcode == 543
(1 real change made)
. assert inlist(race,1,2,3)
.
. * correct age and then check age
. replace age = 38 if idcode == 51
(1 real change made)
. replace age = 45 if idcode == 80
(1 real change made)
. assert (age >= 21) & (age <= 50)
.
end of do-file
```

As you can see, this process combines data checking (a process we must perform) with an automated means of repeating the checking via the `assert` commands. The benefit of this process is that the checking becomes automated. Automation is especially useful if you anticipate receiving updated versions of the raw dataset. Rather than having to repeat the process of manually inspecting all the variables for out-of-range values, you can just rerun the do-file with all the `assert` commands. Any observations that do not meet your assertions will cause the do-file to halt, clearly alerting you to problems in your data.

For more information about the `assert` command, see `help assert`. If you would like even more extensive tools than `assert` for checking and validating your data, I would recommend investigating the `ckvar` package. This suite of user-written programs can be located and downloaded with `findit ckvar`.

The next section describes how you can combine do-files to further automate your data management and data analysis.

9.5 Combining do-files

The previous sections have illustrated how you can use do-files to combine commands in a way that is easily repeated, how to save the results in log files, and how to automate data checking with the `assert` command. This section illustrates how you can combine do-files into a master do-file that performs your entire data management and data analysis all from one simple command.

Let's consider a miniature data analysis project. The raw data for this project are stored in wws.csv. The raw data need to be read, checked, corrected, and labeled, and then the analysis needs to be performed. This entire process is performed by issuing one command, do mastermini. This do-file is shown below.

```
. type mastermini.do
do mkwwsmini
do anwwsmini
```

You can see that this do-file first executes mkwwsmini.do, which reads in the raw dataset, checks the data, and saves wwsmini.dta. The mk is short for make, so this program makes wwsmini.dta. Then anwwsmini.do is run. The an stands for analyze, so this program analyzes wwsmini.dta.

The process of going from raw data to final analysis is illustrated in figure 9.1. In this figure, datasets are represented by ovals, do-files are represented by rectangles, and log files are shown in italics. This flow diagram illustrates how we get from the original source (the raw dataset) to the final destination (the final analysis).

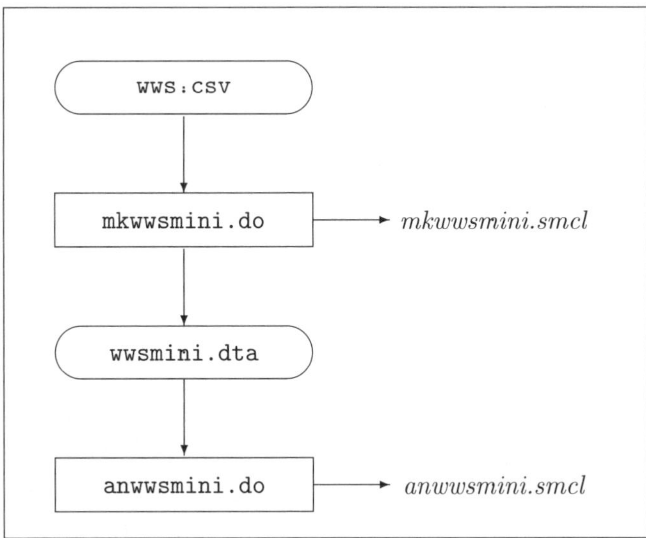

Figure 9.1. Flow diagram for wwsmini project

In this example, the source is the raw dataset wws.csv and that data file is processed by mkwwsmini.do, which produces a Stata dataset called wwsmini.dta. It also produces the log file mkwwsmini.smcl, which we could inspect to verify that the process of creating wwsmini.dta was sound and correct. Then wwsmini.dta is processed by anwwsmini.do, and that produces anwwsmini.smcl, which contains the output of our analysis.

Let's look at the mkwwsmini.do file in more detail. This program reads in the wws.csv raw dataset that was used in section 9.4 and includes the data corrections and data checking commands from that section. The comments include labels for each part of the program (such as [A]). These are used like footnotes below to relate the parts of the program to the explanation given. I would not recommend including such labels in your own programs because of the need to constantly update them as the program changes and evolves.

```
. type mkwwsmini.do
capture log close
log using mkwwsmini, replace
version 11.0
set more off
clear all
set memory 200m
* [A] Read in the raw data file
insheet using wws.csv

* [B] race
* [B1] correct error
replace race = 1 if idcode == 543
* [B2] label variable and values
label variable race "race of woman"
label define racelab 1 "White" 2 "Black" 3 "Other"
label values race racelab
* [B3] double check that race is only 1, 2 or 3
assert inlist(race,1,2,3)

* [C] age
* [C1] correct errors
replace age = 38 if idcode == 51
replace age = 45 if idcode == 80
* [C2] label variable
label variable age "Age of woman"
* [C3] double check that age is from 21 up to 50
assert (age >= 21 & age <= 50)

* [D] save data file
save wwsmini, replace

log close
```

The do-file begins by opening a log file, specifying the `version`, setting `more off`, clearing memory, and allocating enough memory. These are based on the skeleton do-file illustrated in section 9.3. Then part [A] reads in the `wws.csv` dataset.

Part [B] is concerned with the `race` variable. Part [B1] corrects the error previously found, where the race of the person with `idcode` of 543 was coded as a 4 but was supposed to be 1. Part [B2] labels the variable and values of `race`. Then part [B3] verifies that `race` is only 1, 2, or 3.

Part [C] is concerned with the `age` variable. Part [C1] corrects the errors we identified from section 9.4, and then part [C2] labels `age`. Then part [C3] verifies the values of `age`, checking that the ages are all between 21 and 50 (inclusive).

Finally, part [D] saves the dataset, naming it `wwsmini.dta`, and then closes the log file.

Here is the program that analyzes the data:

```
. type anwwsmini.do
capture log close
log using anwwsmini, replace
set more off
version 11.0
clear all
set memory 200m
* [A] read the data
use wwsmini
* [B] run regression predicting age from race
regress age i.race
log close
```

Note how the do-file for the analysis also starts by opening a log file, specifying the `version`, setting `more off`, clearing memory, and allocating enough memory. It then reads the dataset (part `[A]`) and then performs a regression analysis (part `[B]`). Finally, the log file is closed.

All these steps can then be performed with one simple command:

```
. do mastermini
  (output omitted)
```

Executing this one do-file, `mastermini.do`, reads the raw dataset, corrects the errors in the data, labels the data, checks the data, saves the data, and analyzes the data. Note some of the benefits of this strategy:

1. Suppose that someone said that there were errors in `wws.csv` and handed you a revised version. You could simply `use` that revised version and then type `do mastermini`, and it would reproduce the entire analysis with one command.
2. Within `mkwwsmini.do`, the data are checked for errors automatically. If you should be handed a new dataset, the data checking would be performed again, in case errors were introduced in the new file.
3. The file `wwsmini.dta` is the dataset used for analysis, so you can interactively analyze this file, and you can add new analyses that you like to `anwwsmini.do`. If you should need to then run the entire process again, any new analyses you introduce would automatically be performed.
4. Above all else, the process of going from raw data to final analysis is completely explicit and perfectly reproducible. If anyone should inquire about any aspect of how your analysis was performed, you can refer to the steps contained within these do-files to describe exactly what happened. You do not need to rely on your memory or hope that you did something a certain way.

Although this is but a small project, I hope that this section illustrated how this approach can be used and how it can be even more useful for larger and more complex projects.

The next section introduces Stata macros, which can be time saving and useful within do-files.

9.6 Introducing Stata macros

Sometimes in your do-file, you might find yourself referring to the same set of variables over and over again. Perhaps you have a series of regression commands with different outcomes but using the same set of predictors. And if you change the predictors for one model, you would want to change them for all the models. A small example is shown below, in which we run two `regress` commands, each using `age` and `married` as predictors.

```
. use wws2
. regress wage age married
. regress hours age married
```

Imagine that instead of having two `regress` commands, you had 10 or more such `regress` commands. Any time you wanted to add or delete a predictor, you would have to make the change to each of the `regress` commands. Instead, consider below how you could do this using a macro. The command `local preds age married` creates a local macro named `preds`, which contains `age married`. When the following `regress` commands are run, the value of `` `preds' `` is replaced with `age married`, replicating the commands from above.

```
. use wws2
(Working Women Survey w/fixes)
. local preds age married
. regress wage `preds'
      Source |       SS       df       MS              Number of obs =    2244
-------------+------------------------------           F(  2,  2241) =    1.63
       Model |  110.417562     2  55.2087812           Prob > F      =  0.1965
    Residual |   75985.248  2241  33.9068487           R-squared     =  0.0015
-------------+------------------------------           Adj R-squared =  0.0006
       Total |  76095.6655  2243  33.9258429           Root MSE      =   5.823

------------------------------------------------------------------------------
        wage |      Coef.   Std. Err.      t    P>|t|     [95% Conf. Interval]
-------------+----------------------------------------------------------------
         age |  -.0043748   .0230504    -0.19   0.849    -.0495772    .0408276
     married |  -.4619568    .256554    -1.80   0.072    -.9650651    .0411515
       _cons |   8.251724   .8662637     9.53   0.000      6.55296    9.950487
------------------------------------------------------------------------------
```

```
. regress hours `preds´
      Source |       SS       df       MS              Number of obs =    2242
-------------+------------------------------           F(  2,  2239) =   25.66
       Model |  5545.78053     2  2772.89026           Prob > F      =  0.0000
    Residual |  241954.564  2239  108.063673           R-squared     =  0.0224
-------------+------------------------------           Adj R-squared =  0.0215
       Total |  247500.345  2241  110.441921           Root MSE      =  10.395

------------------------------------------------------------------------------
       hours |      Coef.   Std. Err.      t    P>|t|     [95% Conf. Interval]
-------------+----------------------------------------------------------------
         age |  -.0848058    .041155    -2.06   0.039    -.1655117   -.0040999
     married |  -3.179256   .4584825    -6.93   0.000    -4.078351    -2.28016
       _cons |   42.33328   1.546136    27.38   0.000     39.30127    45.36529
------------------------------------------------------------------------------
```

The left and right quotes that hug and surround **preds** asks Stata to replace **'preds'** with the contents of the local macro named **preds**, i.e., **age married**. We can directly display the contents of the local macro **preds**:

```
. display "The contents of preds is `preds´"
The contents of preds is age married
```

Note! Where are these quotes?

It can be tricky to use the correct quotation marks when you want to type **'preds'**. First, I call ' a left quote. On U.S. keyboards, it is usually on the same key along with the tilde (~), often positioned above the **Tab** key. I call ' a right quote and it is located below the double quote (") on your keyboard. The left and right quotes hug the name of the macro, making it clear where the macro begins and ends.

If we want to change the predictors, we only have to do so once, by changing the **local preds** command that defines the macro containing the predictors. Below we add **currexp** to the list of predictors. The following **regress** commands will then use **age married currexp** as the list of predictors.

```
. local preds age married currexp
. regress wage `preds´
. regress hours `preds´
```

You could also use a macro to specify options for the **regress** command. Below the local macro called **regopts** contains **noheader beta** and each of the **regress** commands uses **'regopts'** after the comma to specify those options. As you can see, the output reflects these options, showing the output without a header and showing the standardized regression coefficients.

```
. local preds age married currexp
. local regopts noheader beta
```

```
. regress wage `preds´, `regopts´
```

wage	Coef.	Std. Err.	t	P>|t|	Beta
age	-.0364191	.0231681	-1.57	0.116	-.0332582
married	-.4257074	.2541613	-1.67	0.094	-.0350016
currexp	.1986906	.0244371	8.13	0.000	.1719589
_cons	8.384738	.8600406	9.75	0.000	.

```
. regress hours `preds´, `regopts´
```

hours	Coef.	Std. Err.	t	P>|t|	Beta
age	-.1289253	.0412795	-3.12	0.002	-.0655535
married	-3.199432	.4532619	-7.06	0.000	-.1463295
currexp	.3181376	.0435678	7.30	0.000	.153215
_cons	42.31729	1.531802	27.63	0.000	.

Perhaps you would like to see the results with robust standard errors. You can just add `vce(robust)` to the list of options specified in `regopts`, and then all the results will be shown with robust standard errors.

```
. local preds age married currexp
. local regopts noheader beta vce(robust)
. regress wage `preds´, `regopts´
. regress hours `preds´, `regopts´
```

The power of macros is not limited to analysis commands. Any time you want to make an overall change to the behavior of your do-file, you might want to think whether you can use a local macro to help. It is like being able to throw one light switch that controls a group of lights. For example, in section 9.4, the `assert` command was illustrated for verifying your data by testing assertions. For example, you can use the `assert` command to see if the values for `married` are truly all either 0 or 1.

```
. assert married==0 | married==1
```

Because the command said nothing, the assertion is true. You might also want to assert that if one is currently married (i.e, `if married==1`), then the value of `nevermarried` always contains zero. We check this below.

```
. assert nevermarried == 0 if married == 1
2 contradictions in 1442 observations
assertion is false
r(9);
```

Out of the 1,442 women who are married, this assertion was false for two of them. This command not only told us that this assertion is false but also returned an error code (i.e., `r(9);`). If this command was within a do-file, the do-file would have halted. To avoid halting a do-file on a false assertion, you can add the `rc0`[5] option. An example is shown below.

5. This stands for "return code zero". If a command returns a code of zero, it means that it ran error free, and Stata will continue executing the do-file.

```
. assert nevermarried == 0 if married == 1, rc0
2 contradictions in 1442 observations
assertion is false
```

In this case, a do-file would continue and you could search the log file for the word "false", allowing you to identify multiple false assertions at once. Sometimes you might want your do-file to use the `rc0` option so that you can discover multiple errors in your data. But most of the time, you would want the program to fail if any assertions are false to bluntly indicate than an assertion was found to be false.

Because there is no global setting to change how the `assert` command behaves, you can create a macro that determines how the `assert` command behaves. In the example below, the local macro `myrc` is used to determine whether the program should continue or stop upon encountering a false assertion. The value of `myrc` is set to `rc0`, and each of the `assert` commands is followed by the option `'myrc'`. In this case, a do-file would continue even if there is a false assertion.

```
. local myrc rc0
. assert nevermarried == 0 if married == 1, `myrc´
2 contradictions in 1442 observations
assertion is false
. assert married==0 | married==1, `myrc´
```

By contrast, in the following example, the value of `myrc` is assigned to be nothing, and a do-file would halt if the assertion fails. The second assertion is never tested because the first assertion failed and halted the do-file.

```
. local myrc
. assert nevermarried == 0 if married == 1, `myrc´
2 contradictions in 1442 observations
assertion is false
r(9);
```

This raises an issue, however, about the reason that I keep referring to `myrc` as a *local* macro. I do this because there is another entity called a *global* macro. A local macro is only defined within the current do-file and ceases to exist when a do-file is complete. Also, if a first do-file calls a second do-file, local macros created in the first do-file do not exist within the second do-file. As section 9.5 describes, you might want to create a chain of do-files that perform your data checking and data analysis. If you should want to control the behavior of the `assert` command across this entire chain of do-files, then you would want to use a global macro because a global macro lives and is defined for all do-files until you close your Stata session. Here is an example of how you would repeat the example from above by using a global macro.

```
. global rcopt rc0
. assert nevermarried == 0 if married == 1, $rcopt
2 contradictions in 1442 observations
assertion is false
. assert married==0 | married==1, $rcopt
```

The command `global rcopt rc0` creates a global macro called `rcopt` that has the value of `rc0`. We access the contents of a global macro by prefacing it with a dollar sign, i.e., `$rcopt`, as compared with the way that we accessed the contents of a local macro by specifying it surrounded by quotes, e.g., `‘myrc’`. As described in section 9.5, you can combine your do-files into one master do-file that calls all your other do-files. You could define global macros in your master do-file, which could then be used to control the behavior of all the do-files contained within it. For example, suppose that we had a `master.do` file that looks like this:

```
. global rcopt rc0
. do checkdata
. do modifydata
. do analyzedata
```

The do-files such as `checkdata.do` could use `$rc0` at the end of any `assert` commands. If this was done for all the `assert` commands within all these do-files, then the `global rcopt rc0` command would control the behavior of all the `assert` commands within all the do-files.

The examples from this section illustrated just a couple of different ways that you can use macros as a way to change the behavior of a series of Stata commands. As described earlier, Stata macros can be used like a master light switch, allowing you to flick one switch to control a group of lights. The following sections further expand upon this, illustrating even more that you can do with Stata macros.

9.7 Manipulating Stata macros

This section builds upon the previous section, focusing on the principles of how to manipulate macros. The following sections illustrate practical situations where it is useful to manipulate Stata macros.

Consider this simple example in which the `local` command is used to assign `Hello` to the macro `h`. The next `local` command assigns `world` to the macro `w`. The `display` command shows the contents of these macros.

```
. local h Hello
. local w world
. display "The macro h contains `h´ and the macro w contains `w´"
The macro h contains Hello and the macro w contains world
```

We can combine the local macros `h` and `w` together into a macro called `both`, as shown below. The result is that the macro `both` contains `Hello world`.

```
. local both `h´ `w´
. display "The macro both contains `both´"
The macro both contains Hello world
```

We can also assign an empty string (i.e., nothing) to a macro, as shown below. Previously, the local macro `both` contained `Hello world`, but now it contains nothing; it is empty. This is useful for clearing the contents of a macro.

```
. local both
. display "The macro both now contains `both´"
The macro both now contains
```

Say that the macro `myvars` contains the names of two variables, `age yrschool`. (By the way, Stata does not know or care that these are names of variables.)

```
. local myvars age yrschool
```

As an example of using this macro, we could run a regression predicting `wage` from the variables named in `myvars`.

```
. use wws2
. regress wage `myvars´
```

Below we append the name of another variable, `hours`, to the contents of `myvars`, creating a new macro named `myvars2`. The value of `'myvars'` is `age yrschool`, so the value assigned to `myvars2` is `age yrschool hours`, as shown below. This is followed by showing how you could run a regression predicting `wage` from the variables named in the local macro `myvars2`.

```
. local myvars2 `myvars´ hours
. display "The macro myvars2 contains `myvars2´"
The macro myvars2 contains age yrschool hours
. regress wage `myvars2´
  (output omitted)
```

Applying the same logic as above, we can append the word `hours` to the contents of `myvars`, replacing the existing value of `myvars`, as shown below.

```
. local myvars `myvars´ hours
. display "The macro myvars now contains `myvars´"
The macro myvars now contains age yrschool hours
```

So far, these examples have illustrated macros that contain words or strings. Let's consider the following example:

```
. local x 2+2
. display "The macro x contains `x´"
The macro x contains 2+2
```

Although in our minds, we are tempted to evaluate this (i.e., add the values up) and think that the macro `x` would contain `4`, the macro `x` is treated literally and assigned `2+2`. Contrast this with the example below.

```
. local y = 2+2
. display "The macro y contains `y´"
The macro y contains 4
```

By including the equal sign, we are asking Stata to evaluate the expression to the right of the equal sign. The result is that the contents of y is 4. The equal sign tells Stata to evaluate the expression and then store the result.

Let's consider another example.

```
. local a 5
. local b 10
. local c = `a´ + `b´
. display "The macro c contains `c´"
The macro c contains 15
```

In this case, the local macro c is assigned the value of a (which is 5) plus b (which is 10), which yields 15. This shows that we can perform arithmetic with macros. In fact, you can perform the same kinds of mathematical computations with local macros as you would use when generating a new variable. You can use the operators for addition (+), subtraction (-), multiplication (*), division (/), and exponentiation (^), and you can use parentheses to override (or emphasize) the order or operators. You can also use functions such as exp(), ln(), or sqrt(); see section A.7. The use of some of these operators and functions is illustrated below, creating the nonsense variable d.

```
. local d = (sqrt(`a´) - ln(`b´))/2
. display "The value of d is `d´"
The value of d is -.033258557747128
```

Suppose that we wanted to take the existing value of the local macro a and add 1 to it. We can do so as shown below. On the right side of the equal sign is the expression 'a' + 1, and because the value of a is 5, this becomes 5 + 1. So after this command is issued, the value of a becomes 6.

```
. local a = `a´ + 1
. display "The macro a now contains `a´"
The macro a now contains 6
```

So far, the examples using the equal sign have focused on numeric expressions. You can also specify a string expression after the equal sign when assigning a macro.[6] For example, you might want to convert the contents of the macro myvars into uppercase using the upper() function. We do this as shown below.

```
. local myvarsup = upper("`myvars´")
. display "myvarsup is `myvarsup´"
myvarsup is AGE YRSCHOOL HOURS
```

6. But note that the length of the resulting expression cannot exceed 244 characters.

The equal sign is used to ask Stata to evaluate the contents to the right of the equal sign, and the result is that `myvarsup` is assigned `AGE YRSCHOOL HOURS`.

Although there are many more ways we could explore the manipulation of Stata macros, this section built a sufficient foundation for the following sections. In fact, the next section will illustrate ways in which macros can be used in combination with `foreach` loops to provide shortcuts for performing repetitive commands. For more information about Stata macros, see `help macro`.

9.8 Repeating commands by looping over variables

This section illustrates how you can use `foreach` loops to run one or more commands by cycling through a set of variables you specify. For example, suppose that we would like to run six different regressions predicting the outcomes `wage`, `hours`, `prevexp`, `currexp`, `yrschool`, and `uniondues` from the variables `age` and `married`. We could type these six commands as shown below.

```
. regress wage age married
. regress hours age married
. regress prevexp age married
. regress currexp age married
. regress yrschool age married
. regress uniondues age married
```

Or we could save typing and time by using a `foreach` loop as shown below.

```
. foreach y of varlist wage hours prevexp currexp yrschool uniondues {
.   regress `y´ age married
. }
```

The `foreach` command above cycles (iterates) through each of the variable names supplied after the keyword `varlist`. For the first iteration, the local macro `y` is assigned the value of `wage` and the commands between the braces (`{` and `}`) are executed. This executes the command `regress wage age married`. For the next iteration, the local macro `y` is assigned the value of the next variable name (`hours`), and commands within the braces are executed, executing the command `regress hours age married`. This process is repeated for the remaining variables specified after the `varlist` keyword.

Let's execute the commands shown in the `foreach` loop above but, to save space, show just the output from the first and last iteration of the loop. Note how the Stata log inserts numbers below the `foreach` loop. You do not type these numbers; this is what Stata shows to indicate the nesting of the commands within the `foreach` loop to emphasize that lines 2 and 3 are governed by the `foreach` loop.

(Continued on next page)

```
. use wws2
(Working Women Survey w/fixes)

. foreach y of varlist wage hours prevexp currexp yrschool uniondues {
  2.   regress `y´ age married
  3. }

      Source |       SS       df       MS              Number of obs =    2244
-------------+------------------------------           F(  2,  2241) =    1.63
       Model |  110.417562     2  55.2087812           Prob > F      =  0.1965
    Residual |   75985.248  2241  33.9068487           R-squared     =  0.0015
-------------+------------------------------           Adj R-squared =  0.0006
       Total |  76095.6655  2243  33.9258429           Root MSE      =   5.823

------------------------------------------------------------------------------
        wage |      Coef.   Std. Err.      t    P>|t|     [95% Conf. Interval]
-------------+----------------------------------------------------------------
         age |  -.0043748   .0230504    -0.19   0.849    -.0495772    .0408276
     married |  -.4619568    .256554    -1.80   0.072    -.9650651    .0411515
       _cons |   8.251724   .8662637     9.53   0.000      6.55296    9.950487
------------------------------------------------------------------------------
```

Output for hours omitted to save space
Output for prevexp omitted to save space
Output for currexp omitted to save space
Output for yrschool omitted to save space

```
      Source |       SS       df       MS              Number of obs =    2242
-------------+------------------------------           F(  2,  2239) =    3.84
       Model |   623.75874     2   311.87937           Prob > F      =  0.0217
    Residual |  182070.734  2239  81.3178804           R-squared     =  0.0034
-------------+------------------------------           Adj R-squared =  0.0025
       Total |  182694.493  2241   81.523647           Root MSE      =  9.0176

------------------------------------------------------------------------------
   uniondues |      Coef.   Std. Err.      t    P>|t|     [95% Conf. Interval]
-------------+----------------------------------------------------------------
         age |   -.003459   .0356981    -0.10   0.923    -.0734639    .0665458
     married |  -1.100682   .3974222    -2.77   0.006    -1.880036   -.3213274
       _cons |   6.434747   1.341576     4.80   0.000     3.803884     9.06561
------------------------------------------------------------------------------
```

Suppose that you wanted to run a regression predicting each outcome (from a list of outcomes) from each predictor (from a list of predictors). You could include two `foreach` loops, one for the outcomes and one for the predictors. This is shown below for six outcomes (`wage`, `hours`, `prevexp`, `currexp`, `yrschool`, and `uniondues`) and four predictors (`age`, `married`, `south`, and `metro`).

```
. use wws2
. foreach y of varlist wage hours prevexp currexp yrschool uniondues {
.   foreach x of varlist age married south metro {
.     regress `y´ `x´
.   }
. }
```

The `regress` commands that would be executed from these two nested `foreach` loops are shown below.

```
. regress wage age
. regress wage married
. regress wage south
. regress wage metro
```

```
. regress hours age
. regress hours married
. regress hours south
. regress hours metro
```
commands for prevexp, currexp, yrschool omitted to save space
```
. regress uniondues age
. regress uniondues married
. regress uniondues south
. regress uniondues metro
```

The `regress` command is executed for each variable specified in the first `foreach` loop in combination with each variable specified in the second loop. The first loop contains six variables, and the second loop contains four variables, so this leads to 24 iterations (or 24 `regress` commands). That saves us quite a bit of typing.

The examples so far have illustrated the use of `foreach` for performing a series of `regress` commands. Let's turn our attention to the use of `foreach` with commands related to data management. These examples will use `cardio1.dta`, shown below.

```
. use cardio1
. describe
Contains data from cardio1.dta
  obs:             5
 vars:            12                          22 Dec 2009 19:50
 size:           120 (99.9% of memory free)
-------------------------------------------------------------------------------
              storage  display     value
variable name   type   format      label      variable label
-------------------------------------------------------------------------------
id              byte   %3.0f                  Identification variable
age             byte   %3.0f                  Age of person
bp1             int    %3.0f                  Systolic BP: Trial 1
bp2             int    %3.0f                  Systolic BP: Trial 2
bp3             int    %3.0f                  Systolic BP: Trial 3
bp4             int    %3.0f                  Systolic BP: Trial 4
bp5             int    %3.0f                  Systolic BP: Trial 5
pl1             int    %3.0f                  Pulse: Trial 1
pl2             byte   %3.0f                  Pulse: Trial 2
pl3             int    %3.0f                  Pulse: Trial 3
pl4             int    %3.0f                  Pulse: Trial 4
pl5             byte   %3.0f                  Pulse: Trial 5
-------------------------------------------------------------------------------
Sorted by:
```

This dataset contains five measurements of systolic blood pressure named `bp1`–`bp5` and five measurements of pulse rate named `pl1`–`pl5`. Suppose that you wanted to create a dummy variable that indicated whether a blood pressure measure was high (130 or over). This could be done as a series of `recode` commands, as shown below. This would create five variables named `hibp1`–`hibp5` that would be 0 if the blood pressure was 129 or less and 1 if the blood pressure was 130 or more.

(Continued on next page)

```
. recode bp1 (min/129=0) (130/max=1), generate(hipb1)
. recode bp2 (min/129=0) (130/max=1), generate(hipb2)
. recode bp3 (min/129=0) (130/max=1), generate(hipb3)
. recode bp4 (min/129=0) (130/max=1), generate(hipb4)
. recode bp5 (min/129=0) (130/max=1), generate(hipb5)
```

Instead of typing all those commands, we could use a `foreach` loop as shown below. In comparing the `recode` command from above with the one below, note that '`v`' is inserted in place of bp1–bp5.

```
. foreach v of varlist bp1 bp2 bp3 bp4 bp5 {
.   recode `v´ (min/129=0) (130/max=1), generate(hi`v´)
. }
```

In the first iteration of the loop, the value of '`v`' will be `bp1`; thus

```
.   recode `v´ (min/129=0) (130/max=1), generate(hi`v´)
```

will be replaced by the following command:

```
.   recode bp1 (min/129=0) (130/max=1), generate(hibp1)
```

This process is repeated for `bp2`, `bp3`, `bp4`, and `bp5`. Below we see the listing of the variables after executing the `foreach` loop.

```
. list id bp1 bp2 bp3 bp4 bp5 hibp1 hibp2 hibp3 hibp4 hibp5, noobs
```

id	bp1	bp2	bp3	bp4	bp5	hibp1	hibp2	hibp3	hibp4	hibp5
1	115	86	129	105	127	0	0	0	0	0
2	123	136	107	111	120	0	1	0	0	0
3	124	122	101	109	112	0	0	0	0	0
4	105	115	121	129	137	0	0	0	0	1
5	116	128	112	125	111	0	0	0	0	0

Suppose that we wanted to create the same kind of dummy variables associated with the pulse scores, indicating whether a pulse measurement is high (with high being defined as 90 or over). Instead of doing this with the `recode` command, this could be done with the `generate` command as shown below. (For more on the logic of this statement, see section A.6.)

```
. foreach v of varlist pl1-pl5 {
.   generate hi`v´ = (`v´ >= 90) if ! missing(`v´)
. }
```

This `foreach` loop took advantage of the fact that the variables `pl1`–`pl5` are positioned next to each other, referring to the series of variables as `pl1-pl5` (see section A.11). The first time the loop is executed, the value of '`v`' will be `pl1`, so the `generate` command becomes

```
.   generate hipl1 = (pl1 >= 90) if ! missing(pl1)
```

The result is that hipl1 is 1 if pl1 is 90 or above, 0 if pl1 is 89 or below, and missing if pl1 is missing. The loop is then repeated for pl2–pl5. The results are shown below.

```
. list id pl* hipl*, noobs
```

id	pl1	pl2	pl3	pl4	pl5	hipl1	hipl2	hipl3	hipl4	hipl5
1	54	87	93	81	92	0	0	1	0	1
2	92	88	125	87	58	1	0	1	0	0
3	105	97	128	57	68	1	1	1	0	0
4	52	79	71	106	39	0	0	0	1	0
5	70	64	52	68	59	0	0	0	0	0

Let's extend the power of the foreach loop by combining it with what we learned in section 9.7 about how to append the contents of Stata macros. This will allow us to run a series of regress commands that add one predictor at a time. Suppose that you wanted to perform a series of regress commands like the ones shown below.

```
. regress wage age
. regress wage age yrschool
. regress wage age yrschool hours
```

Note how the outcome is the same, and one predictor is added at a time. With only three predictors, this is not much work, but as the number of predictors grows, so would the amount of our typing. Each command takes the list of predictors from the previous command and appends on a new predictor. The second command appends yrschool to age and the third command appends hours to age yrschool. Instead of manually typing each of these regress commands, we could use a foreach loop, as shown below.

```
. use wws2
. local myvars
. foreach v of varlist age yrschool hours {
.   local myvars `myvars´ `v´
.   regress wage `myvars´, noheader
. }
```

The foreach loop cycles through the variable names one at a time, and the local command within the foreach loop cumulatively gathers the names of each variable. The result is that each time the regress command is executed, it includes not only the current variable from the foreach loop but also the previous variables.

Below we execute these commands and can see that it worked. It ran the regression first using age as a predictor, then using age yrschool as predictors, and then using age yrschool hours as predictors.

(*Continued on next page*)

```
. local myvars
. foreach v of varlist age yrschool hours {
  2.    local myvars `myvars´ `v´
  3.    regress wage `myvars´, noheader
  4. }
```

wage	Coef.	Std. Err.	t	P>|t|	[95% Conf.	Interval]
age	-.0027502	.0230443	-0.12	0.905	-.0479406	.0424401
_cons	7.896423	.8439127	9.36	0.000	6.241491	9.551355

wage	Coef.	Std. Err.	t	P>|t|	[95% Conf.	Interval]
age	.0033213	.0218308	0.15	0.879	-.0394894	.046132
yrschool	.7806215	.0481182	16.22	0.000	.6862606	.8749824
_cons	-2.572595	1.026822	-2.51	0.012	-4.586219	-.5589714

wage	Coef.	Std. Err.	t	P>|t|	[95% Conf.	Interval]
age	.0096621	.021655	0.45	0.656	-.032804	.0521282
yrschool	.7534783	.0479363	15.72	0.000	.6594738	.8474827
hours	.0736386	.0110344	6.67	0.000	.0519999	.0952773
_cons	-5.179999	1.088302	-4.76	0.000	-7.314189	-3.04581

Let's walk through this again more slowly, showing the values of the macros for each iteration of the `foreach` loop.

First, here are the commands:

```
use wws2
local myvars
foreach v of varlist age yrschool hours {
  local myvars `myvars´ `v´
  regress wage `myvars´, noheader
}
```

After reading in `wws2.dta`, the local macro `myvars` is created; it is empty at this time. When the `local` command after `foreach` is executed the first time, the macro `myvars` is empty and the macro `v` contains `age`, so the command

```
local myvars `myvars´ `v´
```

translates into

```
local myvars  age
```

The next line is the `regress` command, which reads as

```
regress wage `myvars´, noheader
```

which is translated into

```
regress wage age, noheader
```

resulting in the output below.

wage	Coef.	Std. Err.	t	P>|t|	[95% Conf.	Interval]
age	-.0027502	.0230443	-0.12	0.905	-.0479406	.0424401
_cons	7.896423	.8439127	9.36	0.000	6.241491	9.551355

Now we repeat the `foreach` loop. The second time through this loop, `myvars` contains the value of `age` and `v` contains `yrschool`. So when the `local` command

```
local myvars `myvars´ `v´
```

is encountered, it translates into

```
local myvars age yrschool
```

And then the `regress` command, which reads as

```
regress wage `myvars´, noheader
```

is translated into

```
regress wage age yrschool, noheader
```

resulting in the following output.

wage	Coef.	Std. Err.	t	P>|t|	[95% Conf.	Interval]
age	.0033213	.0218308	0.15	0.879	-.0394894	.046132
yrschool	.7806215	.0481182	16.22	0.000	.6862606	.8749824
_cons	-2.572595	1.026822	-2.51	0.012	-4.586219	-.5589714

The third time through the `foreach` loop, the macro `myvars` contains `age yrschool` and the macro `v` contains `hours`. The command

```
local myvars `myvars´ `v´
```

then translates into

```
local myvars age yrschool hours
```

And then the `regress` command, which reads as

```
regress wage `myvars´, noheader
```

translates into

```
regress wage age yrschool hours, noheader
```

resulting in the output below.

```
        wage |      Coef.   Std. Err.      t    P>|t|     [95% Conf. Interval]
-------------+----------------------------------------------------------------
         age |   .0096621    .021655     0.45   0.656    -.032804    .0521282
    yrschool |   .7534783   .0479363    15.72   0.000    .6594738    .8474827
       hours |   .0736386   .0110344     6.67   0.000    .0519999    .0952773
       _cons |  -5.179999   1.088302    -4.76   0.000   -7.314189    -3.04581
```

The foreach loop is now complete. If I had only three variables, I would have probably just manually typed in the three regress commands. Using the foreach loop becomes more convenient as the number of variables grows. It becomes even more useful if the variables might change, saving you the hassle of retyping many regress commands just because you wanted to add or delete one variable from this process.

The next section expands on this discussion of the foreach command, showing how to loop over numbers.

9.9 Repeating commands by looping over numbers

The previous section illustrated how the foreach command could be used to repeat a command such as recode or generate across a set of variables. This section extends upon those examples. Consider gaswide.dta.

```
. use gaswide
. list
```

	ctry	gas1974	gas1975	gas1976	inf1974	inf1975	inf1976
1.	1	0.78	0.83	0.99	2.64	2.80	3.10
2.	2	0.69	0.69	0.89	2.30	2.30	2.58
3.	3	0.42	0.48	0.59	2.28	2.44	2.64
4.	4	0.82	0.94	1.09	2.28	2.36	3.00

This file has data from four countries with the price of gas per gallon from 1974 to 1976, along with inflation factors to bring those prices into current dollars. Say that we wanted to make gascur1974, which would be the price of gas in today's dollars. gascur1974 could be created like this:

```
. generate gascur1974 = gas1974 * inf1974
```

We could repeat this command for the years 1975 and 1976, as shown below.

```
. generate gascur1975 = gas1975 * inf1975
. generate gascur1976 = gas1976 * inf1976
```

But if there were many years of data, say, 20 years' worth of data, typing this over and over would soon become cumbersome. When you look at these generate commands, everything is the same except for the year, which changes from `1974` to `1975` and then to `1976`. Contrast this with the examples from the previous section in which the entire variable name changed over each command. In this case, we need to build a `foreach` loop that loops just over the years ranging from 1974, 1975, and 1976.

The example below uses a `foreach` loop but specifies that we are looping over a number list (because the keyword `numlist` was specified) followed by the number list `1974/1976`, which expands to `1974 1975 1976`.

```
. foreach yr of numlist 1974/1976 {
.   generate gascur`yr´ = gas`yr´ * inf`yr´
. }
```

So the first time this loop is executed, the value of `'yr'` is replaced with `1974`, making the command read

```
.   generate gascur1974 = gas1974 * inf1974
```

The second time the loop executes, the value of `'yr'` is replaced with 1975, and then in the final iteration, the value of `'yr'` is replaced with 1976. The resulting variables are shown below.

```
. list ctry gascur1974 gascur1975 gascur1976, abb(10)
```

	ctry	gascur1974	gascur1975	gascur1976
1.	1	2.0592	2.324	3.069
2.	2	1.587	1.587	2.2962
3.	3	.9575999	1.1712	1.5576
4.	4	1.8696	2.2184	3.27

Tip! Reshape long, make variables, reshape wide

In the above strategy, a separate `generate` command was needed to create the gas price in current dollars for each year. Suppose, instead, that we first reshape the data into a long format in which there would be one observation per country per year. Then only one command is needed to create the gas price in current dollars. Then the data can be reshaped back into their original wide format. The commands to do this for this example are the following:

```
. use gaswide
. reshape long gas inf, i(ctry) j(time)
. generate gascur = gas*inf
. reshape wide gas inf gascur, i(ctry) j(time)
```

Either way, you can pick the strategy that makes the most sense to you.

Perhaps you want to make a new variable that is the percent change in gas price from year to year. We could do this for the year 1975 (compared with 1974) like this:

```
. generate gaschg1975 = 100 * ((gas1975 - gas1974)/gas1974)
```

Again imagine that we had 20 such years of data. Rather than repeating and changing the above command 19 times, a `foreach` loop could be used as a shortcut. Note that we need not only a variable that represents the current year (e.g., `1975`) but also a variable that represents the previous year (e.g., `1974`). Using what we learned from section 9.7 about manipulating macros, we use the `local` command to create the variable `lastyr`, which is the value of the current year minus 1.

```
. foreach yr of numlist 1975/1976 {
.   local lastyr = `yr'-1
.   generate gaschg`yr' = 100 * ((gas`yr' - gas`lastyr')/gas`lastyr')
. }
```

The `foreach` command loops over the years 1975 and 1976. These represent the current year, named `yr`. Then the `local` command is used to create `lastyr`, which is `yr` minus 1. The first time the loop is executed, the value of `yr` is `1975` and the value of `lastyr` is `1974`, so the `generate` command becomes

```
.   generate gaschg1975 = 100 * ((gas1975 - gas1974)/gas1974)
```

After the entire set of commands is executed, we get the following results.

```
. list  ctry  gas1974 gas1975 gas1976 gaschg1975 gaschg1976, abb(10)
```

	ctry	gas1974	gas1975	gas1976	gaschg1975	gaschg1976
1.	1	0.78	0.83	0.99	6.410258	19.27711
2.	2	0.69	0.69	0.89	0	28.98551
3.	3	0.42	0.48	0.59	14.28572	22.91666
4.	4	0.82	0.94	1.09	14.63415	15.95745

As illustrated in this section, the `foreach` command combined with the `numlist` keyword can be used to loop over a series of numbers. The next section provides additional examples about how you can use `foreach` to loop across any arbitrary list. For more information, see `help foreach`.

9.10 Repeating commands by looping over anything

The previous two sections have illustrated how `foreach` loops can be used as a shortcut to loop across variables and loop over numbers. Let's explore other ways that loops can be used as a shortcut for data-management tasks. Suppose that we have several comma-separated files that we want to read into Stata. For example, we have three such files, shown below.

```
. dir br_*.csv
   0.1k   9/14/09 18:28  br_clarence.csv
   0.1k   9/14/09 18:29  br_isaac.csv
   0.1k   9/14/09 18:29  br_sally.csv
```

These files each contain book review information from three different people: Clarence, Isaac, and Sally. Below we can see the book reviews from Clarence.

```
. type br_clarence.csv
booknum,book,rating
1,"A Fistful of Significance",5
2,"For Whom the Null Hypothesis is Rejected",10
3,"Journey to the Center of the Normal Curve",6
```

The file contains the book number, the name of the book, and the rating of the book (on a scale from 1 to 10). Suppose that we want to read all these files into Stata and save each one as a Stata dataset. We could do this manually like this:

```
. insheet using br_clarence.csv, clear
. save br_clarence

. insheet using br_isaac.csv, clear
. save br_isaac

. insheet using br_sally.csv, clear
. save br_sally
```

Because there are only three such comma-separated files, this is not too bad, but if there were many such files, repeating these commands could get tedious.

The repetitive nature of these commands is a clue that a `foreach` loop could be used as a shortcut. Everything is the same in these commands except for the name of the reviewer. Because the list of reviewer names is just an arbitrary list (not a variable list or a number list), we can specify the `foreach` command with the `in` qualifier, which permits us to supply any arbitrary list, as shown below.

```
. foreach revname in clarence isaac sally {
.   insheet using br_`revname'.csv, clear
.   save br_`revname'
. }
```

In the first iteration of the loop, each instance of `` `revname' `` is replaced with `clarence`, so the commands translate into the following:

```
.   insheet using br_clarence.csv, clear
.   save br_clarence
```

The same kind of translation process happens in the second iteration, replacing each instance of `` `revname' `` with `isaac`. In the third and last iteration, `` `revname' `` is replaced with `sally`. After executing this loop, we can use the `dir` command to see that there are three datasets for the three reviewers.

```
. dir br_*.dta
  0.8k   2/02/10 18:55  br_clarence.dta
  0.8k   2/02/10 18:55  br_isaac.dta
  0.8k   2/02/10 18:55  br_sally.dta
```

And, for example, we can inspect the dataset with the information from Clarence and see that it corresponds to the information from the comma-separated file that we saw earlier.

```
. use br_clarence
. list
```

	booknum	book	rating
1.	1	A Fistful of Significance	5
2.	2	For Whom the Null Hypothesis is Rejected	10
3.	3	Journey to the Center of the Normal Curve	6

Using the `foreach` loop with the `in` keyword allows you to loop over any arbitrary list. For more information, see `help foreach`.

9.11 Accessing results saved from Stata commands

Nearly all Stata commands produce some kind of output that is displayed in the Results window. In addition, many commands create *saved results*, which contain information related to the results of the command. This section illustrates how you can access and use these saved results.

Let's use `wws2.dta` and issue the `summarize` command.

```
. use wws2
(Working Women Survey w/fixes)
. summarize wage
```

Variable	Obs	Mean	Std. Dev.	Min	Max
wage	2244	7.796781	5.82459	0	40.74659

Suppose that you want to obtain the coefficient of variation for wages. Even if we use the `detail` option, the coefficient of variation is not among the statistics displayed. To compute this value ourselves, we would take the sample standard deviation divided by the sample mean all multiplied by 100. We could do this by typing in the values and using the `display` command to do the computations, as shown below.

```
. display (5.82459/7.796781) * 100
74.705061
```

Instead of manually typing in the numbers for the mean and standard deviation, we can access the values of the mean and standard deviation using the saved results from the `summarize` command. The `return list` command shows the saved results that are available after the `summarize` command.

```
. summarize wage
  (output omitted)
. return list
scalars:
                  r(N) =  2244
              r(sum_w) =  2244
               r(mean) =  7.796780732119998
                r(Var) =  33.92584286115162
                 r(sd) =  5.824589501514388
                r(min) =  0
                r(max) =  40.74658966064453
                r(sum) =  17495.97596287727
```

The saved results for a general command like `summarize` are named `r(`*something*`)`. For example, we can see that after the `summarize` command, the saved result containing the mean is named `r(mean)`, and the saved result containing the standard deviation is named `r(sd)`. We can use these saved results to compute the coefficient of variation as shown below.

```
. display (r(sd)/r(mean)) * 100
74.705057
```

The saved results are transitory. They exist until you run another command that would generate and overwrite the current set of saved results. The help file and reference manual entry indicates, for each command, whether it creates saved results and gives the names and descriptions of the results that are saved.

Let's consider another example. Sometimes we want to center variables; that is, we want to take the variable and subtract its mean. Taking advantage of the saved results, we could center the variable `wage` using the commands below.

```
. summarize wage
    Variable |       Obs        Mean    Std. Dev.       Min        Max
-------------+--------------------------------------------------------
        wage |      2244    7.796781     5.82459          0   40.74659
. generate cwage = wage - r(mean)
(2 missing values generated)
```

Say that you want to center several variables in your dataset. You could combine a `foreach` loop with the ability to access the saved results after `summarize` to create a series of mean-centered variables, illustrated below for the variables `age`, `yrschool`, `hours`, and `currexp`. Because we do not need to view the output of the `summarize` command, it is prefaced by `quietly` to suppress the output and save space.

```
. foreach myvar of varlist age yrschool hours currexp {
  2.   quietly summarize `myvar´
  3.   generate c`myvar´ = `myvar´ - r(mean)
  4. }
(4 missing values generated)
(4 missing values generated)
(15 missing values generated)
```

To check the centering, we can summarize the centered variables to verify that indeed their mean is approximately 0.

```
. summarize cage-ccurrexp
    Variable |       Obs        Mean    Std. Dev.       Min        Max
-------------+--------------------------------------------------------
        cage |      2246   -5.31e-08    5.337859  -15.22707   11.77293
   cyrschool |      2242   -2.09e-08    2.422114  -5.138269   4.861731
      chours |      2242   -9.40e-08    10.50914  -36.21811   42.78189
    ccurrexp |      2231    8.06e-08    5.048073  -5.185567   20.81443
```

Estimation commands (such as **regress**) also create saved results that reflect the results of the estimation. You can access these saved results to perform further computations. For example, you might want to run two different regression models and obtain the difference in the R-squared value.

First, let's consider the following regression analysis.

```
. regress hours currexp prevexp
      Source |       SS       df       MS              Number of obs =    2227
-------------+------------------------------           F(  2,  2224) =   40.49
       Model |  8590.92673     2  4295.46336           Prob > F      =  0.0000
    Residual |  235929.028  2224  106.083196           R-squared     =  0.0351
-------------+------------------------------           Adj R-squared =  0.0343
       Total |  244519.955  2226  109.847239           Root MSE      =    10.3

------------------------------------------------------------------------------
       hours |      Coef.   Std. Err.      t    P>|t|     [95% Conf. Interval]
-------------+----------------------------------------------------------------
     currexp |   .4386054   .0494109     8.88   0.000     .3417092    .5355017
     prevexp |   .3108159   .0556315     5.59   0.000     .2017209     .419911
       _cons |   33.07441   .5582478    59.25   0.000     31.97967    34.16916
------------------------------------------------------------------------------
```

Because this is an estimation command, the saved results can be displayed using the **ereturn list** command (*e* for estimation).

```
. ereturn list
scalars:
                  e(N) =  2227
               e(df_m) =  2
               e(df_r) =  2224
                  e(F) =  40.49145875966638
                 e(r2) =  .0351338471551399
               e(rmse) =  10.29966971356573
                e(mss) =  8590.926728743652
                e(rss) =  235929.0283678002
               e(r2_a) =  .0342661617658909
                 e(ll) =  -8352.088143338988
               e(ll_0) =  -8391.913460716318
               e(rank) =  3
```

```
macros:
              e(cmdline) : "regress hours currexp prevexp"
                e(title) : "Linear regression"
            e(marginsok) : "XB default"
                  e(vce) : "ols"
               e(depvar) : "hours"
                  e(cmd) : "regress"
           e(properties) : "b V"
              e(predict) : "regres_p"
                e(model) : "ols"
            e(estat_cmd) : "regress_estat"

matrices:
                    e(b) :  1 x 3
                    e(V) :  3 x 3

functions:
               e(sample)
```

Notice how the saved results for estimation commands are all named e(*something*); for example, the sample size is stored as e(N), the F statistic is stored as e(F), the residual degrees of freedom is stored as e(df_r), and the R-squared is stored as e(r2).

These saved results persist until we execute another estimation command. Let's save the R-squared value from this model as a local macro called rsq1.

```
. local rsq1 = e(r2)
```

Let's run another model where we add age as a predictor.

```
. regress hours currexp prevexp age
      Source |       SS       df       MS              Number of obs =    2227
-------------+------------------------------           F(  3,  2223) =   34.44
       Model |  10860.1551     3  3620.0517           Prob > F      =  0.0000
    Residual |    233659.8  2223  105.110121           R-squared     =  0.0444
-------------+------------------------------           Adj R-squared =  0.0431
       Total |  244519.955  2226  109.847239           Root MSE      =  10.252

------------------------------------------------------------------------------
       hours |      Coef.   Std. Err.      t    P>|t|     [95% Conf. Interval]
-------------+----------------------------------------------------------------
     currexp |   .5045227   .0511889     9.86   0.000     .4041397    .6049057
     prevexp |   .3861528   .0577007     6.69   0.000     .2729999    .4993056
         age |  -.1999601   .0430355    -4.65   0.000     -.284354   -.1155662
       _cons |   39.52059   1.494495    26.44   0.000     36.58984    42.45134
------------------------------------------------------------------------------
```

Let's save the R-squared for this model as a local macro called rsq2.

```
. local rsq2 = e(r2)
```

We can now compute and display the change in R-squared between the two models, as shown below.

```
. local rsqchange = `rsq2´ - `rsq1´
. display "R-squared for model 1 is `rsq1´"
R-squared for model 1 is .0351338471551399
```

```
. display "R-squared for model 2 is `rsq2´"
R-squared for model 2 is .0444141873437368

. display "Change in R-squared is `rsqchange´"
Change in R-squared is .0092803401885969
```

Nearly all Stata commands will create saved results. The help file (or manual entry) for each command that produces saved results will have a section named *Saved results* that lists the names and descriptions of the saved results. Accessing these saved results allows you to build upon the work of others and feed the saved results into your do-files or ado-files. Also, as illustrated in the next section, the saved results can be captured and stored as Stata datasets.

9.12 Saving results of estimation commands as data

The previous section illustrated how Stata commands create saved results and how you can perform computations using these values. But sometimes you might wish to save the results of a command into a Stata dataset and then further process the results. This section shows how you can save the results of a command with the `statsby` command. Consider this regression command based on `wws2.dta`.

```
. use wws2
(Working Women Survey w/fixes)

. regress hours currexp prevexp
      Source |       SS       df       MS              Number of obs =    2227
-------------+------------------------------           F(  2,  2224) =   40.49
       Model |  8590.92673     2  4295.46336           Prob > F      =  0.0000
    Residual |  235929.028  2224  106.083196           R-squared     =  0.0351
-------------+------------------------------           Adj R-squared =  0.0343
       Total |  244519.955  2226  109.847239           Root MSE      =    10.3

------------------------------------------------------------------------------
       hours |      Coef.   Std. Err.      t    P>|t|     [95% Conf. Interval]
-------------+----------------------------------------------------------------
     currexp |   .4386054   .0494109     8.88   0.000     .3417092    .5355017
     prevexp |   .3108159   .0556315     5.59   0.000     .2017209     .419911
       _cons |   33.07441   .5582478    59.25   0.000     31.97967    34.16916
------------------------------------------------------------------------------
```

Below we use `statsby` to run this same command. Note how the `regress` command comes after the colon after the `statsby` command.

```
. statsby: regress hours currexp prevexp
(running regress on estimation sample)

      command:  regress hours currexp prevexp
           by:  <none>

Statsby groups
----+--- 1 ---+--- 2 ---+--- 3 ---+--- 4 ---+--- 5
```

The `list` command below shows us that the current dataset has been replaced with variables containing the regression coefficients from the above `regress` command.

```
. list, abb(20)

     +------------------------------------+
     | _b_currexp   _b_prevexp   _b_cons  |
     |------------------------------------|
  1. |   .4386055     .3108159   33.07441 |
     +------------------------------------+
```

You can choose other statistics to save in addition to the regression coefficients. In the example below, we save the standard errors as well. Note that the `_b` indicates to save the regression coefficients and the `_se` indicates to save the standard errors.

```
. use wws2
(Working Women Survey w/fixes)
. statsby _b _se: regress hours currexp prevexp
(running regress on estimation sample)
      command:  regress hours currexp prevexp
           by:  <none>
Statsby groups
----+--- 1 ---+--- 2 ---+--- 3 ---+--- 4 ---+--- 5
```

The current dataset now contains variables with the coefficients and standard errors from this regression analysis.

```
. list, abb(20) noobs
```

_b_currexp	_b_prevexp	_b_cons	_se_currexp	_se_prevexp	_se_cons
.4386055	.3108159	33.07441	.0494109	.0556315	.5582478

Stata gives you the ability to save more than just the regression coefficient and the standard error. As we saw in section 9.11, the **ereturn list** command lists the saved results created by an estimation command. The **statsby** command can save these values as well. We saw that the residual degrees of freedom is saved as `e(df_r)`. Below we save the residual degrees of freedom, naming it `dfr` (which Stata converts to `_eq2_dfr`). If we did not give it a name, it would have been named `_eq2_stat_1`.

```
. use wws2, clear
(Working Women Survey w/fixes)
. statsby _b _se dfr=e(df_r): regress hours currexp prevexp
(running regress on estimation sample)
      command:  regress hours currexp prevexp
     _eq2_dfr:  e(df_r)
           by:  <none>
Statsby groups
----+--- 1 ---+--- 2 ---+--- 3 ---+--- 4 ---+--- 5
. list _b_currexp _b_prevexp _se_currexp _se_prevexp _eq2_dfr, abb(20)
```

	_b_currexp	_b_prevexp	_se_currexp	_se_prevexp	_eq2_dfr
1.	.4386055	.3108159	.0494109	.0556315	2224

We can obtain these results separately for each of the 12 levels of industry by adding the by(industry) option, as shown below. Also, the saving() option is used to save the results into the file named wwsreg.dta.[7]

```
. use wws2, clear
(Working Women Survey w/fixes)
. statsby _b _se dfr=e(df_r), by(industry) saving(wwsreg): regress hours
> currexp prevexp
(running regress on estimation sample)

      command:  regress hours currexp prevexp
     _eq2_dfr:  e(df_r)
           by:  industry

Statsby groups
----+--- 1 ---+--- 2 ---+--- 3 ---+--- 4 ---+--- 5
............
```

Let's now use wwsreg.dta, which was created by the statsby command.

```
. use wwsreg, clear
(statsby: regress)
```

The following listing shows, for each of the 12 industries, the regression coefficients and standard errors for currexp and prevexp along with the residual degrees of freedom for the model.

```
. list industry _b_currexp _b_prevexp _se_currexp _se_prevexp _eq2_dfr, abb(20)
> sep(0) noobs
```

industry	_b_currexp	_b_prevexp	_se_currexp	_se_prevexp	_eq2_dfr
1	.5512784	-.5114645	.7195972	1.032018	14
2	0	0	0	0	1
3	.9094636	.1749561	.6948095	.5545271	26
4	.0850188	.0866459	.0729664	.0838895	363
5	.1011588	.4665331	.1564988	.219154	87
6	.1786376	.3468918	.1802799	.1549291	327
7	.3407777	.4575056	.171842	.1849252	188
8	.605177	-.1271947	.3178068	.2920739	82
9	.7354487	.8282823	.4758221	.3549415	91
10	.3959756	-.1265796	.579298	.6179624	14
11	.5019768	.2692812	.0900196	.1030056	812
12	.294879	.1325803	.0968519	.1201941	172

The results for the second industry seem worrisome. Let's return to wws2.dta and manually perform the regress command just for the second industry.

7. If you want to overwrite this file, you need to specify saving(wwsreg, replace).

```
. use wws2, clear
(Working Women Survey w/fixes)

. regress hours currexp prevexp if industry == 2

      Source |       SS       df       MS              Number of obs =       4
-------------+------------------------------           F(  2,      1) =       .
       Model |          0     2           0           Prob > F      =       .
    Residual |          0     1           0           R-squared     =       .
-------------+------------------------------           Adj R-squared =       .
       Total |          0     3           0           Root MSE      =       0

------------------------------------------------------------------------------
       hours |      Coef.   Std. Err.      t    P>|t|     [95% Conf. Interval]
-------------+----------------------------------------------------------------
     currexp |  (omitted)
     prevexp |  (omitted)
       _cons |         40          .        .       .            .           .
------------------------------------------------------------------------------
```

There is a problem with this regression model when run just for the second industry. Let's look at the outcome and predictor variables for this industry.

```
. list hours currexp prevexp if industry == 2

      +---------------------------+
      | hours   currexp   prevexp |
      |---------------------------|
  14. |    40         2         7 |
 498. |    40         9         8 |
1070. |    40         2        10 |
1249. |    40         5         7 |
      +---------------------------+
```

Now we can see the problem. The outcome variable (**hours**) is the same (40) for all observations. This raises an important point. By using commands like **statsby**, it is easy to get so distanced from the analyses that you do not see obvious problems like this.

Bearing in mind that the results for the second industry are problematic, let's return to **wwsreg.dta** and see how we can compute t-values and p-values for the coefficients of **currexp** and **prevexp**. These values are not among the values that can be saved, but we can compute them based on the information we have saved. The t-values are obtained by dividing the coefficient by the standard error, as shown below.

```
. use wwsreg, clear
(statsby: regress)

. generate t_currexp = _b_currexp/_se_currexp
(1 missing value generated)

. generate t_prevexp = _b_prevexp/_se_prevexp
(1 missing value generated)
```

Below we see the t-values for the 12 regression models for the 12 industries. As expected, the value for the second industry is missing (because its standard error was 0).

```
. list industry t_currexp t_prevexp, abb(20) sep(0)

     +-----------------------------------+
     | industry   t_currexp   t_prevexp  |
     |-----------------------------------|
  1. |        1     .766093   -.4955963  |
  2. |        2           .           .  |
  3. |        3    1.308939     .315505  |
  4. |        4    1.165178    1.032858  |
  5. |        5    .6463869    2.128792  |
  6. |        6    .9908907    2.239035  |
  7. |        7    1.983088    2.474004  |
  8. |        8     1.90423   -.4354882  |
  9. |        9    1.545638    2.333574  |
 10. |       10    .6835439   -.2048338  |
 11. |       11    5.576303    2.614238  |
 12. |       12    3.044639    1.103051  |
     +-----------------------------------+
```

The *p*-values are obtained by determining the area under the tail of the *t* distribution that exceeds the absolute value of the *t*-value based on the residual degrees of freedom. By using the `ttail()` function, we can get the *p*-value based on the absolute value of the *t*-value and the residual degrees of freedom (which we saved and named `_eq2_dfr`). We multiply the value by 2 to obtain two-tailed *p*-values.

```
. generate p_currexp = ttail(_eq2_dfr,abs(t_currexp))*2
(1 missing value generated)
. generate p_prevexp = ttail(_eq2_dfr,abs(t_prevexp))*2
(1 missing value generated)
```

The *t*-values and *p*-values are shown below.

```
. list industry t_currexp t_prevexp p_currexp p_prevexp, abb(20) sep(0) noobs

  +---------------------------------------------------------------+
  | industry   t_currexp   t_prevexp   p_currexp   p_prevexp      |
  |---------------------------------------------------------------|
  |        1     .766093   -.4955963    .4563444    .6278678      |
  |        2           .           .           .           .      |
  |        3    1.308939     .315505    .2020064    .7548969      |
  |        4    1.165178    1.032858    .2447122    .3023579      |
  |        5    .6463869    2.128792    .5197304    .0360981      |
  |        6    .9908907    2.239035    .3224716    .0258258      |
  |        7    1.983088    2.474004    .0488136    .0142475      |
  |        8     1.90423   -.4354882    .0603874    .6643523      |
  |        9    1.545638    2.333574    .1256628    .0218208      |
  |       10    .6835439   -.2048338    .5054197    .8406501      |
  |       11    5.576303    2.614238    3.35e-08    .0091084      |
  |       12    3.044639    1.103051     .002696    .2715458      |
  +---------------------------------------------------------------+
```

This section has illustrated how the `statsby` command can take lengthy output and condense it into a simple and easy-to-read report. For more information, see `help statsby`.

9.13 Writing Stata programs

In section 9.3, we saw how you can combine a series of Stata commands together to create a do-file. This section takes that idea, combined with some of the other things we have learned in this chapter, to show you how to write your own Stata programs. Consider this very trivial program called `hello`.

```
. program hello
.   display "Hello world"
. end
```

One unconventional way to define this program is to type it right into the Stata Command window. Note that this merely defines (but does not execute) the program. By defining the program, it is placed into memory ready to be executed. To execute the program, we can type `hello` in the Stata Command window and the program executes, displaying `Hello world` in the Results window.

```
. hello
Hello world
```

It would be more conventional to define the `hello` program by typing it into the Do-file Editor and saving it as a file named `hello.ado` in your current working directory. Be sure that you save the file with the `.ado` extension! The `.ado` extension stands for "automatic do-file", telling Stata that this file contains a program that it can automatically load and execute.

Consider what happens when we type `hello` either in the Stata Command window or in a do-file. Stata first looks to see if this is a built-in Stata program (like `list`). If it is not, Stata then searches to see if `hello` is a program already stored in memory. If it is, Stata executes the copy of `hello` stored in memory. If it is not found in memory, Stata looks for `hello.ado` in a variety of places. One such place is the current working directory. If we stored a copy of `hello.ado` in our current working directory, Stata would find this file and load the program `hello` into memory and then execute it. If we typed `hello` again, this program would already be loaded in memory, so Stata would directly execute the copy stored in memory (and bypass the process of looking for it elsewhere).

Suppose that I have just invoked Stata and I type the command `hello`. Stata searches for and finds `hello.ado` stored in the current working directory. Stata loads the program into memory and then executes it.

```
. hello
Hello world
```

Say that I modify and save `hello.ado` to make it look like this.

```
. type hello.ado
program hello
  display "Good Morning and Hello world"
end
```

When I now type `hello`, I get the following result, making it appear that my changes were not saved.

```
. hello
Hello world
```

Although the changes were saved in `hello.ado`, they were not changed in the version of `hello` stored in memory. We can see the current version in memory by using the command `program list hello`, as shown below.

```
. program list hello
hello:
  1.   display "Hello world"
```

We can make Stata forget the version of `hello` in memory by using the `program drop` command, as shown below.

```
. program drop hello
```

Now when we type `hello`, Stata cannot find `hello` in memory, so it looks for it on disk and loads and executes the updated `hello.ado`.

```
. hello
Good Morning and Hello world
```

Now that we know how to write a basic Stata program, let's write a program that is useful. Sometimes you might want to run a series of regression commands on several outcome variables using the same set of predictors. For example, you might want to predict `wage` and `hours` from the variables `age` and `married`. A simple way of doing this would be using two `regress` commands.

```
. use wws2, clear
(Working Women Survey w/fixes)
. regress wage age married
  (output omitted)
. regress hours age married
  (output omitted)
```

The more outcome variables you have, the less appealing this strategy becomes. This is especially true if you may want to modify the predictors, because for each `regress` command, you would need to change the predictors. As we saw in section 9.8, you could use the `foreach` command to loop across different outcome variables, as shown below.

```
. foreach outcome in wage hours {
.   regress `outcome´ age married, noheader
. }
```

Consider a minor variation of the above commands where we would define a local macro `y` that contains the list of the outcome variables and a local macro `x` that contains the set of predictors. By changing the local macro `y`, you change the outcomes, and by changing the local macro `x`, you change the predictors.

```
local y wage hours
local x age married
foreach outcome of varlist `y´ {
  regress `outcome´ `x´, noheader
}
```

It is not too difficult to extend this strategy one step further to create a Stata program that will do this task for us. Consider this simple program below, called myreg.[8]

```
. type myreg.ado
program myreg
  syntax, y(varlist) x(varlist)

  foreach outcome of local y {
    regress `outcome´ `x´, noheader
  }
end
```

Before explaining the details of how this works, let's try running **myreg** as shown below. This runs a regression predicting **wage** from **age married** and then runs a regression predicting **hours** from **age married**.

```
. myreg, y(wage hours) x(age married)
```

wage	Coef.	Std. Err.	t	P>|t|	[95% Conf.	Interval]
age	-.0043748	.0230504	-0.19	0.849	-.0495772	.0408276
married	-.4619568	.256554	-1.80	0.072	-.9650651	.0411515
_cons	8.251724	.8662637	9.53	0.000	6.55296	9.950487

hours	Coef.	Std. Err.	t	P>|t|	[95% Conf.	Interval]
age	-.0848058	.041155	-2.06	0.039	-.1655117	-.0040999
married	-3.179256	.4584825	-6.93	0.000	-4.078351	-2.28016
_cons	42.33328	1.546136	27.38	0.000	39.30127	45.36529

At the heart of this program is the same **foreach** loop that we used before, which expects the outcome variables to be stored in the local macro y and the predictors to be stored in the local macro x. The **syntax** command tells Stata that the **myreg** program expects an option called y() and an option called x(), which both contain the names of one or more variables. When we ran **myreg**, Stata took the **y(wage hours)** option and stored the contents in the local macro y. Likewise, the **x(age married)** option caused the macro x to contain **age married**. By the time the program reached the **foreach** loop, the local macro y and x had the information to run the desired **regress** commands.

8. If you want to execute this yourself, type the program into the Do-file Editor and save it in your current working directory, naming it **myreg.ado**. This applies for the rest of the programs shown in this chapter. To execute them, you would need to save them as an ado-file with the same name as the program (e.g., **myreg.ado**).

To save space, I have used the **noheader** option, which suppresses the ANOVA table and other summary statistics for the model. Perhaps we might want to see the R-squared for each model. Below is a modified version of this program called myreg2, which adds a **display** command to display a line that provides the R-squared value.

As we saw in section 9.11, we can type **ereturn list** after any estimation command (like **regress**) to see the saved results that are available. One of the saved results is e(r2), which contains the R-squared value for the model. In the program below, the **display** command is used to display the R-squared value for the model.

```
. type myreg2.ado
program myreg2
  syntax, y(varlist) x(varlist)

  foreach outcome of local y {
    regress `outcome´ `x´, noheader
    display as text "R-squared is " as result e(r2)
  }
end
```

We can now execute this program.

```
. myreg2, y(wage hours) x(age married)

        wage        Coef.   Std. Err.      t    P>|t|     [95% Conf. Interval]

         age    -.0043748   .0230504    -0.19   0.849    -.0495772    .0408276
     married    -.4619568    .256554    -1.80   0.072    -.9650651    .0411515
       _cons     8.251724   .8662637     9.53   0.000      6.55296    9.950487

R-squared is .00145104

       hours        Coef.   Std. Err.      t    P>|t|     [95% Conf. Interval]

         age    -.0848058    .041155    -2.06   0.039    -.1655117   -.0040999
     married    -3.179256   .4584825    -6.93   0.000    -4.078351    -2.28016
       _cons     42.33328   1.546136    27.38   0.000     39.30127    45.36529

R-squared is .02240716
```

Let's consider a variation of the **myreg** program. Perhaps you want to perform a multiple regression by entering the first predictor into the model, then the first and second predictor, and then the first, second, and third predictor (and so forth), until all predictors are entered into the model. For simplicity, let's assume this is done just for one outcome variable.

The program **myhier**, below, can help automate running such models.

```
. type myhier.ado
program myhier
  syntax, y(varname) x(varlist)

  local xlist
  foreach xcurr of local x {
    local xlist `xlist´ `xcurr´
    regress `y´ `xlist´, noheader
  }
end
```

This program is like the myreg program that looped across outcome variables, but in myhier, we are looping across the predictor values (the values of x). We are not only looping across those values but also collecting the values in the local macro xlist, drawing upon what we learned about manipulating string macros in section 9.7. At first, the local xlist command is used to assign nothing to xlist. Inside the foreach loop, the command local xlist 'xlist' 'xcurr' is used to accumulate the name of any previous predictors xlist with the current predictor xcurr and assign that to a variable with the running list of predictors xlist.

When the program is run using the command below, the first time through the loop, the value of xlist starts as nothing and xcurr has the value of age, so xlist is assigned age.

The next (and last) time through the foreach loop, xlist contains age and xcurr contains married, so the value of xlist then becomes age married.

```
. myhier, y(wage) x(age married)
```

wage	Coef.	Std. Err.	t	P>|t|	[95% Conf.	Interval]
age	-.0027502	.0230443	-0.12	0.905	-.0479406	.0424401
_cons	7.896423	.8439127	9.36	0.000	6.241491	9.551355

wage	Coef.	Std. Err.	t	P>|t|	[95% Conf.	Interval]
age	-.0043748	.0230504	-0.19	0.849	-.0495772	.0408276
married	-.4619568	.256554	-1.80	0.072	-.9650651	.0411515
_cons	8.251724	.8662637	9.53	0.000	6.55296	9.950487

When running myhier, you might ask about the change in the R-squared at each stage in the model. Maybe we could automatically compute and display the change in R-squared after each regression model is displayed.

The program myhier2 extends myhier by computing and displaying the change in R-squared after each model.

```
. type myhier2.ado
program myhier2
  syntax, y(varname) x(varlist)

  local xlist
  local lastr2 0
  foreach xcurr of local x {
    local xlist `xlist´ `xcurr´
    regress `y´ `xlist´, noheader

    local changer2 = e(r2) - `lastr2´
    display as text "Change in R-squared " as result `changer2´
    local lastr2 = e(r2)
  }
end
```

Before entering the `foreach` loop, we create a variable called `lastr2`, which stands for the R-squared from the previous model. This is initially assigned a value of 0, so the first model is compared with an empty model (with no predictors) that would have an R-squared of 0. The command `local changer2 = e(r2) - 'lastr2'` computes the change in R-squared by taking the R-squared from the current model and subtracting the R-squared from the previous model. The `display` command is used to display the change in R-squared, and then the command `local lastr2 = e(r2)` stores the current R-squared in `lastr2` so that it can be used for comparison with the next model.

An example of using this program is shown below.

```
. myhier2, y(wage) x(age married)
```

wage	Coef.	Std. Err.	t	P>|t|	[95% Conf.	Interval]
age	-.0027502	.0230443	-0.12	0.905	-.0479406	.0424401
_cons	7.896423	.8439127	9.36	0.000	6.241491	9.551355

```
Change in R-squared 6.353e-06
```

wage	Coef.	Std. Err.	t	P>|t|	[95% Conf.	Interval]
age	-.0043748	.0230504	-0.19	0.849	-.0495772	.0408276
married	-.4619568	.256554	-1.80	0.072	-.9650651	.0411515
_cons	8.251724	.8662637	9.53	0.000	6.55296	9.950487

```
Change in R-squared .00144468
```

The `myhier` and `myhier2` programs combine a handful of techniques illustrated in this chapter, yet are powerful and handy. Although we have just scratched the surface of what you can do with Stata programming, I hope that this chapter gave you some useful tools that you can put to good use and gave you a taste of how powerful these tools can be if you should want to learn more about Stata programming. For more information, see `help adofile`.

10 Additional resources

Science has, as its whole purpose, the rendering of the physical world understandable and beautiful. Without this you have only tables and statistics.

—J. R. Oppenheimer

10.1 Online resources for this book

The online resources for this book can be found at the book's web site:

http://www.stata-press.com/books/dmus.html

Resources you will find there include...

- All the datasets used in the book. I encourage you to download the datasets used in this book, reproduce the examples, and try variations on your own. You can download all the datasets into you current working directory from within Stata by typing

```
. net from http://www.stata-press.com/data/dmus
. net get dmus1
. net get dmus2
```

- Errata (which I hope will be short or empty). Although I have tried hard to make this book error free, I know that some errors will be found, and they will be listed in the errata.
- Other resources that may be placed on the site after this goes to press. Be sure to visit the site to see what else may appear there.

10.2 Finding and installing additional programs

I believe that one of the greatest virtues of Stata is the way it combines ease of developing add-on programs with a great support structure for finding and downloading these programs. This virtue has led to a rich and diverse network of user-written Stata programs that extend the capabilities of Stata. This section provides information and tips to help you tap into the network of ever-growing user-written Stata programs.

The `findit` command is the easiest way to find and install user-written Stata programs. The `findit` command connects to Stata's own search engine, which indexes user-written Stata programs from all around the world. Typing, for example, `findit regression` searches for and displays Stata resources associated with the keyword `regression`. The resources searched include the Stata online help, Stata frequently asked questions (FAQs), the *Stata Journal* (SJ) and its predecessor, the *Stata Technical Bulletin* (STB), as well as programs posted on the web sites of Stata users from around the world. All these results are culled together and displayed in the Stata Viewer window.

I will be honest here. The results from `findit` can be a little bit overwhelming at first. The results are organized in a logical and consistent way that can quickly become familiar and easy to navigate. To that end, let's take a little time to look at the search results that you get when you run the command `findit regression`. Below I show an excerpted version of the results, keeping representative entries from each category.

If you do this search for yourself, your results may (and likely will) vary, because the contents being searched change over time. Each chunk of the results is preceded by a description of its contents and tips on how to read it.

```
. findit regression
```

The results (below) begin with a header showing that this is a search for the keyword `regression` and that it is going to provide you with two sets of search results. The first set of search results will be based on the official help files, FAQs, examples, SJs (*Stata Journal* articles and programs), and STBs (*Stata Technical Bulletin* articles and programs). The second set of search results will include web resources from Stata (which includes most, if not all, of the items in the first group) as well as web resources from other users (which does not include items from the first group).

```
Keywords:  regression
  Search:  (1) Official help files, FAQs, Examples, SJs, and STBs
           (2) Web resources from Stata and from other users
```

The heading for the first group of resources that will be searched is

```
Search of official help files, FAQs, Examples, SJs, and STBs
```

The first section of results starts with the search of the official help files, showing relevant help files with links to the help files, as well as relevant entries in the Stata manuals.

For example, the first entry refers to chapter 20 of the *Stata User's Guide* on the topic of *Estimation and postestimation commands*. For that entry, you could either flip to the chapter the physical manual or you could click on the links, which take you to the help file that then contains a link to the online manual.

```
[U]     Chapter 20  . . . . . . . . . . Estimation and postestimation commands
        (help estcom, postest, weights)

[U]     Chapter 25  . . . . Working with categorical data and factor variables
        (help generate, fvvarlist)

[U]     Chapter 26  . . . . . . . . . .  Overview of Stata estimation commands
        (help estcom)

[R]     anova . . . . . . . . . . . . . .  Analysis of variance and covariance
        (help anova)

[R]     areg  . . . . . . .  Linear regression with a large dummy-variable set
        (help areg)
```

The next section includes references to a Stata NetCourse and books related to the keyword `regression`.

```
NC461    . . . NetCourse 461: Introduction to Univariate Time Series with Stata
         http://www.stata.com/netcourse/nc461.html

Book     . . . . . . . . . . . . . . A Gentle Introduction to Stata, 2nd Edition
         . . . . . . . . . . . . . . . . . . . . . . . . . . . . . . Alan C. Acock
         http://www.stata.com/bookstore/acock2.html

Book     . . . . . . . . . . . An Introduction to Modern Econometrics Using Stata
         . . . . . . . . . . . . . . . . . . . . . . . . . . Christopher F. Baum
         http://www.stata.com/bookstore/imeus.html

Book     . . . .  An Introduction to Survival Analysis Using Stata, 2nd Edition
         . . . . . . . . . . . Mario Cleves, William Gould, and Roberto Gutierrez
         http://www.stata.com/bookstore/saus.html
```

The following section shows FAQs from the Stata web site and FAQs from other web sites. The web addresses are links that, when clicked on, will start your web browser and show the web page. Some selected entries are shown below.

```
FAQ      . . . . . . . . . . . . . . . . . . . . . . . . . . Chow and Wald tests
         . . . . . . . . . . . . . . . . . . . . . . . . . . . . . . . . . W. Gould
         8/07    How can I do a Chow test with the robust variance
                 estimates, that is, after estimating with
                 regress, vce(robust)?
                 http://www.stata.com/support/faqs/stat/chow2.html

FAQ      . . . . . .  How can I get an R-squared with robust regression (rreg)?
         . . . . . . . . . . . . . . . . . . . . . UCLA Academic Technology Services
         10/08   http://www.ats.ucla.edu/stat/stata/faq/rregr2.htm
```

The next section shows web pages with examples illustrating regression. These include examples from the Stata web site and other web sites. As with the FAQs, clicking on the web links will open your web browser and take you to the web address shown. A few examples are shown below.

```
Example . . . . . . . . . . . . . Capabilities:  linear regression and influence
         http://www.stata.com/capabilities/fit.html

Example .  Applied Longitudinal Data Anal.: Modeling Change & Event Occurrence
         . . . . . . . . . . . . . . . . . . . . . UCLA Academic Technology Services
         2/08    examples from the book Applied Longitudinal Data
                 Analysis: Modeling Change and Event Occurrence
                 by Judith D. Singer and John B. Willett
                 http://www.ats.ucla.edu/stat/stata/examples/alda/

Example . . . . . . . . Data analysis examples: Multivariate regression analysis
         . . . . . . . . . . . . . . . . . . . . . UCLA Academic Technology Services
         7/08    http://www.ats.ucla.edu/stat/stata/dae/mvreg.htm
```

The next set of results come from the *Stata Journal*. Three selected entries are shown below.

```
SJ-8-3  st0149  . . .  Implementing double-robust estimators of causal effects
        (help dr if installed)  .  R. Emsley, M. Lunt, A. Pickles, and G. Dunn
        Q3/08   SJ 8(3):334--353
        presents a double-robust estimator for pretest-posttest
        studies
```

```
SJ-7-4  st0067_3  . . . . Multiple imputation of missing values: Update of ice
        (help ice, ice_reformat, micombine, uvis if installed)  . . P. Royston
        Q4/07   SJ 7(4):445--464
        update of ice allowing imputation of left-, right-, or
        interval-censored observations

SJ-4-3  st0069  . . . . . Understanding the multinomial-Poisson transformation
        . . . . . . . . . . . . . . . . . . . . . . . . . . . . . . P. Guimaraes
        Q3/04   SJ 4(3):265--273                                (no commands)
        discusses the data transformations required to transform
        a Poisson regression into a logit model and vice versa
```

The first entry refers to a command named `dr` that accompanied the article written by R. Emsley et al. in the *Stata Journal* volume 8, number 3 on pages 334–353. You can click on the link for `st0149` to learn more about the commands or download them. You can also click on the link for `SJ 8(3):334--353` to take your web browser to the online version of the article (older articles are available for free, while newer articles are available for purchase).

The second entry refers to a *Stata Journal* article that provides a third set of updates[1] for programs that P. Royston wrote regarding multiple imputation of missing values. It is common to see such programs updated over time.

The last entry contains no commands, but the link `SJ 4(3):265--273` connects your web browser to the free online PDF version of the article.

This next section refers to programs from the *Stata Technical Bulletin* with links to download the programs. For example, you can download the `outreg` command, described in the first entry, by clicking on the link for `sg97.3`.

```
STB-59  sg97.3  . . . . . . . . . . . . . Update to formatting regression output
        (help outreg if installed)  . . . . . . . . . . . . . . . . J. L. Gallup
        1/01    p.23; STB Reprints Vol 10, p.143
        small bug fixes

STB-59  sg158 . . . . . . . . . . . . . . . . . .  Random-effects ordered probit
        (help reoprob, ghquadm if installed)  . . . . . . . . .  G. R. Frechette
        1/01    pp.23--27; STB Reprints Vol 10, pp.261--266
        estimates a random-effects ordered probit model
```

This next heading indicates that the second group of resources will be displayed. A search engine on the Stata site is used to search the world for resources related to the keyword(s) specified. When I did the search, it found 365 packages (but your results will likely differ). The search results first show entries from the *Stata Journal* and the *Stata Technical Bulletin*.

```
Web resources from Stata and other users

(contacting http://www.stata.com)

365 packages found (Stata Journal and STB listed first)
-------------------------------------------------------
```

1. The article number `st0067_3` indicates this is the third update.

These are references from the *Stata Journal* and *Stata Technical Bulletin*. The *Stata Journal* entries can be identified because they are from http://www.stata-journal.com/, while the *Stata Technical Bulletin* entries are from http://www.stata.com/stb/.

```
st0163 from http://www.stata-journal.com/software/sj9-2
    SJ9-2 st0163.  metandi: Meta-analysis of diagnostic... / metandi:
    Meta-analysis of diagnostic accuracy using / hierarchical logistic
    regression / by Roger Harbord, University of Bristol / Penny Whiting,
    University of Bristol / Support:  roger.harbord@bristol.ac.uk / After

sbe23_1 from http://www.stata-journal.com/software/sj8-4
    SJ8-4 sbe23_1.  Update: Meta-regression in Stata (revised) / Update:
    Meta-regression in Stata (revised) / by Roger Harbord, Department of
    Social Medicine, / University of Bristol, UK / Julian Higgins, MRC
    Biostatistics Unit, Cambridge, UK / Support:  roger.harbord@bristol.ac.uk

sg163 from http://www.stata.com/stb/stb61
    STB-61 sg163.  Stereotype Ordinal Regression / STB insert by Mark Lunt,
    ARC Epidemiology Unit, / University of Manchester, UK / Support:
    mdeasml2@fs1.ser.man.ac.uk / After installation, see help soreg

sg97_3 from http://www.stata.com/stb/stb59
    STB-59 sg97_3.  Update to formatting regression output / STB insert by
    John Luke Gallup, developIT.org / Support:
    john_gallup@alum.swarthmore.edu / After installation, see help outreg
```

These are selected entries from various people around the world. The web address gives some indication of where the information is from, and clicking on the package name will give you more information as well as the ability to download the programs. We can identify packages from employees at Stata because they are from http://www.stata.com/users/. These packages are not official parts of Stata. Employees of Stata tinker with creating programs just like other people do, and they post their programs in their user directory.

```
regh from http://www.fss.uu.nl/soc/iscore/stata
    regh.  heteroscedastic regression / regh computes maximum-likelihood
    estimates for the regression model / with multiplicative
    heteroscedasticity. It also estimates a model in / which the mean is
    included among the predictors for the log-variance. / Author:  Jeroen

regeffectsize from http://www.ats.ucla.edu/stat/stata/ado/analysis
    regeffectsize.  Computes effect size for regression models. / Philip B.
    Ender / UCLA Statistical Consulting / ender@ucla.edu / STATA ado and hlp
    files for simple main effects program / distribution-date: 20090122

spost9_ado from http://www.indiana.edu/~jslsoc/stata
    Distribution-date: 31Jul2009 / spost9_ado Stata 9 & 10 commands for the
    post-estimation interpretation / of regression models. Use package
    spostado.pkg for Stata 8. / Based on Long & Freese - Regression Models for
    Categorical Dependent / Variables Using Stata. Second Edition. / Support

stcstat from http://www.stata.com/users/wgould
    stcstat.  ROC curves after Cox regression / Program by William Gould,
    Stata Corp <wgouldstata.com>. / Statalist distribution, 04 December 2001.
    / / cmd:stcstat calculates the area under the ROC curve based on the /
    last model estimated by help:stcox.
```

The following packages are from the Statistical Software Components (SSC) archive (also often called the Boston College archive). The SSC archive is the premier download site for storing and retrieving Stata programs developed by users from all over the world. The packages can be identified as coming from SSC because the URL for them starts with http://fmwww.bc.edu/RePEc/bocode/.

```
avplot3 from http://fmwww.bc.edu/RePEc/bocode/a
    `AVPLOT3': module to generate partial regression plots for subsamples /
    avplot3 generates "partial regression plots" from an analysis of /
    covariance model, where a category variable has been included in /
    dummy-variable form among the regressors along with a constant / term

gologit2 from http://fmwww.bc.edu/RePEc/bocode/g
    `GOLOGIT2': module to estimate generalized logistic regression models for
    ordinal dependent variables / gologit2 estimates generalized ordered logit
    models for ordinal / dependent variables. A major strength of gologit2 is
    that it can / also estimate three special cases of the generalized model:

outreg from http://fmwww.bc.edu/RePEc/bocode/o
    `OUTREG': module to format regression output for published tables / This
    is a revision of outreg (as published in STB-46, updated in / STB-49,
    STB-56, STB-58, STB-59) which supersedes the latest / version available
    from the STB. The outreg ado file takes output / from any estimation
```

Having searched the packages for the keyword we specified (i.e., `regression`), the following results show any `table of contents` entries that matched the keyword `regression`. As you might surmise, the following section probably overlaps considerably with the previous section; however, it is possible that a table of contents entry may use the word regression even if the contents of the package do not. In such cases, the table of contents entries below may not be contained within the list above. My search found 95 references, but your search may find a different number.

```
95 references found in tables of contents
-----------------------------------------

http://www.stata.com/stb/stb61/
    STB-61 May 2001 / Contrasts for categorical variables: update / Patterns
    of missing values / Simulating disease status and censored age / Violin
    plots for Stata 6 and 7 / Quantile plots, generalized: update to Stata 7.0
    / Update to metabias to work under version 7 / Update of metatrim to work

http://www.indiana.edu/~jslsoc/stata/
    SPost: commands for the post-estimation interpretation of regression
    models / for categorical dependent variables. Based on Long, 1997,
    Regression Models / for Categorical and Limited Dependent Variables and
    Long & Freese, 2003, / Regression Models for Categorical Dependent

http://www.stata.com/users/mcleves/
    Materials by Mario A. Cleves / Mario A. Cleves -- packages created while
    working as a senior / biostatistician at Stata Corp. / Median test for K
    independent samples / Robust test for the equality of variances / Graph
    median and mean survival times / Logit reg. when outcome is measured with
```

This concludes our exploration of the results from the `findit regression` command. Here is more about the `findit` command in a question-and-answer format.

Why are there so many programs from http://fmwww.bc.edu/RePEc/bocode/?

This is the SSC archive, often called the Boston College archive. This is the premier location for storing Stata programs so that they can be retrieved by other users. In fact, Stata has an official program called `ssc` to help people describe and install packages from the SSC archive, as well as see what is new (see `help ssc` for more information).

How do I read an entry from the Stata Journal?

Here is a sample entry.

```
SJ-4-3  st0067 . . . . . . . . . . . . .  Multiple imputation of missing values
        (help micombine, mijoin, mvis if installed) . . . . . . . . P. Royston
        Q3/04   SJ 4(3):227--241
        implementation of the MICE (multivariate imputation by
        chained equations) method of multiple multivariate data
        imputation
```

This entry refers to an article appearing in the *Stata Journal*, volume 4, number 3 on pages 227–241 and was published in the third quarter of 2004. The article concerns multiple imputation of missing values and introduces three commands—`micombine`, `mijoin`, and `mvis`—and these names are linked to their corresponding help files. The article and the package of commands is given the tag `st0067`, meaning that this is the 67th *Stata Journal* program that fell into the "statistics" category (abbreviated as `st`). The tag uniquely identifies every article in the *Stata Journal*. For your information, other category names used for tags include `an` for announcements, `dm` for data management, `ds` for datasets, `gn` for general, `gr` for graphics, `pr` for programming and utilities, and `up` for updates. Here is another example of a *Stata Journal* entry from `findit`.

```
SJ-7-4  st0067_3  . . . . Multiple imputation of missing values: Update of ice
        (help ice, ice_reformat, micombine, uvis if installed)  . . P. Royston
        Q4/07   SJ 7(4):445--464
        update of ice allowing imputation of left-, right-, or
        interval-censored observations
```

No, you are not experiencing déjà vu, this is an entry that shows an update (in fact, the third update) with respect to `st0067`. We could type `findit st0067` to find information about the original article as well as all the updates.

Why do I see multiple entries for a program?

I typed `findit format regression output`, and the output included many different programs, including multiple entries referring to a program that sounded promising called `outreg`. To focus just on that particular program, I ran `findit outreg`. The output included the following entries:

```
STB-59  sg97.3 . . . . . . . . . . . . . Update to formatting regression output
        (help outreg if installed)  . . . . . . . . . . . . . . . . J. L. Gallup
        1/01    p.23; STB Reprints Vol 10, p.143
        small bug fixes

STB-58  sg97.2 . . . . . . . . . . . . . Update to formatting regression output
        (help outreg if installed)  . . . . . . . . . . . . . . . . J. L. Gallup
        11/00   pp.9--13; STB Reprints Vol 10, pp.137--143
        update allowing user-specified statistics and notes, 10%
        asterisks, table and column titles, scientific notation for
        coefficient estimates, and reporting of confidence interval
        and marginal effects

STB-49  sg97.1 . . . . . . . . . . . . . . . . . . . . . . . . Revision of outreg
        (help outreg if installed)  . . . . . . . . . . . . . . . . J. L. Gallup
        5/99    p.23; STB Reprints Vol 9, pp.170--171
        updated for Stata 6 and improved

STB-46  sg97  . . . . . . .  Formatting regression output for published tables
        (help outreg if installed)  . . . . . . . . . . . . . . . . J. L. Gallup
        11/98   pp.28--30; STB Reprints Vol 8, pp.200--202
        takes output from any estimation command and formats it as
        in journal articles
```

The output also included this entry from the SSC archive:

```
outreg from http://fmwww.bc.edu/RePEc/bocode/o
    'OUTREG': module to format regression output for published tables / This
    is a revision of outreg (as published in STB-46, updated in / STB-49,
    STB-56, STB-58, STB-59) which supersedes the latest / version available
    from the STB. The outreg ado file takes output / from any estimation
```

To understand this output, let's follow the evolution of this program. In November 1998, the program **outreg** appeared in the *Stata Technical Bulletin* number 46 (STB-46). The program was assigned the tag **sg97**, being the 97th program that dealt with the topic of general statistics.[2] In May 1999, a revision was published in STB-49, named **sg97.1** (the first revision to sg97), and updates also appeared in STB-58 in November 2000 and again in STB-59 in January 2001. It might appear that **sg97.3** is the most up-to-date version of the program, but people may still continue to update their programs and have sites where the latest versions can be accessed. The entry **outreg from http://fmwww.bc.edu/RePEc/bocode/o** is from the SSC archive and may have even newer versions of programs than the latest version available from the *Stata Journal* or the *Stata Technical Bulletin*. Clicking on the link for **outreg** from the SSC archive showed me a description with a distribution date of 5/14/2002, an even more recent version than the latest STB version. Because this seemed a bit old, I searched for the keyword **outreg** to see if I could find an even newer version and found an entry for **outreg2**; see below.

2. All the STB tags are shown at http://www.stata.com/products/stb/journals/stb1.pdf.

```
outreg2 from http://fmwww.bc.edu/RePEc/bocode/o
    'OUTREG2': module to arrange regression outputs into an illustrative table
    / outreg2 provides a fast and easy way to produce an illustrative / table
    of regression outputs. The regression outputs are produced / piecemeal and
    are difficult to compare without some type of / rearrangement. outreg2
```

This is a related program that was inspired by `outreg` and has been updated recently (August 2009 when I looked) and would merit further investigation. It is not uncommon for user-written programs to inspire other user-written programs.

Why did findit show a program that duplicates a Stata command?

I wanted to see what Stata had to help me convert a string variable into a numeric variable, so I typed `findit string numeric`. The output included the following, which describes a built-in Stata command called `destring`.

```
[R]     destring . . . . . . . . . . . . . . Change string variables to numeric
        (help destring)
```

But the output also included this:

```
dm45_2 from http://www.stata.com/stb/stb52
    STB-52 dm45_2. Changing string variables to numeric: correction / STB
    insert by Nicholas J. Cox, University of Durham, UK / Support:
    n.j.cox@durham.ac.uk / After installation, see help destring

dm45_1 from http://www.stata.com/stb/stb49
    STB-49 dm45_1.  Changing string variables to numeric: update. / STB insert
    by Nicholas J. Cox, University of Durham, UK.  / Support:
    n.j.cox@durham.ac.uk / After installation, see help destring.

dm45 from http://www.stata.com/stb/stb37
    STB-37 dm45.  Changing string variables to numeric. / STB insert by /
    Nicholas J. Cox, University of Durham, UK; / William Gould, Stata
    Corporation. / Support:  n.j.cox@durham.ac.uk and wgould@stata.com /
    After installation, see help destring.
```

The `destring` command was published in STB-37, and then revised in STB-49 and again in STB-52. Then, at some point, this handy program was incorporated into official Stata with the same name `destring`. When an official command shares a name with a user-written command, Stata will load the official command. So even if you downloaded the user-written version of `destring`, the official version would take precedence.

Are there multiple programs that do the same thing?

I wanted to find a program to perform White's test regarding homogeneity of variance in regression, so I typed `findit white test`. The output included these two programs that both appear to do the same thing:

```
white from http://fmwww.bc.edu/RePEc/bocode/w
    'WHITE': module to perform White's test for heteroscedasticity / htest,
    szroeter, and white provide tests for the assumption of / the linear
    regression model that the residuals e are / homoscedastic, i.e., have
    constant variance. The tests differ / with respect to the specification of

whitetst from http://fmwww.bc.edu/RePEc/bocode/w
    'WHITETST': module to perform White's test for heteroskedasticity /
    whitetst computes White's test for heteroskedasticity following / regress
    or cnsreg. This test is a special case of the / Breusch-Pagan test (q.v.
    bpagan). The White test does not require / specification of a list of
```

This is not an uncommon occurrence, and it means that you can try out both programs to see which one you think is most suitable for you.

Who, besides the authors, verifies that the programs are right?

No one does. This is an important point to consider. Even the most diligent programmers make mistakes, and the wonderful and skilled group of people who donate their Stata programs to the Stata community are not exempt from this. Before you rely on a program that you download, you may want to take some steps to establish its accuracy. This might mean running the program against known results or checking with others (or the authors) to see what has been done to validate the program.

Does Nick Cox do anything but write Stata programs?

Yes, he also edits the *Stata Journal*, writes a *Speaking Stata* column for the *Stata Journal*, and answers many questions on Statalist.

10.3 More online resources

There are many online resources to help you learn and use Stata. Here are some additional resources I would highly recommend:

- The Stata *Resources and support* page provides a comprehensive list of online resources that are available for Stata. It lists official resources that are available from StataCorp as well as resources from the Stata community. See http://www.stata.com/support/.
- The Stata *Resource links* page provides a list of resources created by the Stata community to help you learn and use Stata; see http://www.stata.com/links/. Among the links included there, I would highly recommend the UCLA ATS Stata web resources at http://www.ats.ucla.edu/stat/stata/, which include FAQs, annotated Stata output, textbook examples solved in Stata, and online classes and seminars about Stata.
- The *Stata Frequently Asked Questions* page is special because it not only contains many frequently asked questions but also includes answers! The FAQs cover

common questions (e.g., How do I export tables from Stata?) as well as esoteric (e.g., How are estimates of rho outside the bounds [-1,1] handled in the two-step Heckman estimator?). You can search the FAQs using keywords, or you can browse the FAQs by topic. See http://www.stata.com/support/faqs/.

- *Statalist* is an independently operated listserver that connects over 3,000 Stata users from all over the world. I can say from personal experience that the community is both extremely knowledgeable and friendly, welcoming questions from newbies and experts alike. Even if you never post a question of your own, you can learn quite a bit from subscribing to this list or reading the archives (which go all the way back to 2002). You can learn more about Statalist at http://www.stata.com/statalist/.
- The *Stata Journal* is published quarterly with articles that integrate various aspects of statistical practice with Stata. Although current issues and articles are available by subscription, articles over three years old are available for free online as PDF files. See http://www.stata-journal.com/.
- The *Stata Technical Bulletin* is the predecessor of the *Stata Journal*. All these issues are available for free online. Although many of the articles may be out of date, there are many gems that contain timeless information. For more information, see http://www.stata.com/bookstore/stbj.html.

A Common elements

You can't fix by analysis what you bungled by design.

—Richard Light, Judith Singer, and John Willett

A.1 Introduction

This appendix covers topics that are common to many Stata commands or are an inherent part of the structure of Stata. These topics are gathered together and illustrated in this appendix so that the earlier chapters can reference the appendix without repetitively covering these topics.

The appendix begins with an overview of the general structure (syntax) of Stata commands as described in section A.2. Next section A.3 illustrates the `by` prefix. Most commands allow you to use the `by` prefix to repeat the command across different subgroups of observations. Stata comments are useful for adding documentation to your commands and do-files as described in section A.4. Although it is not obvious, every variable in Stata is assigned a data storage type; these data types are described in section A.5. Logical expressions (see section A.6) are most commonly used after the `if` qualifier, where we can specify which observations will be included in the command. Stata functions (see section A.7) can be used in a variety of contexts, but arise most commonly with the `generate` and `replace` commands. The `if` and `in` qualifiers (see section A.8) are permitted by most Stata commands for specifying a subset of observations on which the command operates. The `keep` command can be used for specifying observations or variables that should be retained in the dataset, and the `drop` commands can be used to specify observations or variables that should be eliminated from the dataset (see section A.9). The way that Stata defines and handles missing values is described in section A.10. The appendix concludes with section A.11, illustrating different ways that Stata permits you to specify variable lists, including time-saving shortcuts.

A.2 Overview of Stata syntax

Most Stata commands, including user-written commands (described in section 10.2), follow the same general Stata syntax. If I told you that there was a new Stata command called `snorf`, you would probably be able to guess how the command might work because it probably works according to the same general syntax that most Stata commands follow. This section illustrates the general rules of Stata syntax.

After using `wws2.dta`, issuing the `summarize` command without anything else summarizes all the variables in the dataset.

```
. use wws2
(Working Women Survey w/fixes)

. summarize
  (output omitted)
```

We can specify one or more variables (i.e., a *varlist*) to specify which variables we want summarized (see section A.11 for more about variable lists).

```
. summarize age wage

    Variable |       Obs        Mean    Std. Dev.       Min        Max
-------------+--------------------------------------------------------
         age |      2246    36.22707    5.337859         21         48
        wage |      2244    7.796781     5.82459          0   40.74659
```

Stata supports four different kinds of weights—`pweights`, `fweights`, `aweights`, and `iweights`. Depending on the command, one or more of these weights may be permitted. Below we use the `summarize` command and the variable `fwt` as a frequency weight.

```
. summarize age wage [fweight=fwt]

    Variable |       Obs        Mean    Std. Dev.       Min        Max
-------------+--------------------------------------------------------
         age |      9965    36.21686    5.402677         21         48
        wage |      9950    7.935715    6.020044          0   40.19808
```

You can add an `if` qualifier to summarize just some of the observations. Below we get the summary statistics for `wage` and `age` just for those who work in a union.

```
. summarize age wage if union == 1

    Variable |       Obs        Mean    Std. Dev.       Min        Max
-------------+--------------------------------------------------------
         age |       461    36.46421    5.197964         22         46
        wage |       460     8.70157    4.311632          0   39.23074
```

You can use an `in` qualifier to specify certain observations the command should work on. Below we obtain the summary statistics for the first 200 observations in the dataset.

```
. summarize age wage in 1/200

    Variable |       Obs        Mean    Std. Dev.       Min        Max
-------------+--------------------------------------------------------
         age |       200       35.71    5.210725         22         44
        wage |       200    6.600294    5.538314   1.561996   38.70926
```

Options can be specified by placing a comma at the end of the standard part of the command and then listing the options. For example, the `summarize` command includes an option called `detail` that provides detailed summary statistics. In the example below, detailed summary statistics are requested for all variables.

```
. summarize, detail
  (output omitted)
```

Detailed summary statistics can be requested for specific variables, e.g., `age` and `wage`.

```
. summarize age wage, detail
  (output omitted)
```

When you want to specify both an `if` qualifier and options, be sure to place the comma after the `if` qualifier (as shown below). The `if` qualifier is considered a standard part of Stata commands, as is the `in` qualifier.

```
. summarize age wage if union==1, detail
  (output omitted)
```

We can preface the summarize command with bysort married:, which executes the command once for every level of married, i.e., for those who are not married and then for those who are married.

```
. bysort married: summarize age wage

-------------------------------------------------------------------------------
-> married = 0

    Variable |       Obs        Mean    Std. Dev.       Min        Max
-------------+--------------------------------------------------------
         age |       804    36.50995    5.203328         22         47
        wage |       804    8.092001    6.354849          0   40.19808

-------------------------------------------------------------------------------
-> married = 1

    Variable |       Obs        Mean    Std. Dev.       Min        Max
-------------+--------------------------------------------------------
         age |      1442    36.06935    5.406775         21         48
        wage |      1440     7.63195    5.501786   1.004952   40.74659
```

We can also include the if qualifier and options when using bysort married:, as shown below.

```
. bysort married: summarize age wage if union==1, detail
  (output omitted)
```

bysort married: is an example of a prefix command, in particular, the by prefix. There are several other prefix commands in Stata; see help prefix for more information about them.

This summarizes the major elements of the overall syntax of Stata commands. For more information about Stata syntax, see help language.

A.3 Working across groups of observations with by

Sometimes you might want to run a command separately for each group of observations in your dataset. For example, gasctrysmall.dta contains information on gas prices and inflation rates from four different countries, numbered 1, 2, 3, and 4.

```
. use gasctrysmall
. list, sepby(ctry)

     +--------------------------+
     | ctry   year   gas   infl |
     |--------------------------|
  1. |    1   1974   .78   1.32 |
  2. |    1   1975   .83    1.4 |
     |--------------------------|
  3. |    2   1971   .69   1.15 |
  4. |    2   1971   .77   1.15 |
  5. |    2   1973   .89   1.29 |
     |--------------------------|
  6. |    3   1974   .42   1.14 |
     |--------------------------|
  7. |    4   1974   .82   1.12 |
  8. |    4   1975   .94   1.18 |
     +--------------------------+
```

We can use the command **summarize gas** to obtain summary statistics for the variable **gas** for the entire dataset, as shown below.

```
. summarize gas
    Variable |       Obs        Mean    Std. Dev.       Min        Max
-------------+--------------------------------------------------------
         gas |         8       .7675    .1596201        .42        .94
```

But suppose that we want these results separated by each country. One option would be to use the `if` qualifier, as illustrated below for the first country.

```
. summarize gas if ctry == 1
    Variable |       Obs        Mean    Std. Dev.       Min        Max
-------------+--------------------------------------------------------
         gas |         2        .805    .0353553        .78        .83
```

Rather than repeating this command over and over with an `if` qualifier, we can prefix this command with `by ctry:`, which tells Stata that we would like it to run the **summarize gas** command once for each level of `ctry`. We first need to sort the data by `ctry` using the `sort ctry` command.

```
. sort ctry
. by ctry: summarize gas

------------------------------------------------------------------------------
-> ctry = 1
    Variable |       Obs        Mean    Std. Dev.       Min        Max
-------------+--------------------------------------------------------
         gas |         2        .805    .0353553        .78        .83

------------------------------------------------------------------------------
-> ctry = 2
    Variable |       Obs        Mean    Std. Dev.       Min        Max
-------------+--------------------------------------------------------
         gas |         3    .7833333    .1006645        .69        .89
```

```
-> ctry = 3
    Variable |       Obs        Mean    Std. Dev.       Min        Max
-------------+--------------------------------------------------------
         gas |         1         .42           .        .42        .42

-> ctry = 4
    Variable |       Obs        Mean    Std. Dev.       Min        Max
-------------+--------------------------------------------------------
         gas |         2         .88    .0848528        .82        .94
```

If you prefer, you can combine the `sort` and `by` steps into one step by using the `bysort` command, as shown below.

```
. bysort ctry: summarize gas
  (output omitted)
```

`bysort` can be further abbreviated to `bys`, as shown below.

```
. bys ctry: summarize gas
  (output omitted)
```

Most Stata commands permit you to use the `by` prefix, including descriptive commands (e.g., `summarize` or `tabulate`) and estimation commands (e.g., `regress` or `logistic`). For more information about using the `by` prefix, see chapter 7 and `help by`.

A.4 Comments

When you create a do-file, you might think that you will remember everything that you did and why you did it. For me, my memories fade. When I look at my do-files later, I am grateful when I find notes to myself explaining what I was doing and why. That is where comments can help.

The most common kind of Stata comment begins with an asterisk (*). Anything following the asterisk (on that line) is treated as a comment, as in the example below. This kind of comment can be used at the command line or in a do-file.

```
. use wws2
(Working Women Survey w/fixes)
. * get summary statistics for age
. summarize age
    Variable |       Obs        Mean    Std. Dev.       Min        Max
-------------+--------------------------------------------------------
         age |      2246    36.22707    5.337859         21         48
```

Within a do-file, you can also add a double slash, //, at the end of a command to treat the rest of the line as a comment.

```
. summarize age // get summary stats for age
    Variable |       Obs        Mean    Std. Dev.       Min        Max
-------------+--------------------------------------------------------
         age |      2246    36.22707    5.337859         21         48
```

It can be hard to read long commands that run off the right side of your screen. Instead, you can use continuation comments, symbolized as ///, to indicate that a command continues on the next line. The `recode` command below uses continuation comments to make the command easier to read.

```
. recode occupation (1/3=1 "White Collar") ///
>                   (5/8=2 "Blue Collar")  ///
>                   (4 9/13=3 "Other"), generate(occ3)
(1918 differences between occupation and occ3)
```

Within do-files, comments can be enclosed between the delimiters /* and */. Anything that is enclosed between the /* and */ is ignored. This can be used within a command, as illustrated below.

```
. summarize age /* wage */
    Variable |       Obs        Mean    Std. Dev.       Min        Max
-------------+--------------------------------------------------------
         age |      2246    36.22707    5.337859         21         48
```

You can also span multiple lines with /* and */. This can be used to comment out a block of commands that you do not wish to be executed, or it can be useful for writing multiple-line comments. However, be careful to remember that after you start a comment with /*, you end it with */; otherwise, everything thereafter will be treated as a comment. You can find more information about comments by typing `help comments`.

A.5 Data types

This section describes the different ways that Stata stores variables, technically referred to as *data types* or *storage types*. Let's consider `cardio3.dta` as an example.

(*Continued on next page*)

```
. use cardio3

. describe

Contains data from cardio3.dta
  obs:             5
 vars:            26                          23 Dec 2009 15:12
 size:           660 (99.9% of memory free)
-------------------------------------------------------------------------------
              storage  display     value
variable name   type   format      label      variable label
-------------------------------------------------------------------------------
id              long   %10.0f                 Identification variable
fname           str15  %15s                   First name
lname           str10  %10s                   Last name
bp1             int    %3.0f                  Systolic BP: Trial 1
pl1             int    %3.0f                  Pulse: Trial 1
bp2             int    %3.0f                  Systolic BP: Trial 2
pl2             int    %3.0f                  Pulse: Trial 2
bp3             int    %3.0f                  Systolic BP: Trial 3
pl3             int    %3.0f                  Pulse: Trial 3
bpmean          float  %9.0g                  Mean blood pressure
plmean          float  %9.0g                  Mean pulse
gender          str6   %9s                    Gender of person
bmo             float  %4.0f                  Birth month
bda             float  %4.0f                  Birth day
byr             float  %4.0f                  Birth year
bhr             double %4.0f                  Birth hour
bmin            double %4.0f                  Birth minute
bsec            double %4.0f                  Birth second
age             byte   %3.0f                  Age of person
weight          float  %9.0g                  Weight (in pounds)
famhist         long   %12.0g      famhistl   Family history of heart disease
income          double %10.2f                 Income
zipcode         long   %12.0g                 Zip Code (5 digit)
heart_attack_~t float  %9.0g                  Risk of heart attack from
                                                treadmill test
bdate           float  %td                    Birth date
bdatetime       double %tc                    Birth date and time
-------------------------------------------------------------------------------
Sorted by:
```

Notice the different values in the column labeled “storage type”. These fall into two general types, string types and numeric types.

A string variable, sometimes referred to as a character variable, permits you to store any combination of numbers and characters. String variables can be as short as 1 (i.e., `str1`) and as wide as 244 (i.e., `str244`). This dataset contains three string variables—`fname`, `lname`, and `gender`. `fname` is stored as `str15`, which means that it is a string variable and can hold names that are up to 15 characters wide. `lname` is stored as `str10`, which can hold up to 10 characters, and `gender` is stored as `str6`, which can hold up to 6 characters.

Let's look at the contents of these three string variables:

```
. list fname lname gender

     +------------------------------+
     |    fname        lname gender |
     |------------------------------|
  1. |     Fred      Canning   male |
  2. |    Mario   Washington   male |
  3. |     Hong          Sun   male |
  4. | Salvador     Riopelle   male |
  5. |    Sonia        Yosef female |
     +------------------------------+
```

You might notice that the longest `fname`, `Salvador`, is only eight characters wide, but `fname` is stored as a `str15`. We can use the command `compress fname` to ask Stata to inspect `fname` and change its storage type to the most frugal size possible, as shown below. The result is that `fname` is now stored as a `str8`. If we had a dataset with many observations, this could yield a substantial savings of space.

```
. compress fname
fname was str15 now str8

. describe fname

              storage  display     value
variable name   type   format      label      variable label
-------------------------------------------------------------------------------
fname           str8   %9s                    First name
```

When creating new variables, Stata automatically chooses the storage type for us. Let's illustrate this by creating a variable called `fullname` that combines `fname` and `lname`.

```
. generate fullname = fname + " " + lname
```

In the output below, we can see that the longest name belongs to Salvador, whose full name is 17 characters long.

```
. list fullname
```

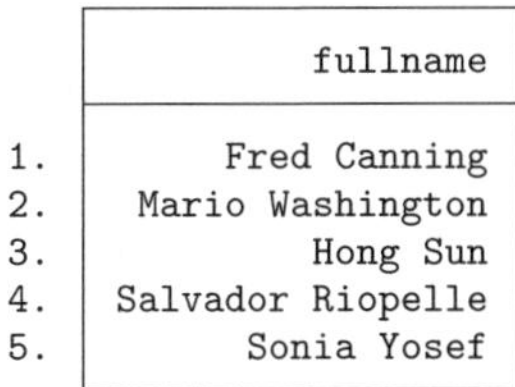

```
     +-------------------+
     |          fullname |
     |-------------------|
  1. |      Fred Canning |
  2. |  Mario Washington |
  3. |          Hong Sun |
  4. | Salvador Riopelle |
  5. |       Sonia Yosef |
     +-------------------+
```

The `describe` command shows that Stata created this variable using the `str17` storage type. Without any extra effort on our part, Stata chose an appropriate length for this new variable.[1]

1. If, for some reason, you needed to specify a different length for `fullname` when creating it (say, `str25`), you could specify `generate str25 fullname = fname + " " + lname`.

```
. describe fullname

              storage  display     value
variable name   type   format      label      variable label
-------------------------------------------------------------------------
fullname        str17  %17s
```

Let's now turn our attention to numeric variables. Stata has five different numeric data storage types: `byte`, `int`, `long`, `float`, and `double`. The first three types are for storing whole numbers (such as age in whole years). They differ based on the largest numbers they can store. The last two types, `float` and `double`, can store nonwhole numbers, such as income (measured to the penny) or weight (measured to the nearest tenth of a pound). The `double` type can store numbers with greater precision (after the decimal) and can store larger numbers (before the decimal) than can the `float` data type.

At the end of `cardio3.dta`, the variables `bdate` and `bdatetime` are special kinds of numeric variables containing date and time values. The `bdate` variable contains date values (birth date), which have the display format of `%td`. The `bdatetime` variable is a date-and-time variable that contains the date and time of birth, and hence, is displayed using a `%tc` format. These kinds of variables are discussed at the end of this section, as well as in section 5.8 and section 5.9.

Let's consider each of these numeric types in more detail. First, consider the storage type called `byte`. This type can store whole numbers between −127 and 100. The variable `age` is stored as a `byte`, and the values of age are all within this range.

```
. describe age

              storage  display     value
variable name   type   format      label      variable label
-------------------------------------------------------------------------
age             byte   %3.0f                  Age of person
. summarize age

    Variable |       Obs        Mean    Std. Dev.       Min        Max
-------------+--------------------------------------------------------
         age |         5        25.4    9.787747         16         40
```

The `int` storage type can hold whole numbers that range from −32,767 to 32,740. Note that the variables `bp1`–`bp3` and `pl1`–`pl3` are stored as `int`. This makes sense because they all can exceed 100 (see below), the highest value for a `byte`. (The variable `pl2` has a maximum value of 97, so it could have been stored as a `byte`.)

```
. describe bp1 bp2 bp3 pl1 pl2 pl3

              storage  display     value
variable name   type   format      label      variable label
-------------------------------------------------------------------------
bp1             int    %3.0f                  Systolic BP: Trial 1
bp2             int    %3.0f                  Systolic BP: Trial 2
bp3             int    %3.0f                  Systolic BP: Trial 3
pl1             int    %3.0f                  Pulse: Trial 1
pl2             int    %3.0f                  Pulse: Trial 2
pl3             int    %3.0f                  Pulse: Trial 3
```

```
. summarize bp1 bp2 bp3 pl1 pl2 pl3

    Variable |       Obs        Mean    Std. Dev.       Min        Max
-------------+--------------------------------------------------------
         bp1 |         5       116.6    7.635444        105        124
         bp2 |         5       117.4    19.17811         86        136
         bp3 |         5         114    11.13553        101        129
         pl1 |         5        74.6    23.36236         52        105
         pl2 |         5          83    12.38951         64         97
-------------+--------------------------------------------------------
         pl3 |         5        93.8    33.20693         52        128
```

The **long** storage type can hold whole numbers that range from −2,147,483,647 to 2,147,483,620. This is a useful storage type when you need to accurately store large whole numbers, like the variable **id** in this dataset.

```
. describe id

              storage  display     value
variable name   type   format      label      variable label
-------------------------------------------------------------------------
id              long   %10.0f                 Identification variable

. list id

     +-----------+
     |        id |
     |-----------|
  1. | 133520121 |
  2. | 275031298 |
  3. | 345821920 |
  4. |  29303092 |
  5. | 938329302 |
     +-----------+
```

Although the **byte**, **int**, and **long** storage types are adequate for storing whole numbers, they cannot store nonwhole numbers (numbers that include fractional values). Stata has the **float** and **double** types for such numbers.

A variable stored as a **float** has approximately seven digits of accuracy. As shown using the **describe** command below, the person's weight (**weight**) measured to the nearest tenth of a pound is stored as a **float** type. This type can hold even the largest weights that we could encounter.

```
. describe weight

              storage  display     value
variable name   type   format      label      variable label
-------------------------------------------------------------------------
weight          float  %9.0g                  Weight (in pounds)

. summarize weight

    Variable |       Obs        Mean    Std. Dev.       Min        Max
-------------+--------------------------------------------------------
      weight |         5      157.88    30.87915      109.9      186.3
```

Note! Floats losing accuracy

The id variable is a nine-digit number and is stored in this dataset as a long type. If id was stored as a float, it would have been stored with only seven digits of accuracy, leading to imprecise storage.

As you might have guessed, the double type has the highest level of precision, storing values with up to 16 digits of accuracy. As shown below, the variable income is stored as a double.

```
. describe income

              storage  display     value
variable name   type   format      label      variable label
-------------------------------------------------------------------

income          double %10.2f                 Income
```

Looking at the values of income (below), it is a good thing that the incomes were stored using the double storage type because the highest income contains nine total digits (seven digits before the decimal and two digits after the decimal). If these values were stored as float, some information could be lost.

```
. list income

     +------------+
     |     income |
     |------------|
  1. | 5987435.32 |
  2. | 1784327.58 |
  3. |  987628.32 |
  4. | 3272828.43 |
  5. | 8229292.21 |
     +------------+
```

When you use generate to make new numeric variables, they are stored as float by default. Consider the example below, where we make income2 to be an exact copy of income. However, it is not an exact copy because the default storage type is float. You can see the discrepancies that result below.

```
. generate income2 = income

. list income income2

     +------------------------+
     |     income     income2 |
     |------------------------|
  1. | 5987435.32     5987436 |
  2. | 1784327.58     1784328 |
  3. |  987628.32    987628.3 |
  4. | 3272828.43     3272829 |
  5. | 8229292.21     8229292 |
     +------------------------+
```

Likewise, if we make a copy of id, we also lose information because the id variable has nine digits in it but the newly generated variable is stored as a float with only seven digits of accuracy.

```
. generate id2 = id
. format id2 %9.0f
. list id2 id

     +-----------------------+
     |       id2          id |
     |-----------------------|
  1. | 133520120   133520121 |
  2. | 275031296   275031298 |
  3. | 345821920   345821920 |
  4. |  29303092    29303092 |
  5. | 938329280   938329302 |
     +-----------------------+
```

Whenever you create a variable in Stata, you can manually select the type you want to store the variable as. For example, let's make a copy of income, calling it incomed, and manually store it as double. This allows us to avoid the loss of accuracy as we saw above.

```
. generate double incomed = income
. format incomed %12.2f
. list income income2 incomed

     +------------------------------------+
     |     income    income2      incomed |
     |------------------------------------|
  1. | 5987435.32    5987436   5987435.32 |
  2. | 1784327.58    1784328   1784327.58 |
  3. |  987628.32   987628.3    987628.32 |
  4. | 3272828.43    3272829   3272828.43 |
  5. | 8229292.21    8229292   8229292.21 |
     +------------------------------------+
```

Let's create idlong, explicitly specifying that it should be created as a type long. The results below show that idlong is the same as id:

```
. generate long idlong = id
. format idlong %9.0f
. list id id2 idlong

     +-----------------------------------+
     |        id         id2      idlong |
     |-----------------------------------|
  1. | 133520121   133520120   133520121 |
  2. | 275031298   275031296   275031298 |
  3. | 345821920   345821920   345821920 |
  4. |  29303092    29303092    29303092 |
  5. | 938329302   938329280   938329302 |
     +-----------------------------------+
```

To avoid accuracy problems, you can instruct Stata to use double as the default type for creating new variables. You can do this by typing

```
. set type double
```

and subsequent numeric variables created using generate will be stored as double for the duration of your Stata session. Or you can type

```
. set type double, permanently
```

and Stata will also save this change for future Stata sessions. These newly created variables will take more storage space. But as shown in the examples below, you could later use the compress command to ask Stata to try to store the variables using a more frugal storage type.

Some commands that create variables do not permit you to control the storage type of a variable. For example, consider the recode command, below, where we create a dummy variable mil to indicate if someone is a millionaire. Because the original variable was stored as double, the generated variable (mil) is also stored as double. Because mil is just a 0/1 variable, we could save space by storing it as byte.

```
. recode income (min/999999.99=0) (1000000/max=1), gen(mil)
(5 differences between income and mil)
. desc income mil
              storage  display     value
variable name   type   format      label      variable label
-----------------------------------------------------------------------
income          double %10.2f                 Income
mil             double %9.0g                  RECODE of income (Income)
```

For one variable, this is not a big problem. But you might have a very large dataset with many such variables stored as double that could be stored as byte.

We can use the compress mil command to ask Stata to inspect the mil variable and select the most frugal storage type that would not result in any loss of information. As we can see below, Stata converted the variable type for mil from double to byte.

```
. compress mil
mil was double now byte
. describe mil
              storage  display     value
variable name   type   format      label      variable label
-----------------------------------------------------------------------
mil             byte   %9.0g                  RECODE of income (Income)
```

Multiple variables can be specified on the compress command. Or you can enter the compress command without specifying any variables to apply it to all variables. It will inspect each variable and select the most efficient storage type for each variable, as illustrated below.

```
. compress
pl2 was int now byte
bmo was float now byte
bda was float now byte
byr was float now int
famhist was long now byte
bdate was float now int
bhr was double now byte
bmin was double now byte
bsec was double now byte
```

Notice that `pl2` is now a `byte`. This is because all the pulse values were under 100.

```
. list pl2

     +-----+
     | pl2 |
     |-----|
  1. |  87 |
  2. |  88 |
  3. |  97 |
  4. |  79 |
  5. |  64 |
     +-----+
```

Say that we use the `replace` command to make `pl2` to be 120 for the first observation. As you can see below, the `replace` command detected that change and promoted `pl2` to type `int`.

```
. replace pl2 = 120 in 1
pl2 was byte now int
(1 real change made)

. desc pl2

              storage  display     value
variable name   type   format      label      variable label
-------------------------------------------------------------------
pl2             int    %3.0f                  Pulse: Trial 2
```

Finally, let's return to the original `cardio3.dta` and consider special issues that arise when using numeric variables that represent dates. The variable `bdate` contains the person's birth date. Below we can see the variables containing the month, day, and year as well as the `bdate` variable, which contains the birth date stored in one variable. (As described in section 5.8, it is the `%td` format that yields the nicely formatted display of `bdate`.)

```
. use cardio3

. list bmo bda byr bdate

     +--------------------------------+
     | bmo   bda    byr         bdate |
     |--------------------------------|
  1. |   7     6   1989     06jul1989 |
  2. |  11    12   1987     12nov1987 |
  3. |   3    10   1981     10mar1981 |
  4. |   6     5   1981     05jun1981 |
  5. |   2     1   1982     01feb1982 |
     +--------------------------------+
```

The describe command shows that the bdate variable is stored using the float storage type, which is sufficient for a variable that stores a date.

```
. describe bdate

              storage  display     value
variable name   type   format      label      variable label
-------------------------------------------------------------------------------
bdate           float  %td                    Birth date
```

But consider the variable bdatetime. This contains both the date and the time of birth in one variable (see section 5.9 for more about date-and-time variables). Let's list the individual variables that contain the day, month, year, hour, minute, and second of birth, as well as the combined bdatetime variable, which stores and displays the date and time of birth as a date-and-time value.

```
. list bmo bda byr bhr bmin bsec bdatetime

     +---------------------------------------------------------------+
     | bmo   bda    byr   bhr   bmin   bsec            bdatetime   |
     |---------------------------------------------------------------|
  1. |   7     6   1989    10      6      7   06jul1989 10:06:07   |
  2. |  11    12   1987    14     11     22   12nov1987 14:11:22   |
  3. |   3    10   1981     5     45     55   10mar1981 05:45:55   |
  4. |   6     5   1981     2     23     25   05jun1981 02:23:25   |
  5. |   2     1   1982     3     11     33   01feb1982 03:11:33   |
     +---------------------------------------------------------------+
```

In contrast with the bdate variable, the bdatetime variable is stored using the storage type double.

```
. describe bdatetime

              storage  display     value
variable name   type   format      label      variable label
-------------------------------------------------------------------------------
bdatetime       double %tc                    Birth date and time
```

It is essential that variables that contain combined date-and-time values be stored as a double; otherwise, you will lose information. When I created this variable, I did so using the following command:

```
. generate double bdatetime = mdyhms(bmo, bda, byr, bhr, bmin, bsec)
```

Note that I explicitly specified to use the storage type double for bdatetime. Let's repeat this command below to create a variable named bdatetime2 but omit double from the generate command; this omission will cause the variable to be stored using the default type of float.

```
. generate bdatetime2 = mdyhms(bmo, bda, byr, bhr, bmin, bsec)
. format bdatetime2 %tc
. describe bdatetime2

              storage  display     value
variable name   type   format      label      variable label
-------------------------------------------------------------------------------
bdatetime2      float  %tc
```

When bdatetime2 is stored as a float, its values do not exactly match the minutes and seconds when the person was born. This is because of the loss of accuracy when storing this kind of variable as a float and forgetting to explicitly store it as a double.

```
. list bmo bda byr bhr bmin bsec bdatetime2

     +--------------------------------------------------------------------+
     | bmo   bda    byr   bhr   bmin   bsec                   bdatetime2 |
     |--------------------------------------------------------------------|
  1. |   7     6   1989    10      6      7   06jul1989 10:06:31 |
  2. |  11    12   1987    14     11     22   12nov1987 14:11:34 |
  3. |   3    10   1981     5     45     55   10mar1981 05:46:01 |
  4. |   6     5   1981     2     23     25   05jun1981 02:23:40 |
  5. |   2     1   1982     3     11     33   01feb1982 03:11:03 |
     +--------------------------------------------------------------------+
```

In summary, understanding Stata data types not only helps you save space by storing your variables in the most frugal manner possible, it also helps avoid mistakes with the loss of accuracy of variables. For even more information, see help data types.

A.6 Logical expressions

When I think about a logical expression, I think of a statement that is true or false. Consider the statement, "Sally works fewer than 40 hours per week." If Sally works 36 hours, then the statement is true, or if she works 43 hours, then the statement is false. Using wws2.dta, we can count how many women work fewer than 40 hours using the count command followed by if (hours < 40). The expression (hours < 40) is a logical expression that can be either true or false. As you can see below, there are 759 women in this dataset for whom this expression is true.

```
. use wws2
(Working Women Survey w/fixes)

. count if (hours < 40)
  759
```

When performing comparisons, you can use < for less than, <= for less than or equal to, > for greater than, and >= for greater than or equal to. You can use == to test whether two values are equal, and you can use != to test whether two values are not equal. Two examples are shown below.

```
. * how many women work 40 hours per week?
. count if (hours == 40)
 1093

. * how many women work at most 40 hours per week?
. count if (hours <= 40)
 1852
```

Sometimes you want to combine logical expressions. For example, whether a woman works 40 or 41 hours combines the two expressions with *or*, asking whether the first is true or the second is true. The symbol for *or* in Stata is |. Perhaps you want to know whether a woman works at least 40 and no more than 50 hours. Here you want to know if the first expression *and* the second expression are true. The symbol for *and* in Stata is `&`. These are illustrated below.

```
. * how many women work 40 hours or 41 hours?
. count if (hours == 40) | (hours == 41)
1095
. * how many women work at least 40 and at most 50 hours?
. count if (hours >= 40) & (hours <= 50)
1383
```

Some logical expressions can be created using Stata functions. For example, we might want to count how many cases have missing values for the `hours` variable. We can do this with `missing(hours)`, which is true if `hours` is missing and false if it is not missing. But perhaps we want to count how many are not missing. In Stata, the symbol for *not* is `!`. Two examples are shown below.

```
. count
2246
. count if missing(hours)
  4
. count if ! missing(hours)
2242
```

These results show that there are a total of 2,246 women in the dataset, of which 4 had missing values for `hours` and 2,242 had nonmissing values for `hours`.

This raises the issue of missing values. As described in section A.10, missing values are stored as the largest possible values. Previously, we counted how many women worked 40 hours or fewer (repeated below). We might (wrongfully) count the number of women who work over 40 hours by specifying `if hours > 40`. This is wrong because the four women with missing work hours are treated as though they work over 40 hours.

```
. count if hours <= 40
1852
. * WRONG (includes missing values too)
. count if hours > 40
 394
```

Below we repeat the last command and properly exclude the women with missing values for `hours`. This command correctly counts the number of women who work over 40 hours because it also stipulates that `hours` must also be nonmissing.

```
. * Correct (excludes missing values)
. count if (hours > 40) & ! missing(hours)
 390
```

Here is another example where you need to explicitly exclude the missing values from a logical expression. When counting the number of women who do not work 40 hours, we need to exclude the missing values because they are also not 40. When constructing logical expressions like these, you should ask yourself what would happen to missing (positively infinite) values and be sure to account for them.

```
. * how many women do not work 40 hours
. count if (hours != 40) & ! missing(hours)
 1149
```

Each logical statement can be either true or false, and Stata assigns a value of 1 to true statements and a value of 0 to false statements. Say that we wanted to make a variable called **workfull** that would be 1 if you worked 40 or more hours and 0 otherwise. We could create that variable like this:

```
. generate workfull = 1 if (hours >= 40) & ! missing(hours)
(763 missing values generated)
. replace  workfull = 0 if (hours < 40)
(759 real changes made)
```

Or we can simplify these commands into one statement:

```
. generate workfull = (hours >= 40) if ! missing(hours)
(4 missing values generated)
. tab workfull, missing

   workfull |      Freq.     Percent        Cum.
------------+-----------------------------------
          0 |        759       33.79       33.79
          1 |      1,483       66.03       99.82
          . |          4        0.18      100.00
------------+-----------------------------------
      Total |      2,246      100.00
```

When the expression `(hours >= 40)` is true, it evaluates to 1; when it is false, it evaluates to 0, yielding the exact results that we wished. By including `if ! missing(hours)`, this command is only executed when **hours** is nonmissing. When **hours** is missing, the value of **workfull** appropriately remains missing.

Say that you wanted to identify women who worked exactly 35, 40, or 45 hours per week. You could make a variable that identifies these women as shown below.

```
. generate workdummy = (hours==35) | (hours==40) | (hours==45)
> if ! missing(hours)
(4 missing values generated)
```

Another way you could do this would be to use the `inlist()` function. Note how this is easier to read and would be much easier to use than the example above if we had even more levels of work hours that we were selecting.

```
. generate workdummy = inlist(hours,35,40,45) if ! missing(hours)
(4 missing values generated)
```

So far, we have dealt with variables that take on whole number values like **hours**. When you form logical expressions with fractional numbers (numbers with values to the right of the decimal place), you might have problems concerning precision and rounding.

For example, there is a variable named **wage2** in this dataset that contains wages rounded to two digits (i.e., the nearest penny). Say that we would like to list the women who make exactly $8.90 per hour. First, I can show you three examples from the dataset to show that such women do exist.

```
. list idcode wage2 if inlist(idcode, 2231, 1370, 1435)

       ┌────────────────┐
       │ idcode   wage2 │
       ├────────────────┤
1288.  │   2231     8.9 │
1660.  │   1435     8.9 │
1683.  │   1370     8.9 │
       └────────────────┘
```

Now having seen that three cases exist, it would seem logical that we can list all such cases like this:

```
. list idcode wage2 if wage2 == 8.9
```

This is frankly baffling, and if we had not previously looked at the three cases with such wages, we might falsely conclude that there are no such women who make exactly $8.90 per hour. This lack of result raises the issue of the precision with which fractional values are stored using computers. For whole numbers, there is no ambiguity about how a number is stored. But for fractional numbers, the precision of the computer representation can vary. As we saw in section A.5, Stata has the `float` and `double` data types for fractional numbers. As we see below, `wage2` is stored as a data type `float`. But Stata does all internal computations with the highest precision possible (i.e., using `double` precision). So when the `wage2` variable (which is a `float`) is compared with the value 8.9 (which is represented as a `double`), the two values are not found to be exactly equal.

```
. desc wage2

              storage  display     value
variable name   type   format      label      variable label
-----------------------------------------------------------------------
wage2          float   %9.0g                  Wages, rounded to 2 digits
```

The solution is to get these two values to the same level of precision (both at the level of a `float`), as illustrated below. When `wage2` is compared with `float(8.9)`, we see all the cases where the wages are exactly equal to $8.90.

```
. list wage2 if wage2 == float(8.9), sep(0)

       ┌───────┐
       │ wage2 │
       ├───────┤
1112.  │   8.9 │
1288.  │   8.9 │
1487.  │   8.9 │
1572.  │   8.9 │
1660.  │   8.9 │
1683.  │   8.9 │
2001.  │   8.9 │
       └───────┘
```

You might be thinking of solving this by trying to make a double-precision version of `wage2` and then comparing that with 8.9. Just for fun, we give this a try below.

```
. generate double wage2d = wage2
(2 missing values generated)
. list wage2d if wage2d == 8.9
```

This showed no observations.

Below we see why this did not work. You can take a more precise value (like 8.9 stored as a `double`) and convert it into a less precise value (like 8.9 stored as a `float`), but you cannot take a less precise value and recover its precision by storing it as a more precise value (i.e., by converting it from `float` into `double`). As you can see below, there are slight differences in the values for `wage2` and `wage2d`.

```
. list wage2 wage2d if wage2 == float(8.9), sep(0)

       +--------------------+
       | wage2       wage2d |
       |--------------------|
 1112. |   8.9    8.8999996 |
 1288. |   8.9    8.8999996 |
 1487. |   8.9    8.8999996 |
 1572. |   8.9    8.8999996 |
 1660. |   8.9    8.8999996 |
 1683. |   8.9    8.8999996 |
 2001. |   8.9    8.8999996 |
       +--------------------+
```

For more information about logical expressions, see `help exp` and `help if`.

A.7 Functions

A function allows you to pass in one or more values and get a value back in return. For example, when you pass in a value of 4 to the `sqrt()` function, it returns the value of 2. This is illustrated below using `wws2.dta`, where the variable `sqrtwage` is created, which is the square root of `wage`.

```
. use wws2
(Working Women Survey w/fixes)
. generate sqrtwage = sqrt(wage)
(2 missing values generated)
. list wage sqrtwage in 1/5

     +---------------------+
     |     wage   sqrtwage |
     |---------------------|
  1. |  7.15781   2.675408 |
  2. | 2.447664   1.564501 |
  3. | 3.824476   1.955627 |
  4. | 14.32367   3.784662 |
  5. | 5.517124   2.348856 |
     +---------------------+
```

As you might imagine, Stata has many functions. In fact, there are so many functions that `help functions` classifies them into about eight different types. The `sqrt()` function is one example of a mathematical function.

Other commonly used mathematical functions include `abs()` (absolute value), `exp()` (exponential), and `ln()` (natural log). Among the mathematical functions are functions for rounding numbers. For example, the `int()` function is used below to take a woman's net worth and remove the pennies. By contrast, the `round()` function is used to round the woman's net worth to the nearest dollar (using traditional rounding rules).

```
. generate networth1 = int(networth)

. generate networth2 = round(networth)

. list idcode networth networth1 networth2 in 1/5, abb(20)
```

	idcode	networth	networth1	networth2
1.	5159	157.8097	157	158
2.	5157	-4552.336	-4552	-4552
3.	5156	-3175.523	-3175	-3176
4.	5154	7323.667	7323	7324
5.	5153	-1482.876	-1482	-1483

The `round()` function is not just limited to rounding to whole numbers (which it does by default). The variable `networth3` is created to contain the net worth rounded to the nearest tenth (i.e., the nearest dime), and the variable `networth4` contains the net worth rounded to the nearest hundred. For more information on math functions in Stata, see `help math functions`.

```
. generate networth3 = round(networth, 0.1)

. generate networth4 = round(networth, 100)
```

We can see the original and the rounded variables below.

```
. list idcode networth networth3 networth4 in 1/5, abb(20)
```

	idcode	networth	networth3	networth4
1.	5159	157.8097	157.8	200
2.	5157	-4552.336	-4552.3	-4600
3.	5156	-3175.523	-3175.5	-3200
4.	5154	7323.667	7323.7	7300
5.	5153	-1482.876	-1482.9	-1500

Note! Stata Function, what's your function?

Perhaps if you asked Jack Sheldon (http://www.schoolhouserock.tv/Conjunction.html) about Stata functions, he might share this with you:

> Stata Function, what's your function?
> Passing in values, getting back results.
> Stata Function, how's that function?
> I've got eight types of functions, that get most of my job done.
> Stata Function, what's their function?
> See the help for `function`, that will get you pretty far.

Let's next explore some of the functions that Stata classifies as programming functions; see `help programming functions`. For example, below we see how the `missing()` function can be used to get summary statistics for `wage` only if the value of `grade` is not missing.

```
. summarize wage if ! missing(grade)
    Variable |       Obs        Mean    Std. Dev.       Min        Max
-------------+--------------------------------------------------------
        wage |      2240    7.803143    5.827396          0   40.74659
```

Suppose that we wanted to get summary statistics for wages but only for those women whose occupations are coded as 1, 3, 5, 8, or 10. Rather than specifying five `if` qualifiers, we can specify `if inlist(occupation,1,3,5,8,10)`, as shown below.

```
. summarize wage if inlist(occupation,1,3,5,8,10)
    Variable |       Obs        Mean    Std. Dev.       Min        Max
-------------+--------------------------------------------------------
        wage |      1391    7.496547    5.451008   1.032247   40.74659
```

Or you might be interested in the summary statistics of wages just for the people who work from 0 to 40 hours per week (inclusively). Although you could isolate such observations by specifying `if (hours >=0 & hours <=40)`, you can use `if inrange(hours,0,40)` as a shortcut, shown below.

```
. summarize wage if inrange(hours,0,40)
    Variable |       Obs        Mean    Std. Dev.       Min        Max
-------------+--------------------------------------------------------
        wage |      1850     7.45381    5.537428   1.004952   40.74659
```

Random-number functions (see `help random number`) can be useful for selecting observations at random to spot-check data. Below the `set seed` command[2] is used,

2. You can choose any number you like for the `set seed` command. If we skip the `set seed` command, we would get a different sample of cases every time we run these commands. The `set seed` command allows us to obtain the same series of random numbers every time for reproducibility.

followed by generate with the runiform() function to create the random variable rannum, which ranges uniformly from 0 to 1.

```
. * ensure you get the same sample
. set seed 8675309
. * make a random number
. generate rannum = runiform()
```

We can then display approximately 1% of the observations selected at random, as shown below.

```
. list idcode age race if rannum <= 0.01
  (output omitted)
```

Or say that we want to inspect exactly 10 observations. We can sort the data on rannum (which sorts the data into a random order) and then show the first 10 observations.

```
. sort rannum
. list idcode age race in 1/10
  (output omitted)
```

This section has just scratched the surface of the wide variety of functions included in Stata. This section omitted string functions because they were covered in section 5.4, date functions because they were illustrated in section 5.8, and date-and-time functions because they were illustrated in section 5.9. For a comprehensive list of the functions included in Stata, see help functions.

A.8 Subsetting observations with if and in

Nearly all Stata commands allow you to specify an if qualifier, restricting the command to operate on the observations that satisfy a logical qualifier. Likewise, nearly all Stata commands permit you to include an in qualifier, which limits the command to operate on a subset of observations specified by the observation number. if and in are illustrated in this section using wws2.dta. We start by computing summary statistics for currexp for all observations in the dataset.

```
. use wws2
(Working Women Survey w/fixes)
. summarize currexp
    Variable |       Obs        Mean    Std. Dev.       Min        Max
-------------+--------------------------------------------------------
     currexp |      2231    5.185567    5.048073          0         26
```

We can add if married==1 to restrict our analysis just to those who are married.

```
. summarize currexp if married==1
    Variable |       Obs        Mean    Std. Dev.       Min        Max
-------------+--------------------------------------------------------
     currexp |      1432    5.078212    4.934883          0         26
```

Of course, we can specify more complex qualifiers. For example, here we restrict our analysis just to those who are married and under age 40. You can see section A.6 for more information about logical expressions.

```
. summarize currexp if (married==1) & (age <40)
    Variable |       Obs        Mean    Std. Dev.       Min        Max
-------------+--------------------------------------------------------
     currexp |      1016    4.829724    4.585643          0         21
```

If you want to include one or more options, they come after the `if` qualifier. For example, to obtain detailed summary statistics for the previous command, the `detail` option would be supplied.

```
. summarize currexp if (married==1) & (age <40), detail
  (output omitted)
```

The `in` qualifier controls which observations the command includes based on the observation number. By using the `in 1/4` qualifier, we can display just the first four observations.

```
. list idcode age race married in 1/4

     +-------------------------------+
     | idcode   age   race   married |
     |-------------------------------|
  1. |   5159    38      2         0 |
  2. |   5157    24      2         1 |
  3. |   5156    26      1         1 |
  4. |   5154    32      1         1 |
     +-------------------------------+
```

You can show the last four observations by specifying `in -4/L`. The `-4` means the 4th observation from the last and `L` means the last observation.

```
. list idcode age race married in -4/L

        +-------------------------------+
        | idcode   age   race   married |
        |-------------------------------|
  2243. |      4    43      1         1 |
  2244. |      3    42      2         0 |
  2245. |      2    34      2         0 |
  2246. |      1    25      2         0 |
        +-------------------------------+
```

I think `in` is especially useful when reading large raw datasets. Suppose that the file `wws_subset.txt` was a very large raw data file with many observations. To see if I am reading the raw data correctly, I start by reading just the first five observations, as shown below.

```
. infile idcode age race married nevmar using wws_subset.txt in 1/5
(eof not at end of obs)
(5 observations read)
. list
```

	idcode	age	race	married	nevmar
1.	1	37	2	0	0
2.	12	2	37	2	0
3.	0	12	3	42	2
4.	0	1	12	4	43
5.	1	1	0	17	6

As the listing shows, it seems that I am not reading the data correctly. I double-check and realize that I omitted one of the variables. I fix that and then try reading the data again. This would appear to be more promising.

```
. infile idcode age race married nevmar grade using wws_subset.txt in 1/5
(eof not at end of obs)
(5 observations read)
. list
```

	idcode	age	race	married	nevmar	grade
1.	1	37	2	0	0	12
2.	2	37	2	0	0	12
3.	3	42	2	0	1	12
4.	4	43	1	1	0	17
5.	6	42	1	1	0	12

I then read the entire file and list the first five and last five observations. I find this to be a quick and easy check that identifies many (but not all) problems when reading in data.

```
. infile idcode age race married nevmar grade using wws_subset.txt
(2246 observations read)
. list in 1/5
```

	idcode	age	race	married	nevmar	grade
1.	1	37	2	0	0	12
2.	2	37	2	0	0	12
3.	3	42	2	0	1	12
4.	4	43	1	1	0	17
5.	6	42	1	1	0	12

```
. list in -5/1
```

	idcode	age	race	married	nevmar	grade
2242.	5153	35	1	0	1	12
2243.	5154	44	1	1	0	16
2244.	5156	42	1	1	0	12
2245.	5157	38	2	1	0	12
2246.	5159	43	2	0	0	12

It is also possible to use `if` when reading a raw dataset. Say that `wws_subset.txt` is an extremely large raw-data file, and we are only interested in the observations of those who are married. Rather than reading the entire file into memory and then deleting the observations for those who are unmarried, we can simply read in just the 1,442 observations for those who are married, as illustrated below.

```
. infile idcode age race married nevmar grade using wws_subset.txt if married==1
(1442 observations read)
```

For more information, see `help if` and `help in`.

Warning! if is not an option

Sometimes people try putting the `if` qualifier after the comma like it is an option.

```
summarize currexp, if married==1 detail
```

This generates an error message like

```
option if not allowed r(198);
```

This error might look like the command does not support the `if` qualifier, but it is saying that the command does not recognize an option called `if`. Repeating the command with the comma coming after the `if` qualifier produces the desired results.

```
summarize currexp if married==1, detail
```

A.9 Subsetting observations and variables with keep and drop

The `keep` and `drop` commands have two uses: to eliminate variables from the current dataset or to eliminate observations from the current dataset. We will explore each of these in turn using `cardio1.dta`, shown below.

```
. use cardio1
. describe
Contains data from cardio1.dta
  obs:            5
 vars:           12                          22 Dec 2009 19:50
 size:          120 (99.9% of memory free)

              storage  display     value
variable name   type   format      label      variable label

id              byte   %3.0f                  Identification variable
age             byte   %3.0f                  Age of person
bp1             int    %3.0f                  Systolic BP: Trial 1
bp2             int    %3.0f                  Systolic BP: Trial 2
bp3             int    %3.0f                  Systolic BP: Trial 3
bp4             int    %3.0f                  Systolic BP: Trial 4
bp5             int    %3.0f                  Systolic BP: Trial 5
pl1             int    %3.0f                  Pulse: Trial 1
pl2             byte   %3.0f                  Pulse: Trial 2
pl3             int    %3.0f                  Pulse: Trial 3
pl4             int    %3.0f                  Pulse: Trial 4
pl5             byte   %3.0f                  Pulse: Trial 5

Sorted by:
```

This dataset contains five observations, with an ID variable, age, five measures of blood pressure, and five measures of pulse. Let's list the five observations from this dataset.

```
. list
```

	id	age	bp1	bp2	bp3	bp4	bp5	pl1	pl2	pl3	pl4	pl5
1.	1	40	115	86	129	105	127	54	87	93	81	92
2.	2	30	123	136	107	111	120	92	88	125	87	58
3.	3	16	124	122	101	109	112	105	97	128	57	68
4.	4	23	105	115	121	129	137	52	79	71	106	39
5.	5	18	116	128	112	125	111	70	64	52	68	59

To drop the variable **age**, we can type **drop age** as shown below. By using the **list** command, we can see that **age** has been dropped from the working dataset.

```
. drop age
. list
```

	id	bp1	bp2	bp3	bp4	bp5	pl1	pl2	pl3	pl4	pl5
1.	1	115	86	129	105	127	54	87	93	81	92
2.	2	123	136	107	111	120	92	88	125	87	58
3.	3	124	122	101	109	112	105	97	128	57	68
4.	4	105	115	121	129	137	52	79	71	106	39
5.	5	116	128	112	125	111	70	64	52	68	59

You can specify multiple variables after the **drop** command to drop more than one variable. Say that you wanted to drop all the blood pressure variables. You could type **drop bp1 bp2 bp3 bp4 bp5** or you could type **drop bp1-bp5** (as shown below) because all these variables are consecutively positioned in the dataset. The listing of the observations shows that the blood pressure variables have been dropped from the dataset.

```
. drop bp1-bp5
. list

     ┌─────────────────────────────────┐
     │ id   pl1   pl2   pl3   pl4   pl5 │
     ├─────────────────────────────────┤
  1. │  1    54    87    93    81    92 │
  2. │  2    92    88   125    87    58 │
  3. │  3   105    97   128    57    68 │
  4. │  4    52    79    71   106    39 │
  5. │  5    70    64    52    68    59 │
     └─────────────────────────────────┘
```

Let's read in **cardio1.dta** again to illustrate the use of the **keep** command, which removes variables by specifying just the variables you want to keep. So if we wanted to just keep **id** and the blood pressure readings, we could use the **keep** command below.

```
. use cardio1
. keep id bp*
```

As we can see below, the working dataset now contains just the identification variable and the blood pressure measurements.

```
. list

     ┌─────────────────────────────────┐
     │ id   bp1   bp2   bp3   bp4   bp5 │
     ├─────────────────────────────────┤
  1. │  1   115    86   129   105   127 │
  2. │  2   123   136   107   111   120 │
  3. │  3   124   122   101   109   112 │
  4. │  4   105   115   121   129   137 │
  5. │  5   116   128   112   125   111 │
     └─────────────────────────────────┘
```

The **drop** and **keep** commands can also be used to eliminate observations from the current dataset. The **drop** command can be combined with **if** and **in** to specify which observations you would like to eliminate (drop) from the dataset (see section A.8 for more information about **if** and **in**). For example, the command **drop in 1/3** would eliminate the first three observations from the current dataset. Or as shown below, the command **drop if age < 30** drops all people who are under age 30.

```
. use cardio1
. drop if age < 30
(3 observations deleted)
```

```
. list
```

	id	age	bp1	bp2	bp3	bp4	bp5	pl1	pl2	pl3	pl4	pl5
1.	1	40	115	86	129	105	127	54	87	93	81	92
2.	2	30	123	136	107	111	120	92	88	125	87	58

As you might expect, the `keep if` and `keep in` commands specify which observations to keep. Typing `keep in 2/3` would keep the second and third observations. Below we use `cardio1.dta` and then issue the `keep if age <= 20` command to keep just those who are 20 years and younger.

```
. use cardio1

. keep if age <= 20
(3 observations deleted)

. list
```

	id	age	bp1	bp2	bp3	bp4	bp5	pl1	pl2	pl3	pl4	pl5
1.	3	16	124	122	101	109	112	105	97	128	57	68
2.	5	18	116	128	112	125	111	70	64	52	68	59

You can type `help drop` for more information about the `drop` and `keep` commands.

A.10 Missing values

Missing values are the bane of a researcher's existence. If I could, I would outlaw missing values. But for now, we need to contend with them. Fortunately, Stata offers many features and tools to help you deal with missing values. One such feature is the ability to have many different kinds of missing values. In fact, Stata gives you 27 different missing-value codes. The default is `.`, and on top of that you can use `.a`, `.b`, `.c`, up to `.z`. For example, your study might have people who are missing because they refused to answer (and you might assign those missing values a `.r`), and others did not know (and you might give those a missing value of `.d`), and so forth. All these different types of missing values are recognized as being missing in Stata commands, but the different types of codes permit you to distinguish one kind of missing value from another.

Consider the dataset `cardio1amiss` below. In this dataset, there are five measures of blood pressure (`bp1`–`bp5`) and five measures of pulse (`pl1`–`pl5`). In this study, people could have a missing value because of a recording error (which I assigned as a `.a`) or because of withdrawing from the study (which I assigned as a `.b`).

```
. use cardio1amiss
. list
```

	id	age	bp1	bp2	bp3	bp4	bp5	pl1	pl2	pl3	pl4	pl5
1.	1	40	115	86	129	105	.b	54	87	93	81	.b
2.	2	30	123	136	107	111	120	92	88	125	87	58
3.	3	16	.a	122	101	109	112	105	97	128	57	68
4.	4	23	105	115	121	129	137	52	79	71	106	39
5.	5	18	116	128	112	125	.a	70	.a	52	68	59

Although it might not seem relevant, you should know how Stata stores missing values inside the computer. Missing values are stored as numbers larger than any valid values. Below we sort the data on `bp5` and then list the sorted observations. The observations with missing values on `bp5` are at the end of the list because they have the highest values. Missing values are higher than all other valid values. Among the missing values, they are sorted from `.` being the lowest missing value, followed by `.a`, then `.b`, continuing up to `.z` being the highest missing value.

```
. sort bp5
. list
```

	id	age	bp1	bp2	bp3	bp4	bp5	pl1	pl2	pl3	pl4	pl5
1.	3	16	.a	122	101	109	112	105	97	128	57	68
2.	2	30	123	136	107	111	120	92	88	125	87	58
3.	4	23	105	115	121	129	137	52	79	71	106	39
4.	5	18	116	128	112	125	.a	70	.a	52	68	59
5.	1	40	115	86	129	105	.b	54	87	93	81	.b

Say that we wanted to view just the cases where the blood pressure at time 5 was over 115. We might be tempted to try this but see what happens.

```
. list if bp5 > 115
```

	id	age	bp1	bp2	bp3	bp4	bp5	pl1	pl2	pl3	pl4	pl5
2.	2	30	123	136	107	111	120	92	88	125	87	58
3.	4	23	105	115	121	129	137	52	79	71	106	39
4.	5	18	116	128	112	125	.a	70	.a	52	68	59
5.	1	40	115	86	129	105	.b	54	87	93	81	.b

The above command displayed the observations where `bp5` was missing because when `bp5` was missing, it was over 115. To exclude the missing values from this listing, we could add `& ! missing(bp5)` to the `if` qualifier, which excludes missing values on `bp5`.

```
. list if (bp5 > 115) & ! missing(bp5)
```

	id	age	bp1	bp2	bp3	bp4	bp5	pl1	pl2	pl3	pl4	pl5
2.	2	30	123	136	107	111	120	92	88	125	87	58
3.	4	23	105	115	121	129	137	52	79	71	106	39

Another usage you might see is specifying `& (bp5 < .)`, which says that `bp5` should be less than the smallest missing value. This is just another way of saying that `bp5` should not be missing.

```
. list if (bp5 > 115) & (bp5 < .)
```

	id	age	bp1	bp2	bp3	bp4	bp5	pl1	pl2	pl3	pl4	pl5
2.	2	30	123	136	107	111	120	92	88	125	87	58
3.	4	23	105	115	121	129	137	52	79	71	106	39

If you wanted to display the cases where `pl5` was missing, you could use the following command:

```
. list if missing(pl5)
```

	id	age	bp1	bp2	bp3	bp4	bp5	pl1	pl2	pl3	pl4	pl5
5.	1	40	115	86	129	105	.b	54	87	93	81	.b

If you wanted to see just the observations where the fifth pulse was missing because of the participant withdrawing (coded as `.b`), you could type

```
. list if pl5==.b
```

	id	age	bp1	bp2	bp3	bp4	bp5	pl1	pl2	pl3	pl4	pl5
5.	1	40	115	86	129	105	.b	54	87	93	81	.b

You can supply a list of expressions (separated by commas) in the `missing()` function. If any of the expressions is missing, then the `missing()` function is true. This is used below to identify observations that are missing for any blood pressure or pulse measure.

```
. list if missing(bp1,bp2,bp3,bp4,bp4,pl1,pl2,pl3,pl4,pl5)
```

	id	age	bp1	bp2	bp3	bp4	bp5	pl1	pl2	pl3	pl4	pl5
1.	3	16	.a	122	101	109	112	105	97	128	57	68
4.	5	18	116	128	112	125	.a	70	.a	52	68	59
5.	1	40	115	86	129	105	.b	54	87	93	81	.b

Note that the `missing()` function takes a list of expressions, not a varlist; see section A.11. If you specify, for example, `missing(bp*)`, you will get an error. But consider the command below. You might think that it worked, because it ran, but it does not yield the same results as above. What happened? Because the `missing()` function is expecting an expression, it interpreted `bp1-pl5` to be `bp1` minus `pl5`, and thus listed all the observations where the difference in these two variables was missing.

```
. list if missing(bp1-pl5)
```

	id	age	bp1	bp2	bp3	bp4	bp5	pl1	pl2	pl3	pl4	pl5
1.	3	16	.a	122	101	109	112	105	97	128	57	68
5.	1	40	115	86	129	105	.b	54	87	93	81	.b

You might want to get the average pulse for the five trials. Here is one way. Notice that when `id` is 5 or 1, the average is missing. This is because any arithmetic operation on a missing value yields a missing value.

```
. generate plavg = (pl1 + pl2 + pl3 + pl4 + pl5)/5
(2 missing values generated)
. list id pl*
```

	id	pl1	pl2	pl3	pl4	pl5	plavg
1.	3	105	97	128	57	68	91
2.	2	92	88	125	87	58	90
3.	4	52	79	71	106	39	69.4
4.	5	70	.a	52	68	59	.
5.	1	54	87	93	81	.b	.

Perhaps you want to get the average of the observations that are present. Then you can use `egen` as shown below.

```
. egen plavg2 = rowmean(pl1 pl2 pl3 pl4 pl5)
. list id pl*
```

	id	pl1	pl2	pl3	pl4	pl5	plavg	plavg2
1.	3	105	97	128	57	68	91	91
2.	2	92	88	125	87	58	90	90
3.	4	52	79	71	106	39	69.4	69.4
4.	5	70	.a	52	68	59	.	62.25
5.	1	54	87	93	81	.b	.	78.75

You can also use the `rowmiss()` function with `egen` to compute the number of missing values among specified variables and the `rownonmiss()` function to compute the number of nonmissing values among the variables specified, as shown below.

```
. egen missbp = rowmiss(bp1 bp2 bp3 bp4 bp5)
. egen nonmissbp = rownonmiss(bp1 bp2 bp3 bp4 bp5)
```

```
. list id bp1-bp5 missbp nonmissbp
```

	id	bp1	bp2	bp3	bp4	bp5	missbp	nonmis~p
1.	3	.a	122	101	109	112	1	4
2.	2	123	136	107	111	120	0	5
3.	4	105	115	121	129	137	0	5
4.	5	116	128	112	125	.a	1	4
5.	1	115	86	129	105	.b	1	4

We can assign value labels to missing values just as we do for any other value (see section 4.3). Below we create a value label named `plbpmiss`, indicating that `.a` represents a recoding error and `.b` represents that the participant dropped out.

```
. label define plbpmiss .a "RecErr" .b "Dropout"
```

Now we can label the blood pressure and pulse variables with `plbpmiss`. We can see the labels in the listing of the five pulse variables shown below.

```
. label values pl1-pl5 bp1-bp5 plbpmiss
. list id pl1 pl2 pl3 pl4 pl5
```

	id	pl1	pl2	pl3	pl4	pl5
1.	3	105	97	128	57	68
2.	2	92	88	125	87	58
3.	4	52	79	71	106	39
4.	5	70	RecErr	52	68	59
5.	1	54	87	93	81	Dropout

For more information about how to code missing values, see section 5.6. You can also see `help missing` for more information about missing values in Stata.

A.11 Referring to variable lists

The Stata help files and manuals refer to a concept called a *varlist*. A simple example of a *varlist* is a list of one or more variables, separated by spaces. Stata supports several shorthand conventions for referring to variable lists, specifically `*`, `?`, `-`, `_all`, and `~`. Let's illustrate these shorthands using `cardio3.dta`, which is used and described below.

```
. use cardio3

. describe

Contains data from cardio3.dta
  obs:             5
 vars:            26                          23 Dec 2009 15:12
 size:           660 (99.9% of memory free)
-------------------------------------------------------------------------------
              storage  display     value
variable name   type   format      label      variable label
-------------------------------------------------------------------------------
id              long   %10.0f                 Identification variable
fname           str15  %15s                   First name
lname           str10  %10s                   Last name
bp1             int    %3.0f                  Systolic BP: Trial 1
pl1             int    %3.0f                  Pulse: Trial 1
bp2             int    %3.0f                  Systolic BP: Trial 2
pl2             int    %3.0f                  Pulse: Trial 2
bp3             int    %3.0f                  Systolic BP: Trial 3
pl3             int    %3.0f                  Pulse: Trial 3
bpmean          float  %9.0g                  Mean blood pressure
plmean          float  %9.0g                  Mean pulse
gender          str6   %9s                    Gender of person
bmo             float  %4.0f                  Birth month
bda             float  %4.0f                  Birth day
byr             float  %4.0f                  Birth year
bhr             double %4.0f                  Birth hour
bmin            double %4.0f                  Birth minute
bsec            double %4.0f                  Birth second
age             byte   %3.0f                  Age of person
weight          float  %9.0g                  Weight (in pounds)
famhist         long   %12.0g      famhistl   Family history of heart disease
income          double %10.2f                 Income
zipcode         long   %12.0g                 Zip Code (5 digit)
heart_attack_~t float  %9.0g                  Risk of heart attack from
                                                treadmill test
bdate           float  %td                    Birth date
bdatetime       double %tc                    Birth date and time
-------------------------------------------------------------------------------
Sorted by:
```

The simplest variable list is one variable that is fully spelled out. For example, below we get summary statistics for the variable `bp1`.

```
. summarize bp1

    Variable |       Obs        Mean    Std. Dev.       Min        Max
-------------+--------------------------------------------------------
         bp1 |         5       116.6    7.635444        105        124
```

Or we can name multiple variables, separated by spaces, as shown below.

```
. summarize bp1 bp2 bp3

    Variable |       Obs        Mean    Std. Dev.       Min        Max
-------------+--------------------------------------------------------
         bp1 |         5       116.6    7.635444        105        124
         bp2 |         5       117.4    19.17811         86        136
         bp3 |         5         114    11.13553        101        129
```

We could save ourselves some typing by using `summarize bp*`, which displays summary statistics for all variables that start with `bp`.

```
. summarize bp*

    Variable |       Obs        Mean    Std. Dev.       Min        Max
-------------+--------------------------------------------------------
         bp1 |         5       116.6    7.635444        105        124
         bp2 |         5       117.4    19.17811         86        136
         bp3 |         5         114    11.13553        101        129
      bpmean |         5         116    4.600725        110        122
```

Oh dear, in the above example, the variable `bpmean` was included as well (because it also starts with `bp`). We can try another trick; we can specify `bp?`, which is a shorthand for any three-letter variable that starts with `bp`, but the third letter can be anything. This excludes `bpmean` because it has six letters.

```
. summarize bp?

    Variable |       Obs        Mean    Std. Dev.       Min        Max
-------------+--------------------------------------------------------
         bp1 |         5       116.6    7.635444        105        124
         bp2 |         5       117.4    19.17811         86        136
         bp3 |         5         114    11.13553        101        129
```

The `?` and `*` can appear anywhere: at the end (as we saw above), squished in the middle (e.g., `b*1`), or even at the beginning. Below we can specify `*2`, which shows summary statistics for all variables ending in `2`.

```
. summarize *2

    Variable |       Obs        Mean    Std. Dev.       Min        Max
-------------+--------------------------------------------------------
         bp2 |         5       117.4    19.17811         86        136
         pl2 |         5          83    12.38951         64         97
```

We can specify a range of variables based on the position in the dataset using the `-`. For example, `x-y` means the variables from `x` to `y`, as they are positioned in the dataset.

Below we try this trick for getting the blood pressure values from 1 to 3, but it does not work because these variables are not positioned next to each other in the dataset. The pulse readings are in the middle.

```
. summarize bp1-bp3

    Variable |       Obs        Mean    Std. Dev.       Min        Max
-------------+--------------------------------------------------------
         bp1 |         5       116.6    7.635444        105        124
         pl1 |         5        74.6    23.36236         52        105
         bp2 |         5       117.4    19.17811         86        136
         pl2 |         5          83    12.38951         64         97
         bp3 |         5         114    11.13553        101        129
```

Let's reorder the variables so the blood pressure and pulse measures are grouped together.

```
. order fname bp1 bp2 bp3 bpmean pl1 pl2 pl3 plmean
```

With the variables ordered this way, we can refer to bp1-bp3 and that refers to the three blood pressure measures as shown below. Section 5.15 shows more details about ordering the variables in your dataset.

```
. summarize bp1-bp3
    Variable |       Obs        Mean    Std. Dev.       Min        Max
-------------+--------------------------------------------------------
         bp1 |         5       116.6    7.635444        105        124
         bp2 |         5       117.4    19.17811         86        136
         bp3 |         5         114    11.13553        101        129
```

What if we need to specify a variable list that contains all the variables in the current dataset? The keyword _all can be used in such cases. The example below uses the order command with the alpha option to alphabetize the order of variables in the dataset. Because _all was specified as the variable list, all the variables in the dataset will be alphabetized.

```
. order _all, alpha
```

Finally, the ~ can be used to abbreviate one (and only one) variable. It works like the *, except that it refers to one variable. If more than one variable is indicated by this shorthand, then Stata returns an error (thus preventing you from referring to more variables than you intended). This is used below to refer to the variable heart_attack_risk_treadmill_test.

```
. summarize heart~test
    Variable |       Obs        Mean    Std. Dev.       Min        Max
-------------+--------------------------------------------------------
heart_atta~t |         5    .4288191    .3189803   .0445188   .8983462
```

For more information on variable lists, see help varlist.

Subject index

C

Q

R

S

T

U

V